Water Treatment Unit Processes

Water Treatment Unit Processes

DAVID G STEVENSON

Consultant, Newbury, UK

Imperial College Press

Published by

Imperial College Press
57 Shelton Street
Covent Garden
London WC2H 9HE

Distributed by

World Scientific Publishing Co. Pte. Ltd.
P O Box 128, Farrer Road, Singapore 912805
USA office: Suite 1B, 1060 Main Street, River Edge, NJ 07661
UK office: 57 Shelton Street, Covent Garden, London WC2H 9HE

Library of Congress Cataloging-in-Publication Data
Stevenson, David G.
    Water treatment unit processes / David G. Stevenson.
       p.   cm.
    ISBN 1-86094-074-9
    1. Water -- Purification   I. Title.
    TD430.S73   1998
    628.1'662--dc21   97-42054
                                   CIP

British Library Cataloguing-in-Publication Data
A catalogue record for this book is available from the British Library.

First published 1997
Reprinted 1999

Printed in Singapore.

Contents

Part A. General Concepts

Part B. Primary Treatments

Part C. Granular Media Filtration

Table of Symbols

a	Surface Area (particles & grains)
A	Area
A_c	Capture cross section around a single grain
A_h	Total capture cross section in a unit area of bed and depth h
A_p	Horizontally projected area of the grains in a bed per unit volume.
B	Width of a trough
C	Concentration (in units consistent with the other concentrations, normally volume fraction)
C^*	Equilibrium concentration
C_d	Coefficient of discharge
C_o	Initial concentration
δC	Incremental change in concentration
d	Floc or particle diameter, or smaller particle.
D	Grain diameter, or larger particle diameter, Impellor diameter, pipe diameter, Dispersivity.
D_m	Hydraulic diameter
D_p	Pore diameter
e	Bed Voidage
e_A	Air Voidage within the bed
e_i	Unexpanded bed voidage (before or after fluidisation)
e_o	Clean bed voidage
E	Bed expansion, energy/unit volume
f	Friction factor
F_D	Force due to drag
F_L	Filterability number
F_w	Force due to weight
g	Gravitational constant
G	Shear rate
h	Height or depth
H	Head (water column)
ΔH	Headloss (difference), (Suffixes described in the text)
J	No. of tanks in series.
K	Roughness factor
K_B	Boltzmann constant

K_D	Diffusion Coefficient
K_F	Floor Headloss Coefficient
K_L	Liquid phase mass transfer coefficient
l	Length
L	Length
L_T	Total length of wall in a set of filters
N	Number concentration
N_o	Initial number concentration
N_p	Power Number
N_Q	Flow Number
p	pressure
Δp	pressure difference
P	Power
Q	Volumetric flow
Q_L	Flow per unit length of launder
Q_w	Flow per unit width of weir
R	Ratio of decline
R'	Average shear stress on the surface of media
S	Specific surface area
S_c	Surface area of capillaries
t	time
T	Temperature (Absolute), tank diameter
u	Velocity
u_a	Air velocity
u_b	Backwash rate or velocity
u_c	Velocity of flow in a capillary
u_l	Velocity of flow in a lamella
u_{max}	The maximum value of the velocity in a manifold or trough in which the velocity varies along the length
u_s	Settling velocity
u_y	Velocity at distance y from a surface
u_w	Effective water velocity when air is present
V	Volume
w	Width of a lamella
W	Width of a filter or impeller blade
y	Distance from a surface

Re	Reynolds No.
Sc	Schmidt No.
Sh	Sherwood No.

α	Angle of friction
β	Partition coefficient
γ	Surface Tension
λ	Filtration coefficient (defined in Equation 3.7), or hydraulic gradient
λ_o	Filtration coefficient in clean media
μ	Viscosity
π	Ratio of circumference to diameter of a circle
ρ	Density, (subscripts S & L for solid & liquid)
σ	Standard deviation, σ^2 = variance
θ	Mixing time, residence time, blade angle

Glossary of Terms Used in Granular Media Filtration

Adhesion Stress
The shear stress that must be overcome to remove a particle from the surface of a grain, or the stress that will prevent deposition.

Aggressive Water
Water that is undersaturated with calcium carbonate and liable to cause corrosion of iron.

Air Blinding
Clogging of filter media with air, which prevents water passing.

Air Scour
The agitation of a bed of media by means of air distributed from below.

Angular Material
(Angularity)
Material with sharp edges produced by crushing massive material such as anthracite, glass, flint etc.

Anoxic
Devoid of dissolved oxygen

A.O.C.
Assimilable (biodegradable) organic carbon.

Aspect Ratio
(Sphericity)
The ratio of the largest to the smallest dimension of a given grain.

AWWA
American Water Works Association

Backwash

The procedure used to flush deposited solids out of the granular material filter at the end of the filtering run. This normally involves feeding clean water in at the base of the granular bed washing in an upwards direction.

Bed Volume

A volume of water corresponding to the overall bulk volume of the bed of media.

BOD

Biological Oxygen Demand.

Breakthrough

The point where suspended solids penetrate through the filter bed to an unacceptable extent.

Bulk Density

The weight of a unit volume of the material in a stated condition, eg tipped or backwashed.

Bulk Material

Loose, unpacked material.

Capture Cross Section

The projected area around the grains in a bed viewed in the direction of flow through which particles have to pass to be captured. (The term is used for a similar purpose in nuclear physics for neutron capture).

Charge

(n)   a) The granular material content of one filter.
        b) Electrical charge on particles which affects the inability to adhere to
           surfaces. (Zeta Potential)
(v)   To fill.

Ciliate (algae), flagellate

Algae having hair like tails.

Clarifier
This term is used in several industries and denotes different things. In the water industry it refers to the primary separation stage upstream of filters, usually sedimentation but sometimes applied to flotation.

Clogging Head
The headloss (measured in metres water gauge) caused by the dirt in a filter. It corresponds to the total headloss minus the clean bed headloss at the same flow rate.

Coagulant (Coagulation)
A chemical that will cause a stable colloidal suspension to start to agglomerate and adhere to surfaces.

Coagulant aid
A chemical, eg. ployelctrolyte, added to improve the action of the above.

Conductance
The inverse of resistance.

Crown (Pipe)
The top of the pipe.

Demand
The amount of a disinfectant consumed by reaction with impurities in the water before any residual is left.

Determinand
A parameter such as the concentration of an element or compound in water which is to be measured and recorded.

Destabilisation
Neutralisation of any electrical charge or other influence which prevents particles agglomerating. Equivalent to coagulation but without any subsequent processing.

Dirt

Loose removable material capable of dispersion (eg clay) which is of substantially smaller particle size than the grains of the filtering material.

Dump (Water)

The discharge of residual supernatant water from a filter (usually preceded by draining down) prior to commencing the wash procedure.

EBCT

Empty bed contact time, ie. bed volume/flow (The converse of bed volumes per hour).

Effective Size

A traditional but meaningless and non-preferred term used widely in the USA to describe the 10 percentile size (the sieve size that the 10% by weight of the material would penetrate).

Effective Specific Gravity

The resultant value of the specific gravity when a porous grain is saturated with water (relevant to its settling behaviour).

Epilimnion

The upper aerobic layer of a reservoir or lake.

Expansion

The ratio of the height of a fluidised bed to the settled height.

Filter

A device, which includes a barrier or medium, for separating solids from liquids.

Filter Medium (Media)

The term 'media' is in common use to describe granular filtering materials. It is grammatically plural and strictly speaking should apply to the bulk material to be installed in several filters. (A single filter has a filter medium). To avoid ambiguity the term 'material' used in the AWWA Specification is preferred for legal documents.

Floc
Coagulated and agglomerated particle in suspension.

Floc Blanket
A fluidised or settling bed of floc particles or a mass of such particles in hindered settlement.

Flocculation
The process of causing small particles in suspension to agglomerate into larger clusters which then separate more readily. The term is especially applied to the use of low speed paddles or energy dissipation in tanks, pipes or capillaries for this purpose. The term is sometimes used to describe coagulation.

Friable Material
Material such as anthracite (coal), granular activated carbon, pumice, etc., which tends to crumble if compressed.

Grain
A single unit indivisible particle of media

Gravel
This is a non-specific term signifying coarse material above about 3mm mean size, mainly silica normally but not exclusively, used in a passive role as a support material.

Gravity Filter
An open filter working at atmospheric pressure where the flow is created by a difference in head between the inlet and outlet.

Hazen
Degrees. A measure of natural colour in water (based on mg/l platinum as platinic chloride stabilised with cobalt).

Headloss, Head
Pressure loss expressed as a height or head of water $1m = 9810$ Pa (Newtons/m^2).

High Lift Pump
Final high pressure pump delivering into supply.

Hydraulic Diameter
The diameter of an equivalent circular pipe which has the same hydraulic characteristics as a specific non-circular pipe, duct or channel. It is calculated as (4 × cross section ÷ wetted perimeter).

Hydraulic Size
The uniform grain size that would produce the same resistance to flow as the material under consideration (at the same voidage). (See Appendix to Chapter 30 for method of calculation). The term is also used by some to describe the size of spherical grain that settles at the same velocity as an arbitrary non-spherical grain.

Hypolimnion
The anaerobic lower layer of a reservoir or lake.

Invert
The bottom of a pipe or channel.

Jar Test
A term employed loosely describing laboratory (bench top) treatment tests.

Laminar Flow
Streamline flow where the pressure loss is linearly related to velocity.

Lateral
A term used in the water and engineering industries to denote a branch or one of a set of branch pipes or ducts connected to a single common manifold.

Launder
A collecting channel or trough usually set at high level to decant water from a tank (i.e. decanting trough).

Low Lift Pump
Usually describes the pump feeding the treatment plant.

Micron
A traditional term for 1μm.

Mld
A unit of flow in common use in the United Kingdom. Megalitres/day=Thousands of cubic metres/day=tcmd (the latter is not used in this book!)

Negative Head
A condition in a filter bed where the hydrostatic pressure is below atmospheric pressure. This can lead to the nucleation of air bubbles.

Nozzle (Filter)
A unit component which is set into a filter floor to retain the media, allow the filtrate to pass through and also to regulate the flow of water and possibly air during backwashing.

NTU
Nephelometric Turbidity Unit, ie based on light scattering using a standard formazin suspension. Numerically the same as Formazine units (FTU) and Jackson units (JTU).

OA, PV, Oxygen Absorbed, Permanganate Value
A measure of the organic content of water capable of reacting with permanganate under standard test conditions.

Packing, Packing Layer
Coarse sand and gravel placed beneath the working layer of sand in a filter. See Support Material.

Plug Flow
Piston type flow where the fluid across the entire cross section moves at the same velocity.

Porosity

In the context of this book, the term is applied to the internal space within the grain (as with activated carbon). Some use the term to describe voidage, see below.

NOTE: Porosity may be 'open' and capable of filling with water or 'closed' and isolated from the surrounding fluid.

Pressure Filter

An enclosed filter which operates at a pressure above atmospheric.

Primary Treatment

Any process applied upstream of filtration (see Clarification).

Quartering

A procedure whereby a pile of material is divided in four by inserting a spade or divider blade across the diameter of the pile and then again at right angles.

Residual

The concentration of active disinfectant left after part of the dose has reacted (see demand).

Reynolds Number

A dimensionless number which characterises the flow regime in a pipe or round a particle.

Riffler

A device consisting of a set of alternating chutes to split a sample of granular material in two without segregation.

Rise Rate

The vertical velocity of fluid in a filter during backwash. Also applied to the flow in settlement tanks. (m^3/m^2-hr=m/hr or mm/s)

Sand

This is a non-specific term. Normally it refers to a silica based granular material smaller than about 3mm mean size of S.G. $2.65 \pm .05$. It is also used non-specifically to describe the material that plays the major role in the filtering action.

Schmutzdecke

German for dirt blanket. Used to describe the surface dirt layer in a slow sand filter.

Sharpness

See Angular Material

Size (Size Range)

Unless otherwise stated filter materials are defined in terms of an upper and lower sieve size based upon BS 410 (1986). The lower size will be the standard sieve mesh which retains 95% by weight or more of the material, and the larger size the standard mesh on which 5% or less is retained, regardless of the size range.

Sparge

Noun. A perforated pipe used to distribute a fluid.
Verb. To discharge a fluid, particularly air, through a sparge pipe.

Specific Gravity

The density of the grain or material divided by the density of water (only applicable to the grains not the material en masse).

Sphericity

See Aspect Ratio.

Strainer

The component of a nozzle which retains the media. A sieve-like component.

Streamline

The path taken by an element of fluid, eg. when passing round an obstacle.

Streamline Flow
Laminar flow, Viscous flow in which the velocity is proportional to the applied pressure.

Submerged Specific Gravity
The difference between the specific gravity of the grain and water.

Support Material/Gravel
A granular material coarser than the working material which is not intended to capture suspended solids and neither is it intended to move or be fluidised in filter washing. Its role is to prevent percolation of the finer material into under-drains or equivalent means of collection of the filtrate, and in some cases to protect the latter from abrasion and reduce the pressure loss of the collection system.

Surface Flush
A procedure whereby unfiltered water is caused to flow across the surface of a filter bed.

Surface Scour
A method of assisting the cleaning of the media involving high velocity water jets impinging on the surface of the bed.

THM
Trihalomethane. A class of chloroform-like chemicals that are produced as a by-product from the reaction of chlorine on natural organic matter.

TOC
Total organic carbon

Transitional Flow
Flow which has characteristics between streamline and turbulent.

Turbidity
A measure of the light scattering ability of the fluid, used in water treatment as an easily measured surrogate for suspended solids concentration. The units of measurement accepted in the water industry are based an arbitrary but widely accepted standard using formazin to an American Public Health Association Standard, with measurement by light scattering (Nephelometry), hence the Formazin Turbidity Unit or Nephelometric Turbidity Unit (FTU and NTU).

Turbulent Flow
Fluid flow involving eddies and inertial effects characterised by the velocity being related to the square root of the applied pressure.

Underdrain
The floor of a filter, particularly with respect to its function in collecting the filtrate.

Uniformity Coefficient
A arbitrary traditional non preferred term with used in the USA and elsewhere to describe the ratio of the 60 percentile to the 10 percentile size (see Effective Size). The size range is preferred in Britain.

Upwash
Backwash.

Voidage (Voids)
The ratio (normally expressed as a percentage) of the volume of the space between the grains to the overall volume of the granular material. This varies with the circumstance, which must be stated eg 'as poured', 'backwashed', 'packed'. Some use the term 'porosity' but this must be defined as bed porosity to distinguish it from grain porosity.

Washout
The outlet channel or pipe for conveying the used wash water.

Washout Weir

A side weir over which the used wash water is decanted.

Working Material

That part of the granular filling that provides the filtering action and from which the deposited solids are backwashed.

Wormhole

A string of clean pores within the clogged section of a granular bed.

Zeta Potential

The electrical charge on a surface immersed in a conducting fluid.

Preface

The need for a book such as this became apparent while working with PWT Ltd. when I found that there were few useful books available which could be recommended to new recruits. Internal design documents did not give the background theory. There was a well known Handbook, which describes another contractor's technology in a qualitative manner, and a number of American books, which tend not to go into detail, and in any case discuss a rather different tradition. With over 25 years of experience in the industry, plus further related experience in chemical engineering and surface chemistry one was reluctant to see the gathered material pass into oblivion, without others benefitting hence this project began. The author's conviction that such a book as the present one was needed was confirmed when attending a presentation of MSc projects in which a horizontal flow clarifier was proposed for a precipitation softening application.

Many books discuss the whole of the water cycle, reservoirs and storage, distribution, and necessarily the discussion of treatment is very much condensed. On the other hand there is a good number of research books and papers which perhaps discuss narrow issues often with equations that leave the ordinary engineer speechless. Although some equations are quoted in this book they are for the most part simple and directly applicable to the problem being discussed. Many references are made to Coulson and Richardson's classic and regularly updated textbook which has served several generations of chemical engineers and fuller explanations of many equations may be found there. (Perhaps one day this will be extended to include water treatment processes as a part of chemical engineering). The are no doubt gaps in this book and it is hoped that these will be filled in due course.

One is conscious of a danger in expressing views on the advantages and disadvantages of any particular system or design. The intention here has been to be impartial, analysing the various systems without promoting or disparaging any particular contractor or manufacturer's technology. Neither is the book exhaustive. The examples chosen were selected to represent types. It may be that this book will be the basis for later additions.Suggestions and additional material for any gaps that become apparent would be welcomed.

The help, encouragement and friendship of past and present colleagues is acknowledged freely. The information presented is the outcome of experience of earlier colleagues, the investigations of many who worked for me, the colleagues who presented me with problems for consideration and the many friends in the industry who were customers, employees of bodies such as the Water Research Centre or members of the Chartered Institute of Water and Environmental Management. Lastly of course the book would not have been possible without the tolerance, encouragement and co-operation of my wife.

D.G.S. 1997

Part A

General Concepts

CHAPTER 1.

INTRODUCTION AND EARLY HISTORY

1.1 Introduction

The term **water treatment** is widely understood as the steps taken to purify water for drinking (potable) or industrial purposes. The processes used are mainly physico-chemical. Wastewater or sewage treatment is mostly biological. While some of the processes are similar the technologies differ in some fundamental ways. This book is specifically directed to the former. The processes used in water treatment are not so widely understood outside the industry. For example many may be unaware that to purify and stabilise water a variety of chemicals are generally added. Such additions facilitate the removal of substances which are not easily removed by physical means alone, to make it wholesome and non corrosive. Also many, including politicians and law makers do not always appreciate the statistical nature of water quality. One can never guarantee anything in this area with 100% certainty, only with a high probability. Raw waters vary with the seasons, with rainfall and unpredictably with algal blooms etc. There are many trace organic materials which interfere with treatment. All that can be done is to build a treatment plant incorporating processes that stand a very good chance of meeting the quality targets practically all of the time at an acceptable cost. With more stringent quality standards enforced by more rigorous inspection arrangements and the threat of penalties utilities are having to resort to more advanced treatment processes. The use of activated carbon has become common in the past 10-15 years and membranes are now being considered as a possible routine unit process. It is debatable whether such expenditure is always the most cost effective way of improving the health of the community.

Most plants treating variable surface waters consist of at least two stages. The first is able to handle the very variable raw water quality, sometimes with high levels of suspended solids, and produce a reasonably consistent intermediate quality not normally good enough for supply but with most of the impurities removed and by specification good enough for filtration. In the industry these primary processes are usually called clarification. The ultimate quality as far as a suspended solids and many other parameters are concerned is achieved by filtration. Most filters for water treatment are based on granular materials such as sand and because the solids being filtered out are retained within the voids in the sand bed they have a limited capacity. They are able to produce a high quality filtrate when correctly operated. The clarification and filtration processes are therefore complementary.

In some cases these stages are not able to remove all the undesirable species and activated carbon may be needed to absorb pesticides for example. The required bacteriological quality of the water is achieved by a combination of the above plus the

1

addition of a specific disinfectant such a chlorine, which is usually applied as the last stage after most of the suspended solids have been removed. A filtered water will not be free of pathogenic organisms unless specifically disinfected in this way or else filtered through a membrane fine enough to remove them.

1.2 Early History

Traditionally settlement has been the main method of clarifying water on a large scale. In some cases this happened naturally as rivers flowed into lakes. Egyptian tomb paintings of 1450 BC depict settlement jars and syphons for clarifying water and wine. Julius Caesar is reputed to have found an extensive series of underground aqueducts in Alexandria which brought water to cisterns in which it was allowed to clarify by settlement. The Egyptians cut off the supply of fresh water and diverted salt water from the Mediterranean. Caesar proceeded to dig new wells. However the simple principles involved have been developed into more complex and efficient systems, and in recent years inverted by the extensive use of dissolved air flotation.

Filtration also was first practised in ancient times. It seems that the earliest examples were infiltration wells dug on river banks and seashores. Riverside wells could be deemed to be an example of cross flow filtration in that the solids filtered out are carried on their way by the flow of the river. Wells on seashores frequently tapped sweet water aquifers flowing from the land into the sea. For many centuries the situation was not appreciated and consequently it was widely assumed that sand was filtering out the salt. Such infiltration wells date back to Roman times. The technology has been revived in recent years with shore wells being used as a source of clean water for reverse osmosis desalination plants.

After the legendary constructions of the Romans, Sir Francis Drake, in the 16th century, was one of the earliest water engineers in Britain. He built the Devonport leat from Dartmoor to bring fresh water to the Naval dockyard and the neighbouring area of Plymouth. He made good use of the gradient and built many mills on the way which were let out to bring him an income. The Dartmoor water however did not receive treatment.

Filtration is defined in dictionaries as a process of solid/liquid separation. It is more specifically applied to the use of some form of physical barrier, in contrast to sedimentation or flotation. It has been borrowed by many other technologies, so that we have optical filters, and electrical ones, as well as biological filters which are really supported growth bioreactors and need a separate solid/liquid separation stage. The first reference to filtration as we now know is has been attributed to Porzio an Italian physician in 1685. The invention arose out of work aimed at conserving the health of soldiers in the Austro-Turkish war of that year.

Water treatment as we know it today in dedicated tanks or containers dates back to the closing years of the 18th Century. Although James Peacock was granted a patent for an upflow filter with graded support gravel and a reverse flow wash in 1790 this was an exceptionally early example of backwashing. There are still examples of such traditional Scottish upflow filters in operation. Early filters relied on the processes now associated with slow sand filtration with purification involving biological processes. The first was possibly that known as a Lancashire filter, which is believed to have been in use as early as 1790, but was not described until 1850.

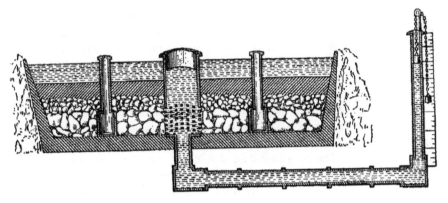

Fig. 1.1 The principle of the Lancashire Filter.

An annular filter with an integral settling tank was built in the early 19th Century at Paisley. However the first fully satisfactory and reliable slow sand filter is generally acknowledged to be the one built by James Simpson at Chelsea in London, which was one acre in size and commissioned in 1829. For many years afterwards, Simpson's filter was the only technology in general use. In 1858 a patent appeared on the use of a rake, presumably to break up the 'schmutzdecke' (surface clogging layer). In 1871, Holly in the USA took out a patent on a reverse flow wash, downflow filter. Cook patented the concept of a battery of filters with one being taken out of service for washing in 1877.

Hyatt, who died in 1885 introduced the concept of a coagulant (ferric chloride was the first) and over this period, the general practice of what we now know as rapid filtration became established initially using pressure filters with perforated floors and rakes. It is not clear whether the principles of fluidisation were understood and it seems likely that cleaning then was by a sub-fluidising flow assisted by raking.

4

Fig. 1.2 The concept of the Paisley Settlement Tank and Filter.

In 1830 Darcy established the principles of laminar or viscous flow in granular bed and noted that the pressure loss was linearly related to the flow rate. The understanding of the effect of voidage and particle size was not established until 1927 when Kozeny published his work on flow in porous media. (Curiously even in some much more recent papers, referred to later, the importance of voidage has been ignored).

It in interesting to note that an evaluation programme on filtration plant and processes was undertaken at Louisville in 1896-7. Many of the conclusions are valid today. For example, it was stated that as much as possible should be removed from the water before filtration and that the water should be properly coagulated just before entering the filter. The coagulant dose should not exceed two grains per gallon (ie 28mg/l) which is a maximum dose that many might still regard as appropriate for direct filtration. The history of water filtration and treatment practice going back to Egyptian times is described fully in a timeless book by Baker (1947) which is well worth reading.

In the early years of the 20th century Sir William Paterson and Frank Candy were prominent in Britain. The former pioneered a filter design using exposed laterals embedded in multiple layers of gravel that was adopted by several other companies. Frank Candy developed a floor using fireclay ducts embedded in concrete, initially with brass nozzles. The basic principles of the process remained much as they were in the latter years of the previous century. More advanced clarification processes started to emerge during the 1930's.

A sand filter is a tank or pressure vessel which contains a bed of sand over a perforated floor. The size may range from 200mm diameter or less for a pilot unit to 200m². Large treatment works have up 10,000m² of such filters or considerably more if slow sand filters are used. As such filters have developed other granular materials such as anthracite, granular activated carbon, pumice, garnet etc. have been introduced for special purposes. Hence the wider term granular media filtration is now more appropriate.

Water treatment unit processes evolved originally by the efforts of civil engineers before the advent of chemical engineering. Thus there has tended to be a blinkered outlook between the two disciplines, and Coulson and Richardson (1990) for example make no reference to sand filtration in their standard textbook on chemical engineering, while at the same time there are many concepts in chemical engineering that have taken many years to be discovered by civil engineers eg. chemical reaction engineering.

Filtration in chemical engineering tends to involve the retention of the particles on a medium such as a cloth, paper, a porous sinter or membrane and the water in the suspension has to pass through the solids already deposited (cake). This is not usually very permeable and relatively high pressures eg. 7 bar may be involved. In granular media filtration the solids collect in the narrower pores and crevices leaving highways (wormholes) for the water to pass through to the layers below. The pressure (or head) is very much less and the majority of such filters operate at up to 2m head of water (0.2 bar) only. The depth of media in conventional filters is usually 600-1000mm, and up to 2000mm in some cases. This produces a situation which has been described as 'in depth' filtration, although this is a much misused term which has often been used to describe new developments in a manner that implies that conventional granular media filtration is otherwise. In all forms of granular media filtration the deposit collects mainly in the bed rather than on top.

Filters have a limited capacity and operate best on low levels of suspended solids because of the limited volume in the pores between the grains in the bed (normally about 40% of the bed volume) and the fact that such filters generally operate batchwise, (ie. they clog and have to be taken out of service to be backwashed). For polishing duty after a clarifier the suspended solids in the feed may be only 5mg/l. For direct filtration levels of 50-100mg/l. are handled by specially designed filters. However the available pretreatment stages of settlement or flotation are truly continuous processes and are better able to handle high and varying concentrations. At the same time they produce a consistent residue, unfortunately higher than that required for potable and most industrial purposes but in the ideal range for granular media filtration, which can then remove a further 80-99%.

Granular media filtration is similar to absorption in some respects. It does not have a defined size cut off point in the same way as a sieve. The concentration of any particular size of particle in the feed decays exponentially with depth at a rate depending on the filtration conditions, including the chemical treatment. The concept of a size rating as used with cartridge filters is regarded generally as irrelevant. With appropriate conditions colloidal particles are removed efficiently but never at 100%.

Although this book has been written in the context of potable and process water treatment starting from natural water sources granular filtration is used widely for many other large scale applications,in water recovery and recycling within industrial processes such as steel making (cooling of machinery and conveying of mill scale), paper making (water recycling and fibre recovery), power stations, food, dairy and brewing etc. Sea water is frequently treated prior to reverse osmosis. Oil droplets behave much as particles and sand filters are used in the petrochemical industries as well as others for oil removal. In considering other applications it must be remembered that the capacity depends on the volume of the solids removed not the weight.

A most interesting review of the situation in 1908 was written by Don (1909) for the Institution of Mechanical Engineers in Britain. The emphasis on water treatment was directed much more to the removal of bacteria rather than aesthetic parameters, such as turbidity and colour. Already at that date, some of the design criteria still used today had become established. Rapid gravity filters at Gloucester built by Jewel had a maximum design rate of 5 m/h. Such filters had rake arms driven by water motors. At this point there is no mention of air scour. Similar filters dating from this period have continued in service to the present day.

Already there were prejudices at that time and in the discussion on Don's paper one speaker commented that mechanical filters should be put side by side with 'some well established sand filters', ie slow sand filters, implying that there were doubts about the effectiveness of 'mechanical' filters. There was concern about the amount of water used for washing rapid gravity filters although this appears to be no more than present practice. It is believed that wash water recovery was the exception even if there were any examples at that date.

It seems that the term 'Polarite', now applied to a catalytic manganese dioxide, was originally applied to an artificially produced material which was a subject of a patent by Spencer in 1858. It was produced by roasting comparatively pure iron ore in intimate contact with carbonaceous material such as sawdust. It was used in the form of coarse sand and was claimed to catalyse the oxidation of organic matter and ammonia as well as providing disinfection, This material was known variously as Polarite, Purite or Oxidium. Candy Filters Ltd. promoted pressure filters in those days using an Oxidium media, in which air was trapped under the top of the shell after each backwash to provide oxygen for the supposed reaction. It is of course conceivable that the material slowly corroded, producing ferric hydroxide which acted as a coagulant. However, this is only speculation.

It is clear from the early records that the failure of filters was mostly associated with either the complete absence of, or an inadequate, backwash and it was Simpson's slow sand filter in which the dirt layer was skimmed off manually which first enabled sand filtration to achieve reliability. Washing still remains the most critical process in filtration after, perhaps, correct chemical treatment.

One has the impression that the basic form of sand filtration was established in the late 19th Century and for very many years little changed. Examples of 8ft steel pressure filters built in banks of 40-80, often with line shafting to drive rakes have persisted into the late 20th Century and many pressure vessels are still in use. Mechanical dosing pumps, often driven from water turbines have been replaced by electric dosing pumps but many works still have a 19th Century steam age appearance and until about 1970 there were still a number of steam driven pumps in service. This background influenced the design and many measures were introduced to minimise the peak power demand. For this reason pressure filters were popular. Repumping was expensive in capital if not in operation. Elevated storage tanks were often used to reduce the power demand for filter backwashing, whereas the low relative cost of electric pumps and the reliable nature of power supplies in the United Kingdom has led to the almost universal use of direct pumping for this purpose.

Lateral filter floors (using perforated parallel pipes) came in even with James Simpson's slow sand filters and continue to the present time. Frank Candy's variant using pottery clay ducts 14 ins long which were laid on the structural floor by bricklayers, carrying brass bushes to take crude strainers also in brass continued into the late 1960's. These arrays of ducts with dry joints were covered with concrete to produce a smooth floor which looked like a suspended floor from above, with only the strainers being visible. A derivative using extruded plastic in place of fireclay and with plastic strainers replaced the latter only about 1970 and is still the basis of cost effective current designs.

Other suppliers tended to use the Paterson exposed perforated laterals (ie. parallel pipes connected to a header or main) surrounded with deep gravel packing layers. The lateral pipes were in cast iron or asbestos cement until the advent of cheap engineering plastics in the 1970s.

It is believed that air scour came into use in Britain in the 1920's and became almost universal on open gravity filters. However this air scour remained as the separate rather than the combined version and the most common working filter media has tended to remain at 0.5-1.0mm or 0.6-1.18mm in Britain. In the USA media has tended to be one sieve size smaller while in Continental Europe the pattern has been towards rather coarser media with combined air and water washing. This pattern has persisted until the present although since 1980 continental influence has increased and even in the USA, where process parameters have been more strongly regulated, air scour is beginning to appear.

Much attention has been focused on the control of filters and many ingenious hydromechanical devices were developed. Quite a number were less than perfect and it was partly for this reason that Cleasby introduced the concept of declining rate filtration, discussed in Chapter 23. Today mechanical control devices have been largely replaced by electrical control systems. The water industry had to pioneer several technologies which are now commonplace in the chemical industry, for example metering pumps.

The main change in filtration practice in the late 20th Century has probably been a reduction in the load carried by granular media filters as a result improvements in clarification ie. settlement with the introduction of solids recirculation and floc blanket clarifiers (particularly in Britain and areas where British companies operated), and since the mid 1970's dissolved air flotation. (Until recently horizontal flow clarifiers with mechanical flocculators tended to be used in the USA, and only in the mid 1990s was dissolved air flotation being considered seriously).

In parallel there have been improvements in coagulation practice, through a better understanding and the introduction of coagulation aids such as activated silica in the 1950s and then organic polyelectrolytes which came into general use in the 1970s. These enabled the settled water quality to be improved although if used unwisely they could create other problems on the filters. This is a separate area of technology and outside the scope of this book. For the most part the design of granular filters themselves remained unchanged.

A mid-century review of filtration, albeit brief, is given in the Institution of Water Engineers' Manual of British Water Supply Practice (1954). In this it is noted that air scour was by no means universal. It was also lamented that there was no theory to back up filtration technology. The main objective in selecting filtration rates was to achieve long filter runs in order to minimise labour for washing. The strike in the industry in the early 1980's greatly accelerated automation and now automatic washing is the rule. Filter optimisation is more a matter of minimising the overall cost of the service.

As in many older industries designs were originally empirical. The past 50 years and particularly the past 25 years have seen a major improvement in the understanding of theoretical basis. Nevertheless there remains an ingrained conservatism and until recently few wished to be the first to employ any new process. The privatisation of the industry in Britain (1989) coupled with pressures to achieve higher qualities has caused some change of attitude but even so there is a considerable element of caution and radical departures from proven practice are rare.

The 1990's have seen two new concerns arise, namely pesticides, where political rather than toxicity based standards have led to the wide introduction of ozonisation followed by activated carbon, and the identification of cryptosporidium in water supplies as a health hazard. Two reports (Anon 1990 and 1995) produced by a committee of experts appointed by the British Government contain advice on good practice in operating water treatment plant that must be taken seriously by operators in a situation where disregard may lead to criminal prosecution. One may speculate on the next problem to face the industry.

1.3 References

Don,J., (1909), "The Filtration and Purification of Water for Public Supply". Proc. Inst. Mech. Eng. (Jan/Feb. 1909),7. London.

Baker,M.N., (1949), The Quest for Pure Water, American Water Works Assn. New York (now Denver)

Institution of Water Engineers, Hobbs, A.T. Ed.(1954) Manual of British Water Supply Practice, (Second Ed.) Heffer, Cambridge

Anon (1990 and 1995), Cryptosporidium in Water Supplies, (First and Second Badenoch Reports). H.M.Stationery Office, London.

Coulson, J.M., Richardson, J.F., Backhurst, J.R. and Harper, J.H. (1991) Chemical Engineering. Pergamon

CHAPTER 2.

TREATMENT PROCESSES

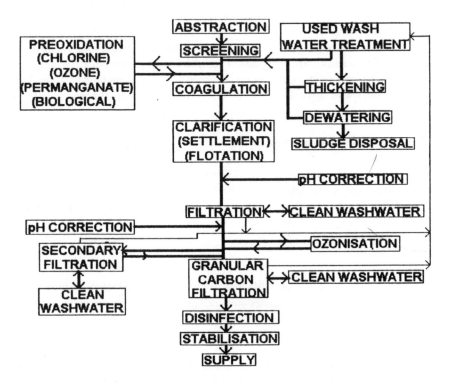

Fig. 2.1 A chart of water treatment processes, showing their position in the treatment sequence. It is unlikely that all of these stages will be included in any one plant. However it illustrates the position and function of the processes discussed in this and later chapters.

2.1 Scope

This chapter is intended to provide a general background to the selection of treatment processes available to the engineer or scientist facing the task of designing a new works or refurbishing an existing works. The way in which the process units work and the detailed design of treatment units is discussed in later chapters. Surface ratings given in the present text are only for guidance. Specialist suppliers with appropriate experience may be able to offer higher rates and if pilot trials can be run the designer will have much greater confidence. Very considerable sums of money can be saved in this way.

It should be borne in mind that there is seldom one single solution to a given treatment problem and alternatives should be explored and costed to arrive at the most cost effective overall solution, taking into account capital and running costs, including manpower and waste disposal. Time in initial trials and experiments usually well spent.

2.2 Raw Water Sources and Data

Water sources used for drinking and public supply divide naturally into surface sources eg. rivers, lakes and reservoirs and ground water sources from wells and boreholes. In rare cases saline water even sea water may be the only alternative.

Possible Contaminants

Rivers

High turbidities, colour, pesticides, bacteria, cryptosporidium, giardia lamblia, ammonia, nitrate, metals, oil, tastes and odours. Rivers can be subject to rapid variations and pollution incidents.

Reservoirs

As rivers, except that turbidity, ammonia and nitrate, bacteria will be much less. Algae blooms, toxins and ease of treatment may be worse due to peptisation by products from algae and loss of weighting solids. Reservoirs are less variable but turnover (caused by wind or suface cooling) can produce ammonia and manganese.

Boreholes

These may produce excess carbon dioxide. Acidic boreholes -Iron and manganese, ammonia, nitrate, hydrogen sulphide, heavy metals. Chalk and limestone boreholes - high hardness, nitrate, pesticides.

Sodium and calcium (occasionally magnesium) may also be unacceptable in any of these waters.

Sea water
This will have high sodium, magnesium and chloride. Boreholes in coastal areas may suffer from ingress of sea water. In arid areas evaporation may exceed rainfall and salts dissolved from the ground may be concentrated leading to brackish or saline water.

Data Collection

When designing a water treatment works it is essential to consider the following:-

> Historical raw water data. Statistics and trends
> Seasonal variation in water demand.
> Current and future demand forecast.
> Future upgrading potential with regard to raw water quality and impending treated water quality requirements.
> Any existing structures that may exist at the site.
> Existing processes at the site concerned or on similar sources in the locality.

Sampling programmes should cover extended periods, preferably over several years and particular attention should be given to extreme conditions regardless of the time at which they occur. It is common for routine sampling to follow a regular time pattern and for random excursions to be overlooked. Special measures need to be taken and samples secured at possibly antisocial hours.

Much data is collected on a routine basis for the purposes of source quality compliance. This is often of little use for process design purposes. Also archive data is frequently summarised in terms of maxima, minima and averages (possibly with standard deviations). Such manipulation hides useful information. One needs to know the full composition on extreme occasions because the maximum pH or minimum pH is unlikely to occur at a time of extreme alkalinity for example. It should also be noted that sulphide and carbon dioxide are often lost in transit and need to be measured on site unless considerable care is exercised in taking and preserving the sample.

To design a treatment purposes the following data, expressed as mg/l or μg/l, must be avaiable. The figures indicate likely ranges.

Primary data.
Turbidity (or suspended solids or both) Nephelometric Turbidimetric Units, (NTU). (0-50,000)
Colour (filtered) °Hazen, or chloro platinic acid standard. (0-500°)
Alkalinity as mg/l $CaCO_3$ (50mg/l CaCO3≡1 meq/l) (0-300mg/l exceptionally 500mg/l as $CaCO_3$)
pH (3-9).

Secondary data.
Iron (filtered) (0-15mg/l) often traces.
Manganese (filtered) (0-2mg/l).
Aluminium (filtered) (0-1mg/l).
N.B. Suspended Fe, Mn and Al count as suspended solids.
Calcium (as Ca or $CaCO_3$)(0-500mg/l as $CaCO_3$).
Magnesium (as Mg or equivalent $CaCO_3$)(0-100mg/l as $CaCO_3$).
Nitrate (as N or NO_3) (0-100mg/l as NO_3).
Ammonia (as N or NH_3) (0-5mg/l)
Pesticides and other listed trace organics if suspected.

Tertiary Data.
Sodium   (0-200mg/l)
Phosphate (as P or PO_4) (0-2mg/l as PO_4).
Silica (as SiO_2) (0-20mg/l as SiO_2).
Boron
Heavy metal contaminants.

There may be other determinands that past history or observation show to be significant such as sulphide. A knowledge of the catchment or source will enable the analyst to decide what can be ignored or what may be expected. The full range of determinations applied to river carrying industrial effluent would be wasted on a chalk borehole or upland coloured water. A utility contemplating a new source should confirm the absence of all the determinands listed in the regulations. A contractor tendering for a new plant will expect to be advised of any contaminants beyond the primary and secondary determinands listed above.
It will normally be assumed that bacteria are present, and surface water is always a potential source of organisms such as giardia lamblia and cryptosporidium.

An investigation of the catchment will give the best indication of the risk of occurrence of these organisms, which will in any case be very intermittent. Treatment will include disinfection.

Reservoirs in particular can generate serious blooms of algae and counts identifying numbers and species will assist the selection of treatment processes. Measurements of chlorophyll are frequently made (being reasonably easy) but give little idea of the algal problem. Algae counts with approximate numbers of each species are much more valuable to the process designer.

2.3 Treated Water Quality

The water treatment works must be designed to ensure that the final water leaving the works complies with the relevant regulations in the territory concerned. Regulations controlling the supply of potable water in the UK are 1991 Water Industries Act and the EC Drinking Water Directive 80/778 EC.

In other territories local regulations may be in operation. These are usually based on the WHO Guidelines, although not necessarily the latest edition. WHO Guidelines were issued in 1971 and 1984 but the latest is dated 1993.

Intermediate Water Quality

The objective of a treatment plant is to provide final water going into supply with an acceptable quality. Sometimes specifications demand that the quality from intermediate stages shall be of a given standard. While this may be of some use with conventional treatment processes it may be counterproductive to the design of the most cost effective process. Thus while it is reasonable to request such data for guidance in evaluation it should not be mandatory unless the contract is for a new process unit which may be required to work in conjunction with an existing unit, eg for a clarifier to work with an existing filter.

Intermediate data can be valuable as a means of controlling and monitoring plant performance. An example is the installation of turbidimeters on filters. However with excessive duplication of instrumentation there can be a temptation not to maintain all of them or to ignore the data.

2.4 Raw Water Storage

This can have a significant effect on the quality of a river water eventually reaching the treatment plant in addition to the more obvious reasons for storage. Any raw water storage facility will depend on the available land, the seasonal variation in demand and the seasonal variation in raw water supply. For example, a water treatment works supplying a seaside town with a high summer population is likely to require an impounding water storage reservoir if the raw water source is a river as the river level is likely to be low in the summer when water demand is high.

Required raw water storage capacity can vary between 12 hours to several months.

Bankside Storage (up to 7 days)

This is normally employed for short term storage of river water either to smooth out rapid quality fluctuations and simplify process control and /or to provide a buffer in the event of pollution of the source. Bankside storage is particularly useful on "flashy" rivers where optimum dosing conditions vary rapidly and where the works serves sensitive or extensive areas.

Reservoirs

These may either be fed by a river, dammed to provide the desired storage (impounded) or they can be filled by pumping from a river in which case the pumps can be shut off in the event of poor river quality. The reservoir can be located several kilometres from the water treatment works. A reservoir may be sized to allow continuous supply during five consecutive years of typical drought. Means of drawing off water from several levels is necessary so that anoxic water can be avoided and also the best water with the lowest algae count can be selected.

Advantages.
The plant output can be maintained in the event of a prolonged drought or when the raw water may be unsuitable for treatment;
Reduction of nitrate levels by biological denitrification.
Oxidation of ammonia.
Smoothing of water quality fluctuations.

Disadvantages.
Thermal stratification with a eutrophic (nutrient rich) water may result in thehypolimnion (lower layer) becoming anaerobic causing an increase in iron, manganese, ammonia, sulphides, phosphorus and silica. In addition, the epilimnion (upper layer) warms, remains oxygenated and is subjected to sun light which may result in algae blooms. Algae can interfere with coagulation and be difficult to remove by sedimentation. They can cause an increased filtration headloss resulting in an increased backwash frequency. Thermal stratification can be prevented by inducing mixing. Techniques include jet induced circulation, pumping and air injection.

2.5 Pretreatment

Pretreatment may be defined as any process applied to the water before coagulation and the main separation stage. The dosing of chemicals which act in the clarification stage, e.g. chlorine and activated carbon is not normally regarded as pretreatment.

2.5.1 Screens

Floating booms and static screens are normally fitted to the works inlet to prevent large solids and oil from entering the treatment works. Screening requirements are governed largely by the process units included in the plant. Pumps must be protected from large debris which could damage impellers. The relevant size depends on the size of the pump. Certain static mixers and clarifiers have relatively small orifices which can block e.g. Pulsators and flat bottomed clarifiers. Upflow filters (seldom used) in particular require fine screening to protect the nozzles.

An extension of the screening principle is the microscreen, which has apertures down to 15 μm and is sometimes used to remove filter blocking algae.

2.5.2. Aeration

Aeration may be required to:-

a) Raise a low dissolved oxygen concentration to any level specified by the customer/operator or to assist oxidation of pollutants in an initially anoxic water.

b) Strip excess dissolved oxygen and other gases which could cause air blinding in filters.

c) Remove carbon dioxide, raise the pH and thereby reduce the aggressiveness of the water and if present to create conditions for the precipitation of iron and manganese.

d) Remove dissolved gases such as hydrogen sulphide or traces of volatile organics.

e) Improve palatability.

f) Remove radioactive radon from ground water.

The six main types of aerator in general use are:-
cascades,
spray aerators,
packed towers (usually forced draft)
submerged injection,
pressure injection,
surface aerators.

These are discussed in more detail in Chapter 6.

2.5.3 Presettlement

On many river sources overseas, particularly in arid areas where storms can occur and also rivers affected by snow melt turbidities may rise to levels up to 10,000 mg/l suspended solids. Certain types of clarifier are also unable to remove gritty solids without silting up. In such situations a pre-settlement stage may be included. This may sometimes have the option of chemical dosing (e.g. polyelectrolyte alone) but often troublesome solids settle fast enough on their own. The aim of presettlement is to remove those solids which can cause problems in subsequent units or overload them. Jar tests sometimes show that waters with high concentrations of solids respond best with presettlement treatment with a polyelectrolyte at the first stage and conventional coagulation at the normal clarification stage. A presettlement stage would normally be expected to yield a product with less than 100-200 mg/l suspended solids or rather less if dissolved air flotation is contemplated.

2.5.4 Oxidation

Chemical oxidation of raw water can be used to precipitate iron and manganese. At one time prechlorination was common, partly because of its role in inactivating ciliate algae and improving their removal; however the problem of trihalo methanes formed by the reaction between chlorine and organic matter has led to a marked decrease in the practice. Potassium permanganate is also useful, particularly with manganese removal, and avoids trihalomethane formation. In recent years pre-ozonisation has become popular and evidence suggests that it assists coagulation as well as improving the clarified water quality. Otherwise oxidation tends to be deferred to later in the process stream. **Biological Oxidation** using natural bacteria supported on natural river silt or added fine sand can be a useful method of oxidising manganese, ammonia and improving tastes. Hopper tank clarifiers are used and rise rates can range from 10 to 20m/h. An example exists at Mythe, Tewkesbury, also data from large scale pilot trials were published by WRc. The dissolved oxygen in the raw water limits the amount of oxidation and pre-aeration may be necessary. For higher levels the flow may be recycled with intermediate re-aeration.

2.6 Coagulation and Precipitation

This process step, which is used in most conventional treatment processes, involves the addition of chemicals to destabilise colloidal suspensions of suspended solids which will not otherwise settle, float or filter, or to precipitate soluble organic matter such as colour. The usual coagulants are aluminium salts, or ferric salts. In addition there are other proprietary inorganic coagulants, many partially polymerised.

The pH of coagulation is critical if the best quality is to be achieved. The pH becomes even more critical as the dissolved solids, particularly the hardness, falls.

In addition to the above mentioned coagulants there are a number of organic polyelectrolytes which can be used as coagulants, particularly on suspended solids. They tend not to be effective on their own against colour but can offer a reduction in the dose of inorganic coagulants. Such materials are generally more useful in direct filtration rather than with settlement or flotation.

In addition the same polyelectrolytes and others are used to strengthen the floc or precipitate particles to make them settle faster or adhere better to the filter media. In this application they are known as coagulant aids. In this case it is usual to allow a delay between true coagulation and the addition of coagulant aids. It is possible to achieve spectacular improvements in settlement rate with a consequent reduction in size and cost of equipment. Some utilities are reluctant to use these materials on grounds of toxicity but in the UK their use is governed by an approval scheme under the Drinking Water Inspectorate, Committee for Chemicals and Materials of Construction for use with Drinking Water and Swimming Pools, who issue lists of acceptable materials and maximum allowable doses. The above comments are applicable to all of the clarification processes listed.

"Flocculation" is a term which is sometimes loosely applied to the coagulation stage. In fact it is the step where the initial precipitate particles start to agglomerate and become separable. It is discussed under the individual unit processes.

2.7 Clarification (Primary Treatment)

While granular media filters are capable, with correct chemical treatment, of giving the desired quality they have a limited capacity for solids and can be used on their own only with waters with low suspended solids and where the coagulant dose is low. In all other cases a clarification stage will be required. The available processes as discussed below are all truly continuous processes capable of handling fairly high and variable concentrations of solids, while producing an effluent suitable for conventional polishing filtration.

Clarification processes in common use in treatment plants include settlement and flotation systems. The process units exploit the specific qualities of the hydroxide floc. Later chapters in this book explore the principles and design of these units.

Sometimes a separate coarse filtration system may be used but centrifuges and other processes used in the chemical industry tend to be uncompetitive at the scale of operation involved in water treatment.

2.8 Filtration

This forms the heart of conventional water treatment and it is at this stage that the final quality should be achieved with respect to suspended solids and other parameter to be removed by precipitation. However while it is able to provide the desired final water quality the dirt or floc removed is retained within the voids between the grains of sand or other media and as this amounts to only 40% or so of the sand bed the capacity is limited. Thus direct filtration without a clarification stage is only possible if the chemical dose or solids in the raw water are low.

The solids removed on a filter are in a delicate state and as discussed later considerable care must be exercised to prevent them being dislodged.

In some cases two stages of filtration are used, particularly where manganese is present. The first filters remove suspended solids and colour at a low pH and the manganese is precipitated at a higher pH on secondary filters. The high pH would tend to redissolve the colour and aluminium and produce a high residual concentration if the stages were combined.

All aspects of filtration are discussed fully later in this book.

2.9 Activated Carbon

Activated carbon is used in water treatment for the absorption of soluble and volatile organic compounds and to provide biological treatment. Compounds removed include those giving rise to taste, either in their own right or after reaction with chlorine, and compounds such as pesticides, hydrocarbons etc. which may exceed the required limits. Biological action enables soluble organic compounds which may provide a source of food to zooplankton and bacteria in the distribution system to be digested biologically within the treatment plant and thereby to reduce the assimilable organic carbon (AOC). which may lead to fouling in the distribution system. Ammonia is also oxidised efficiently by such biological action but this is limited by the available dissolved oxygen .

2.9.1 Powdered Activated Carbon

Powdered activated carbon may be dosed to any of the clarification stages without reasonable limit and at doses below 5-10mg/l direct on to filters. In this latter case no new process facilities are required, only a means of slurrying and dosing.

Powdered activated carbon (sometimes referred to as PAC, but liable to confusion with poly aluminium chloride) is therefore useful in emergency situations eg following a fuel spillage at the source, and with occasional taste problems. The effective dose is greater than with granular carbon and as it is only dosed when required it does not provide the continuous safety of granular carbon. Dosing tends to be started after a problem has been detected. Powdered carbon is not normally an adequate protection against pesticides, nor will it provide any facility for biological action.

Powdered activated carbon will also tend to neutralise ozone and chlorine and it should not be added to process units simultaneously.

2.9.2 Granular Activated Carbon (GAC)

This is a form of filter media and it is used in filters which may be virtually identical to sand filters. Indeed many conventional sand filters have been converted to GAC, in which case they cover both the functions of filtration and absorption. The particle size is then chosen primarily to suit the filtration role. Several grades of carbon are available and the suppliers can even adapt grades to suit special circumstances.

Soluble organic molecules will diffuse through the floc deposit (which tends to be on the leading edges of grains anyway) and the absorption action proceeds independently.

For dedicated carbon absorption, eg. for pesticides or hydrocarbons separate post absorbers are often employed. These may be designed specifically for absorption duty and are usually much deeper than conventional sand filters (eg. 3.0m or more). The basis of design is the empty bed contact time (EBCT) which is the inverse of the number of bed volumes per hour. It is thus cheapest to use higher rates and deeper beds which would not give satisfactory filtration. For taste removal an EBCT of 6 min. is common, 10 min for ammonia oxidation and for pesticide removal 15-30 min. The actual figure is best determined by pilot column trials on the actual water.

GAC filters will also filter out much of any residual turbidity passing the main filters. Also the inherent biological action will allow slimes and fungal hyphae to grow. Regular back washing is therefore necessary, but not at the same frequency as true filters. (1-2 times per week is usual). Often an air scour facility is provided but it may not be used regularly because the air tends to mix the carbon and destroy the 'absorption front', thereby reducing the efficiency of the process. The carbon is normally expanded by 15-20% by the water during washing. Higher figures often recommended by the suppliers stem from American filtration practice and do not appear to be necessary, particularly if air is available to break up aggregates.

Combined air and water washing may abrade the carbon and lead to serious loss (not to mention the difficulty of retaining the carbon within the bed).

The life of GAC for taste removal is typically about 3 years, this being achieved with the aid of resident bacteria. For pesticide removal the life may only be 3 months and means for removing the carbon are needed.

The carbon is regenerated, usually off-site, thermally in special furnaces by specialist contractors, often other than the suppliers of the virgin material. Normally the carbon from a particular works is returned to that works because of concerns about cross contamination. The regenerated carbon can cause problems (eg sulphides and alkalinity) and it has to be well backwashed when returned to service.

Non regenerable, throw away carbons are also available. These may be regarded by the relevant authorities as hazardous for land disposal if pesticides have been absorbed, although the substances absorbed will be more firmly retained than in their original state.

Granular activated carbon will destroy any chlorine or ozone present in the feed water. It will also remove any permanganate present but not colloidal manganese dioxide from upstream reactions. Ammonia is likely to be oxidised biologically on matured carbon beds, but it is not affected chemically. Chloramine is converted to ammonia by GAC. Disinfection must always follow GAC filters to kill off bacteria shed from the filters.

Solutes adsorbed on GAC are firmly fixed and are not easily dislodged as with suspended solids on sand. On the other hand the GAC itself is friable and damaged if handled incorrectly. The design of activated carbon contactors is discussed in more detail in Chapter 28.

2.10 Specific Processes

2.10.1 Ozonisation

Ozone is an unstable, sparingly soluble gas but also the most powerful oxidising agent available under dilute conditions. It is also a powerful disinfection agent although mostly it is used as an oxidising agent. Its power is such that it tends to be considered as a separate process beyond the scope of normal oxidation. Unfortunately it decays rapidly and does not normally persist in the distribution system. Thus chlorine is normally required for final disinfection before the water goes into supply. Ozone has been introduced extensively to assist in the destruction of pesticides and to improve final water quality.

The pure gas is dangerous and it must be generated in a diluted form on site and dissolved immediately. It can be generated electrolytically and some small units are on the market. Electrolytic generation is expensive on power because 6 Faradays are required per mole. High concentrations are produced and the equipment is relatively simple. Ozone is also produced by the action of short wave ultraviolet energy on air or oxygen. Concentrations and efficiency are low and the process is used mainly for air conditioning.

In water treatment ozone is generated mainly from carefully dried air or oxygen using a high voltage electrical discharge, 4000-10,000 volts.The equipment involves electrodes, a gas path and an insulating dielectric, commonly in glass but increasingly a ceramic. The voltage must be alternating to make electrons fly across the gas path in the presence of an insulating dielectric. Traditional devices used 50 Hz, but modern designs use higher frequencies or even shaped pulses.

Ozone is the most expensive disinfectant/ oxidising agent. The cost includes the electrical power for generation, the drying the feed air, and the cost of dissolving the ozonised air plus destruction of any residual undissolved ozone. Ozone doses are typically 1-3 mg/l. The largest component of the cost is the power used. 25kW per kg O_3 is a guide. With pure or enriched oxygen the power consumption is much less but the overall cost is tempered with the cost of the oxygen itself. Costings are specific to the particular site.

The electrical discharge destroys ozone as well as creating it. There is therefore an optimum concentration. Higher concentrations improve the dissolution efficiency but increase the generation power consumption. Also with pure oxygen a considerable proportion is discharged to the atmosphere from the dissolution chambers.

A typical contact condition for the main ozonisation stage is to allow 4 min. with a residual of 0.4mg/l. Like chlorine, ozone reacts with organic matter present and some of the dose is used to satisfy the "demand" before a residue is left. The latter provides the driving force for disinfection. This demand must be measured. It cannot be estimated from other parameters such as Oxygen Absorption (Permanganate Value).

The most common type of ozone contactor is a deep (5m) tank with baffles to provide serpentine flow in 3-4 compartments, with gas diffusers on the bottom of each of them to introduce the ozonised air or oxygen. Such a device will tend to supersaturate the water and a stripping weir is normally installed down stream to prevent gas blinding of downstream filters. Ozone is a toxic gas, and will volatilise off from any water in which there is a residual. Thus water emerging from a contactor must either be passed through a GAC filter (which destroys ozone) without any free water surface being exposed, or it must be dosed with a reducing agent such as sulphur dioxide or sodium bisulphite to reduce the residual without depleting the dissolved oxygen.

Many favour `pre-ozonation' ie. treatment with ozone before coagulation. In this which uses a dose of ozone less than the demand is injected. This is fully utilised so that there is no residual. The contact period is usually about 2min. This can assist coagulation and clarification. However it is wise to undertake trials to check benefits before investing in expensive equipment. Overdosing can be detrimental.

The ozone is not absorbed with 100% efficiency and the residue in to gas stream must be destroyed by absorption, catalytically or thermally.

Ozone does not oxidise organic matter to carbon dioxide. It opens many aromatic rings and will bleach colour. Sometimes this may return. The TOC does not usually fall. The final mixture tends to be more biodegradable than the original and normally a biological stage is included after ozone treatment, eg. granular carbon or slow sand filters, otherwise the assimilable organic carbon (AOC) will rise and problems with after growths in the distribution system are likely to be encountered.

Ozone will oxidise any complexed iron present and manganese is usually oxidised to permanganate. This is removed on GAC filters.

The recent widespread introduction of ozone in the UK is largely due to the requirement for meeting EC limits on pesticides, where a combination of ozone followed by granular activated carbon is effective.

Detailed information will be available from the specialist suppliers. Because of the considerable interest in the subject the technology is advancing fairly rapidly.

2.10.2 Nitrate Removal

Where ever possible problems with excessive nitrate are dealt with by blending or by transferring to alternative sources. In some cases boreholes drawing water from a greater depth may produce less nitrate.

Biological processes for the removal of nitrate use a carbon source such as ethanol, sugar, acetic acid or less desirably methanol. These can be operated in fairly simple equipment such as upflow filters, coarse downflow filters, fluidised beds etc. An aerobic stage may be needed to remove the excess carbon source. One Danish process has been promoted whereby subsidiary bores have been sunk to inject a carbon source to enable denitrification to take place within the aquifer.

Worthwhile removal of nitrate is obtained in reservoirs with a.storage period of several weeks. However experience has been limited to lowland river abstraction where the BOD in the incoming water is significant.

Biological methods for removing nitrate from ground water inevitably involve bacterial contamination and some particulate contamination which is not present in the raw water. It may be possible to use one of the newer non-coagulation filters as a longstop followed by disinfection but mostly non biological treatments are preferred.

Ion exchange has mainly been used for nitrate removal. The specificity is poor although it is better with so called nitrate specific resins which are less affected by sulphate. The stoichiometric efficiency varies with the source but is about 16%, with regeneration with brine. The process replaces all the anions with chloride unless the resin is contacted with sodium bicarbonate after regeneration. Even so there is an increase in chloride which increases the corrosivity of the water. Ion exchange will tend to remove all of the nitrate and the raw water will be partially treated and blended.

The process produces a waste corresponding to about seven times the dissolved solids load (as sodium nitrate removed) as a 2% spent brine effluent which has to be discharged with dilution or tankered away.

Electrodialysis particularly with semi specific membranes is making advances. Unfortunately it removes most of the species present, producing a lower TDS, and still produces a saline effluent although with a lower solids load. The power consumption is an inverse function of the conductivity of the water. It is possible to control the removal of dissolved salts but it may still be more economical to treat part of the flow and blend. Likewise reverse osmosis is applicable but again it is non specific.

2.10.3 Ammonia Removal

Ammonia is removed most easily by biological means although in so doing nitrate is produced. Small amounts can be oxidised by chlorine (but not by ozone or permanganate), some 8-10 parts of chlorine being required by each part of ammonia. Traces of ammonia, as chloramine are usually left.

Any biological process step will remove ammonia, eg. reservoir storage (except during turnover), slow sand filtration, activated carbon filtration (providing there is an excess of dissolved oxygen). Vertical flow hopper tanks with fine suspended sand are simple high rate devices (20m/h) evaluated by WRc as mentioned earlier.

2.10.4 Fluoride Addition/Removal

Fluoridation where required is achieved by dosing fluorosilicic acid, its sodium salt or sodium fluoride. Particular attention is given in the design of the equipment to avoid over dosing.

Fluoride is difficult to remove from water. It can be removed non specifically by any desalination process. It is removable by bone char and activated alumina but these adsorbents have a limited life. In Britain any problems are resolved by blending. In other parts of the world it may have to be tolerated. For very high levels it can be reduced by absorption on magnesium hydroxide (produced by dosing magnesium sulphate and lime). It is not possible to achieve WHO levels of about 1mg/l by such means.

2.10.5 Phosphate Removal

Phosphate is not normally a problem, indeed it is vital to health. It can however cause treatment problems such as eutrophication of reservoirs. It also interferes with precipitation softening reactions. In one large German scheme a specific treatment plant was built to remove phosphorus entering an impounding reservoir. Such specific plants are common for tertiary treatment of sewage effluents in Scandinavia.

Ferric sulphate is most commonly used, eg. by direct dosing of the inflow in pumped storage reservoirs, and as a coagulant in softening clarifiers.

2.11 Softening

Softening implies the removal of calcium and possibly magnesium from water either to make the water more acceptable to consumers or for industrial purposes, including pretreatment for reverse osmosis.

Normally water above about 250-300mg/l $CaCO_3$ hardness will lead to complaints, especially if the bicarbonate alkalinity is similar to the hardness, in which case rapid scaling of kettles can occur.

2.11.1 Ion Exchange

Domestic softeners are usually based on ion exchange using resin beads which act as fixed anions which start in the sodium form. When a typical hard water is passed through the bed of beds the bivalent magnesium and calcium ions are preferentially absorbed and displace the sodium until the bed is exhausted. Strong sodium chloride brine is able to reverse this absorption.

The process gives very low levels of calcium and magnesium in the product and for partial softening a bypass stream is provided. The run terminates fairly abruptly with a breakthrough of hardness.

The spent by product brine is typically 3-5% mixed calcium (and magnesium if present) plus sodium chloride which may present a disposal problem. The efficiency of industrial softeners is nearly 100%. Domestic units, often using a timer or a water meter, and with a crude co-current contact system may be only 50% or less with respect to brine usage.

An alternative which removes bicarbonate hardness only is dealkalisation. This uses a carboxylic resin which is in the acid form initially and removes calcium and magnesium and replaces them with the hydrogen ion. This releases carbon dioxide which must be stripped off in a packed tower aerator. The overall effect is to lower the dissolved solids content and no sodium is added. The regenerant is sulphuric or hydrochloric acid and the waste stream calcium and magnesium sulphate or chloride, according to the acid used, plus a little free acid. The chemical efficiency is fairly high.

Pressure vessels are normally used and process rates are typically 10-30m/hr. Specialist suppliers should be consulted.

2.11.2 Precipitation Softening

Excess magnesium and calcium may be precipitated by the following reactions:-

$$Ca(HCO_3)_2 + Ca(OH)_2 \rightarrow 2\ CaCO_3 + 2\ H_2O$$
$$Ca\ SO_4 + Na_2CO_3 \rightarrow CaCO_3 + Na_2SO_4$$
$$CaSO_4 + Ca(HCO_3) + 2\ NaOH \rightarrow 2\ CaCO_3 + Na_2SO_4 + 2H_2O$$
$$MgSO_4 + 2\ NaOH \rightarrow Mg(OH)_2 + Na_2SO_4$$
$$\text{also}\quad Ca(OH)_2 + Na_2CO_3 \rightarrow 2\ NaOH + CaCO_3$$

Precipitation softening is discussed in detail in Chapter 14.

2.11.3 Membrane Softening

This has become popular in part of the USA. It is a form of reverse osmosis and uses membranes that reject calcium and magnesium but sodium only poorly. The waste stream is hard water and the amount of waste is limited by the saturation limits of calcium carbonate and possibly calcium sulphate. Sequestrant chemicals such as polyphosphates, phosphonates and polyacrylates, are added to delay the precipitation of these salts and concentrations may rise as much as 2-4 times the published solubilities. The percentage rejection can be fairly high (30%). Continuous chemical dosing may be necessary to reduce the waste by inhibiting precipitation on the membranes. Such plants will also remove nitrate but less efficiently.

The product water needs no further filtration.

2.11.4 Water Conditioning

A number of proprietary devices exist on the market which are claimed to prevent scaling by calcium carbonate. These generally employ magnets or electric fields. At the present time no fundamental explanation exists nor any published basis for the process design. Until such a basis has been established they are not recommended for use other than in cases where empirical trials have shown them to work. The action appears to be one of modifying the crystal form of calcium carbonate. The various forms have different solubilities. Also traces of bivalent ions such as zinc have a profound influence on nucleation of calcium and some devices appear to work more on this basis than by electromagnetic means. They cannot soften the water in the sense of removing calcium on magnesium.

2.12 Iron and Manganese Removal

The conditions for removing iron and manganese conflict with those for normal clarification and separate treatment is often necessary.

2.12.1 Borehole Sources

Iron and manganese may occur in borehole sources, in which case organic impurities are likely be absent. Often the water is aerated to raise the pH by stripping carbon dioxide, alkali added if necessary and the water passed through a catalytic filter. Normal practice in the UK is to use manganese dioxide (polarite) as a catalyst additive to ensure a good performance from the start, although the entire bed soon becomes catalytic.

American practice tends to use a manganese zeolite which is dosed with permanganate either continuously or after backwashing or exhaustion.

To achieve low residual levels of manganese some free chlorine may be necessary even with polarite.

pH is important and this will need to be at least 7.0 or as high as the pHs permits. The pH must not exceed pHs or the media will become coated with calcium carbonate and the catalytic action lost. Filtration rates follow conventional practice but can be higher at low input levels.

Iron and manganese can also be removed in biological filters at some very high rates (eg 20m/h). A number of contractors now offer their own proprietary versions. The process depends on the precise composition of the raw water. The necessary nutrients must be present and pilot trials are recommended.

2.12.2 Surface Waters

Often the manganese level in surface water is very variable. This is the result of anoxic layers in the reservoir being disturbed by wind or inversions. It is possible with low alkalinity water to raise the pH with lime or caustic sada, and to add permanganate to oxidise the manganese, then after a short contact period to lower the pH with coagulant and clarify normally.

More often the alkalinity is too high for this and the water is coagulated conventionally at pH5-6, to remove organic matter including colour on a settlement or flotation stage. The pH is then raised before filtration to about 6.7 to reduce the residual aluminium, and then raised again to pHs at which point the water is chlorinated. The manganese is then removed on secondary filters at a high rate (up to 18m/h), on coarse media.

It is possible to avoid the first stage if the coagulation dose is low.

Ozone will oxidise manganese to permanganate and it must then be reduced again. Activated carbon with do this, but it may interfere with the action of the carbon if attempts are made to regenerate it and use it again.

There are other variations. Sometimes permanganate is used between clarification and filtration if the pH can be adjusted to a point at which aluminium does not go into solution. If colour is not excessive it is often possible to coagulate with an iron salt at a low pH and to raise the pH before filtration, adding chlorine or permanganate to remove the manganese on the filters without significant colour going back into solution. An efficient clarification stage is necessary. Jar testing is essential.

2.13 Desalination

As a general principle precipitation processes are preferred as a means of removing substances from water. The resultant solids present less of a disposal problem inland than liquid wastes. However sodium, chloride and sulphate cannot be removed by such means economically and one of the following processes must be employed if the total dissolved solids have to be reduced.

2.13.1 Evaporation

Evaporation is perhaps the easiest to understand and the economics are relatively independent of the solids concentration. Thus it has in the past been used mainly for seawater and other waters with dissolved solids concentrations above 1-2% (Sea water is 3.5%, the Gulf is 4.5%). Pretreatment is required to prevent scale formation. However in recent years evaporation has encountered competition from reverse osmosis. The latter tends to be more competitive in larger installations but even here the two are still competitors. Both need to be evaluated on specific sites. The conversion is limited by the precipitation potential of the residual salts. Specialist contractors should be consulted.

2.13.2 Reverse Osmosis

This has become the work horse of desalination for salinities above about 1500mg/l up to 4.5%. For this duty hollow fibre and spiral wound membranes are used and in both cases the feed water must be cleaned to levels better than potable quality to achieve low "fouling indexes" otherwise membranes will block prematurely. Such pretreatment involves many conventional water treatment processes such as iron removal, precipitation softening and silica removal, and multi media filtration with precise coagulant control.

The conversion efficiency is again limited by the solubility of the residual salts. Calcium carbonate is removed by softening and pH adjustment. Calcium sulphate is commonly limiting, although its solubility is increased by various sequestrant chemicals which delay precipitation. Even barium can be limiting. Silica must also be watched. For brackish water (~2000mg/l) 75-85% conversion is often possible, but on seawater only 30% may be achieved because the osmotic pressure, which acts as a back pressureand greatly reduces the effective driving pressure. Thus a pretreatment plant may have a throughput of 3 times the product flow.

Operating pressures range from 10bar to 60 bar. Power is the main operating cost. The design of the process involves an assessment of the water chemistry in association with published data on membrane fluxes, with an allowance for membrane compression or loss of permeability over the life of the membrane. Membrane life depends on good pretreatment but 3-5 years is now common.

Normally enquiries will be put to specialist contractors representing the membrane manufacturers.

2.13.3 Electrodialysis

This relies on the passage of salts through membranes under the influence of an electric field. As a result the dissolved salts are removed from the water. The latter is not filtered as in reverse osmosis. Monovalent ions are of course more mobile than bivalent ions and these are removed preferentially. The process uses electricity in proportion to the quantity of salts removed (following Faraday's Law), but the voltage depends on the conductivity of the electrolyte as well as the geometry. Thus it is most economical at lower salinities, and where high purity of product is not demanded. An upper limit will usually be around 1500-2000mg/l. salinity. There is only a limited number of suppliers.

2.13.4 Ion Exchange

This is an effective means of achieving low final salinities but the running cost is that of the equivalent quantity of acid and alkali to regenerate the resin. Unless these are available as a by product from another processes ion exchange will not normally be the first choice, indeed reverse osmosis or electrodialysis are often used as pretreatments in front of ion exchange.

The feed water must have low level of organic matter to avoid fouling of resin but otherwise the equipment is relatively cheap and robust. Again specialist suppliers will usually be consulted unless there is an ongoing demand to warrant the accumulation of expertise.

2.13.5 Post treatment

The above processes tend to produce a water with very low dissolved solids which are corrosive, non palatable and too low in calcium for potable use. In many cases it is possible to blend the product with part of the feed, but if this is sea water there will be too much sodium and insufficient calcium.

Recarbonation is used extensively to raise the calcium bicarbonate level to at least 50mg/l (as $CaCO_3$). On distillation plants waste carbon dioxide from the acidification and degassing stages is redissolved in the product and neutralised with lime slurry or by passing through limestone filters. The choice depends on the local geology. Limestone filters are much more expensive but the running cost will be very low if limestone is available locally as is often the case in many arid areas. The filters need regular washing to remove impurities and undersize residues, and to collapse the wormholes that develop as the media dissolves. Such filters are adaptations of standard potable filtration technology.

Disinfection after desalination is more a matter of prevention than of removing bacteria. Any bacteria in the source will have been removed at the pretreatment stage. Bacterial contamination of membranes is a significant hazard, and bisulphite is used to inhibit growth. The normal oxidising disinfection agents attack some membrane materials.

2.14 Specific Removal Processes

To summarise the above the following processes are available to treat the stated determinands:-

Suspended Solids	Coagulation, clarification and filtration.
Colour	as above at the correct pH.
Excessive alkalinity	Acidification and removal of carbon dioxide softening.
Carbon dioxide	Aeration (stripping).
Calcium, Magnesium	Softening, ion exchange, precipitation, membranes.
Iron	Precipitation, clarification and filtration. Use of pH control at coagulation.
Aluminium	Filtration at the correct pH.
Manganese	Discussed in the text above.
Heavy Metals	Traces of most of these are absorbed on ferric floc and are removed by normal coagulation at the correct pH.
Phosphate	Coagulation with ferric or use of ferrous salts.
Ammonia	Biological treatment. Low levels-superchlorination.
Nitrate	see text above.
Nitrite	Chlorination.
Sodium, potassium	Desalination (which removes all dissolved salts).
Sulphate, chloride	as above.
Sulphide	Aeration, biological treatment, oxidation.
Silica	Precipitation softening involving magnesium precipitation, membrane filtration.
Fluoride	Granular activated alumina. Precipitation on magnesium hydroxide.(There is competition from bicarbonate), blending.
Boron	Difficult to remove, desalination, blending.
Detergents	Activated carbon.
Hydrocarbons, PAH	Activated carbon.
Pesticides	Activated carbon preceded by ozone where economical.
Taste and odour	Activated carbon. Sometimes ozone or chlorine dioxide.

2.15 Disinfection

Although the earlier stages of treatment will remove a large percentage of bacteria present in the raw water the overall efficiency necessary to meet bacterial standards is far greater than for solids removal on its own. Thus a separate disinfection stage is necessary. Indeed the conditions at this point usually have a large margin of safety so that the removal of bacteria earlier is not in fact taken into account. On the other hand bacteria are shielded from the disinfectant by suspended solids and the water should have reached the final quality in this respect before disinfection (the WHO require <1NTU).

As a first approximation all disinfection processes have first order decay kinetics, where:-

$$N/N_o = \exp(-kct)$$

N/N_o = ratio of final bacterial concentration to initial concentration.
c= concentration of disinfectant (or light flux in the case of UV)
k= reaction constant which is specific to the organism and the disinfectant.
t= time.
For pasteurisation c is not relevant.
Thus doubling the time will give a squared residual efficiency (90% will become 99%).

A typical river source may have up to 10^4 E. Coli/100ml. For 99% of samples <1 an effective concentration of 0.01/100ml is required, thus a 10^6 fold reduction is necessary. Very polluted water may contain 10^6/100ml.

At this level the design of the contact tank is very important as any bypassing will produce an unacceptable water.

2.15.1 Chlorination

The WHO guidelines require a 30 min. contact with 0.5mg/l of free chlorine at a pH < 8.5 and with a turbidity < 1 NTU. They make no reference to the performance of the contact tank and from published data on the kinetics it appears that this includes a pragmatic margin of safety.

The hypochlorite ion is much less effective than un-ionised hypochlorous acid and as a matter of good practice disinfection by chlorination should occur at as low a pH as possible. pH correction where used should follow disinfection.

Chlorine reacts with ammonia to form chloramine then dichloramine. Dichloramine will tend to react with monochloramine to produce nitrogen and hydrochloric acid (the break point reaction). Thus chlorination will oxidise ammonia. However about 8-10 parts of chlorine are required to remove 1 part of ammonia. Thus chlorine tends to be uneconomical as a disinfectant when appreciable amounts of ammonia are present.

Chloramine itself is a weak disinfectant, and generally regarded as too weak for primary disinfection. Thus if ammonia is present the water may present a high 'chlorine demand' which must be satisfied before sufficient residual free chlorine is left for rapid disinfection.

Chlorine is added either as the dissolved gas, or as hypochlorite solution. The latter may be delivered as a solution of the sodium salt, generated electrolytically on site from brine, or overseas as calcium hypochlorite powder (bleaching powder).

Free chorine reacts with many natural organic compounds to produce undesirable trihalomethanes and normal practice is now to delay contact with chlorine until the level of residual organic matter has been reduced as far as possible. However in some cases chlorination is useful in improving the removal of certain algae, and the risk must be balanced.

2.15.2 Chlorine Dioxide

Chlorine dioxide is an explosive water soluble gas which is always produced on site and dissolved immediately in water. It is generated from sodium chlorite plus an acid. Chlorine may be used as the acid but some chlorine is then mixed with the chlorine dioxide. Several proprietary generator units are available on the market.

Sodium chlorite is a very strong oxidising agent and shipment other than by road or rail may be difficult. Thus chlorine dioxide is an option only in UK and other industrialised countries, where producers are located.

The advantages of chlorine dioxide are :-
1. It does not produce THM's
2. More particularly it destroys many tastes and odours that are resistant to chlorine or which are intensified by chlorine such as phenol (TCP odour).
3. In contrast to chlorine it is not affected by pH and is therefore more effective at high pH. The active agent is chlorine dioxide itself, which is non-ionisable.
4. It does not react with ammonia.

Disadvantages are:-
1. The extra cost
2. In some situations chlorine dioxide may react with other organic vapours to produce new odours, for example chlorine dioxide can vaporise from tap water and produce a mousy odour with volatile organics from household furnishings such as new carpets.

3. Chlorine dioxide is yellow and in swimming pools will give a yellow tinge to the water.

4. It slowly reverts under alkaline conditions to chlorite which is toxic like nitrite. Re-chlorination will convert the chlorite back to chlorine dioxide.

Chlorine dioxide may be regarded as another option in difficult cases. Bench top testing is usually necessary.

2.15.3 Ozone

This is used mainly as an oxidising agent and has been discussed earlier

For disinfection the concentration and contact time (CT value) is less than for other disinfectants, and it will kill many viruses and other stubborn organisms which resist chlorine. (It is not fully affective against cryptosporidium under normal conditions). A typical reference condition for disinfection is 0.4mg/l for 4 min.

As already mentioned ozone is not stable and in most cases any residues are removed within the treatment plant. Thus if a residual disinfection action within the distribution system is required then either chlorine (including chloramine) or chlorine dioxide must be added at the end of the treatment train.

The advantages of ozone are primarily its effectiveness against a very wide range of organisms, and the visual quality of the product water.

The disadvantages are the cost and complexity of generation, the fact that is leaves no residual disinfecting action, and the tendency for after growths to occur if a downstream biological stage is not present.

Disinfection by products are present as with chlorine. These tend to be regarded as less hazardous than those from chlorine but less is known about them.

2.15.4 Ultra violet light

This is a simple process of passing water though a tube or channel in which ultraviolet lamps are installed. Ultraviolet light at a wavelength of 254nm inactivates bacteria. Residence tines are only a few seconds and an energy dose of 20-40 mw-s is normally applied.

To be effective the water must be fairly clean and colourless, in particular it must have a low UV absorption coefficient. Suppliers of equipment usually check this when quoting.

The cost of an installation tends to increase linearly with size because the lamps are small and the number increases linearly with scale.

Like ozone there is no residual disinfection action once the water has left the unit. Thus any distribution system will need careful monitoring and possible routine flushing with hypochlorite.

UV systems tend to be used on small supplies and colourless borehole water, as well as for treating process water for the drinks industries etc.. However because there is no disinfectant demand as with the other options it can be competitive when the oxidant demand is otherwise high, such as with final sewage effluents. There are no disinfection byproducts.

It has the advantage of requiring no attention or chemical supplies apart from an annual lamp change. The proprietary units are fitted with a UV photometer to give an alarm signal when the flux falls below the safe level.

2.15.5 Pasteurisation (Thermal)

This is used widely in the processing of foodstuffs such as milk. It would be applicable to water if the energy costs associated with it were not excessive compared the other options. However it is widely advocated in emergency situations when consumers are advised to boil water before drinking. It has advantages over chlorination and ultraviolet light in that the heat penetrates all organisms present regardless of turbidity. Temperature of about 70°C for a few minutes only are needed. The actual time varies inversely with temperature and the organism to be killed.

Heat exchangers can be designed to operate with differential as low as 2-5°C in which case pasteurisation can be economical for applications such as wash water and sludge treatment if re-cycling of cryptosporidium is intended. Direct contact water heating, either by cascade towers or submerged combustion, is possible (cf. sewage sludge pasteurisation). It is of course used almost universally used in milk treatment. Pasteurisation is more effective than other treatments for killing cryptosporidium.

It is essentially suitable for emergency treatment and for water with suspended solids too high for UV treatment. Nevertheless it should not be overlooked as it forms the backstop for public health when other measures fail.

2.16   pH Correction and Stabilisation

Water ready for distribution must be free from suspended solids, colour , bacteria and the other determinands laid down in the Water Regulations. It must also be non corrosive to common materials of construction. For practical purposes this usually implies that the pH must be close to equilibrium with respect to calcium carbonate. Treatment by conventional means will usually make the water more acid and aggressive. Thus an alkali is often required to raise the final pH (in addition to the alkali required earlier in the treatment process to achieve an optimum for coagulation and the removal of residual metals etc.)

A water with a low bicarbonate alkalinity will have a low buffering index and will tend to be unstable in contact with that anything which effectively adds acid or alkali will cause a large change in pH. Thus for low alkalinity water it is preferable to use sodium carbonate for pH correction. To make the water less corrosive and at the same time improve the ease of treatment (and ease of pH control) and also to bring the hardness into line with preferred levels for good health, carbon dioxide is sometimes added early in the process so that extra lime can be added later on to raise the bicarbonate hardness to at least 50 mg/l.

Lime is preferred as an alkali where a reasonable alkalinity already exists. It is cheapest as a commodity. However on smaller plants the simplicity of liquid caustic soda makes the latter a common choice. Sodium carbonate may be considered if this effectively reduces the number of chemicals to be brought onto the plant. The recent introduction of a lime cream in Britain has given designers a wider choice.

pH correction for stabilisation should preferably follow disinfection so that the pH at the disinfection stage is as low as possible. With alkalinities above about 150mg/l it is less likely that the pH will be sufficiently below the equilibrium value to require additional correction.

In some situations, where post precipitation of manganese or iron, or calcium carbonate can occur, or where old pipes are corroding further corrosion or deposition can be inhibited by the addition of sodium hexametaphosphate at a level of about 1-2 mg/l. This is added after filtration as it can interfere with the efficiency.

Phosphates are also added to reduce plumbosolvency, particularly in softer waters. Any convenient form may be used, but liquid orthophosphoric acid is generally the cheapest and most convenient form. A target of 1mg/l is usual. This is best added after filtration, and preferably before final pH correction. It may be added with the main disinfectant dose. Disodium phosphate is an alternative that does not affect the pH.

2.17 Laboratory Testing

An approximate idea of the best chemical treatment is obtained by a bench top procedure known as the jar test. It involves the addition of the appropriate chemicals to a sample of water, with rapid mixing, followed by slow stirring or flocculation for about 20 min. The sample is then allowed to settle and the turbidity of the supernatant water measured together with the concentration of residual coagulant ions and, if relevant, colour. (Full details are given elsewhere, see Stevenson, 1986).

Usually the supernatant from the jar test is decanted off and filtered. The residual concentration of coagulant can then be measured. A series of such tests is run with various doses to enable the minimum turbidity and residual coagulant to be established. It may be found that manganese is not removed at the coagulation stage. Separate tests can be performed on the filtrate to find a treatment for such residues. For example the pH can be raised, a known dose of chlorine water added, and the sample refiltered.

Settlement rates are determined by timing the fall of the larger floc particles over a fixed distance. This provides a guide which, with experience, can be used to determine the surface rating of settlement tanks. Floc size has been a popular parameter which is estimated using comparator chart, but this is an unreliable indicator because it ignores the density of the floc, and smaller floc may often settle faster than large floc if loaded with clay.

It is possible to assess the best order of addition of chemicals, eg. alkali before coagulant or vice versa. It is necessary in this case to do a dummy run first to establish the alkali dose for a given coagulant dose to achieve a given pH, and then to add the necessary doses in reverse order to a second aliquot of the raw water.

Polyelectrolytes can also be selected by similar tests. It will usually be found that the best performance is achieved if the coagulant is added and allowed to flocculate for a period of several seconds up to a few minutes before the diluted polyelectrolyte is mixed in. Conditions in the jar test cannot easily represent those of a floc blanket or solids contact clarifier, which operates with a high floc concentration. It is usually found that doses needed in jar tests are higher than those found on the plant, but promising materials can be short listed in this way.

The alkali dose for pH correction is readily determined on the supernatant following the jar test. Also after filtration it is normal to determine the chlorine demand by adding a given quantity of chlorine water or hypochlorite and back titrating after a short contact time, ideally 30 min.

Laboratory tests aimed at direct filtration must be modified as settlement data is irrelevant. The flocculation time is reduced eg. 10 min, to simulate the delay time in the supernatant water in the filter. The sample is then filtered through a No.1 Whatman paper and analyzed for turbidity, colour and residual coagulant. Some prefer to use a small sand filter column 25mm diameter by 150mm deep.

Laboratory tests of this type are worth any amount of data on the raw water. They can identify problems at an early stage, are quick to carry out and are suitable in many cases for on-site work.

This is normally the first step in confirming any treatment scheme. The choice of process depends to a fair extent on the dose of coagulant required, and jar tests should be included in any regular sampling programme in aid of a new scheme. If pilot trials are anticipated laboratory jar tests are still worthwhile and may help to identify problems to be investigated on the pilot plant.

2.18 References

Anon. (1980) Council Directive relating to the Quality of Water for Human Consumption. (80/778/EC), H.M.S.O. London.

Anon. (1993) Guidelines for Drinking Water Quality, 2nd Edn. Vol.1. Recommendations. WHO, Geneva. (H.M.Stationery Office, London)

Stevenson, D. G. (1986) Solid/Liquid Separation Equipment Scale Up, Ed.Purchas,D. and Wakeman, R. J., Uplands Press and Filtration Associates.

Part B

Primary Treatments

CHAPTER 3.

THE BEHAVIOUR OF PARTICLES

3.1 Introduction

The operation of all the non-filtration solid/liquid separation processes in water treatment is governed by the settlement or buoyant rise under gravity of particles in water, either individually or collectively. Their behaviour as a mass relates back to the characteristic of the individuals. Flotation is not an exception because the micro bubbles behave as buoyant particles. Adsorbed films impart a rigidity that makes them behave as a solid.

In water treatment the primary particles may be grit, silt or colloidal clays, as well as organic matter ranging from algae to colour colloids. These names are employed non specifically. Grit includes all particles down to about 100-200µm, ie. including fine sands. Silts cover the faster settling particles that are too fine to be identified by the naked eye as individual particles, and go down to, say, 20µm. Clays cover finer materials that do not settle rapidly, ie <3-5m/h. Particle sizes may go down to below 1µm.

In addition there are low density solids such as algae, whose density approaches that of water. They contain much water and may be compared on the macro scale with cucumber and soft fruit. The dry solids content may be low compared with the volume occupied. Their size ranges from 5µm to 100µm or more. Cladophera, for example, forms strings up to 1m or more in length.

Pathogenic organisms such as bacteria, giardia lamblia and cryptosporidium oocysts will have densities similar to water and on their own are unlikely to settle significantly.

A range of techniques is available to the water scientist to modify the particles present in the raw water so that they can separate at a rate which can be exploited. As will be shown below the settlement rate varies with the square of the particle diameter so that even modest agglomerates produced by flocculation give a much higher rate of separation. Such agglomerates are not necessarily permanent but dynamic units which fall apart and reassemble under the resultant shear if settlement is too fast. The agglomerates capture smaller particles as they fall and the process is repeated. Flocculation produces a mottled appearance in the suspension somewhat like cirrocumulus or altocumulus cloud.

Apart from small air bubbles the particles encountered will tend to be irregular and non symmetrical. This will cause them to follow a curved path initially, the curve finally ceasing when the particle has turned into the position of slowest settlement, presenting the largest horizontally projected area. Readers will be familiar with the orientation of a feather when falling. The concave surface is uppermost. An assumption of spherical particles will tend to over estimate settlement rates. A number of definitions of sphericity exist. A simple one commonly used is the ratio of the minimum dimension to the maximum one, which can be measured by the microscope. However for use in hydraulic calculations a conversion of surface area to that of the sphere of the same size as measured by the same procedure (eg. sieve analysis or projected diameter) will be required.

3.2 Settlement

In 1851 Stokes showed that the total drag on a spherical particle streamline flow is defined by:-

$$F = 3\pi\mu du$$

If the particle is falling under its own weight:-

$$F = \frac{\pi}{6} . d^3 (\rho_S - \rho_L) g$$

Hence:-

$$u = d^2 (\rho_S - \rho_L) \frac{g}{18\mu}$$

This has been shown to be valid for particle Reynolds Numbers (Re') up to 0.2 where Re' = $ud\rho/\mu$. Thus for example silica particles with a specific gravity of 2.65 in water at 10°C will reach this limiting Reynolds No. if the diameter is 75μm. The settlement rate will be 16m/h. Smaller particles will follow Stokes law accurately enough for practical purposes. Floc particles produced by coagulation with iron or aluminium salts (discussed later) may settle at up to 10m/h and have a diameter of up to 5mm. These exceed the above limit for Reynolds Number. For higher Reynolds Numbers a drag force per unit area is introduced.

$$R' = \frac{4F}{\pi d^2}$$

Thus:-

$$\frac{R'}{\rho u^2} = \frac{4F}{\pi d^2 \rho u^2}$$

Stokes Law may then be rewritten:-

$$\frac{R'}{\rho u^2} = 12 \frac{\mu}{u d \rho} = 12 Re^{\prime -1}$$

At Reynolds Numbers above 0.2 the situation is complex but Kahn and Richardson (1987) have produced the following equation following an extensive examination of the data.

$$\frac{R'}{\rho U^2} = [1.84 Re^{\prime -0.31} + 0.293 Re^{\prime 0.06}]^{3.45}$$

This provides a convenient basis for calculating settlement rates for particles up to Re = 10^5 which may be used in computer programmes. Plots of settlement rates for silica particles and anthracite with a specific gravity of 1.45 are given in Fig. 3.1.

Above Re = 10^5 the boundary layer separates at an earlier point and the resistance falls considerably:-

$$\frac{R'}{\rho u^2} = 0.05$$

A very full discussion of this subject will be found in Coulson et al. (1990).

(To calculate the settlement rate for a particle, using the above equation, an iterative programme must be used to solve the equation as it has velocity on both sides.)

Naturally occurring particles are very variable and effort on characterising them in order to calculate settlement rates accurately would be misplaced. Direct measurement will eliminate errors. However the above equations have their uses, one being in the design of traps for retaining filter media, which is fairly well defined. In this case the lower size limit will be chosen, because under size material is not required. With air bubbles in clean water and with oil drops, particularly with larger sizes, internal recirculation can occur. This will reduce the viscous drag. Hadamard (1911) has produced the following equation for a factor to be multiplied with the free fall velocity as calculated from the Stokes Equation.

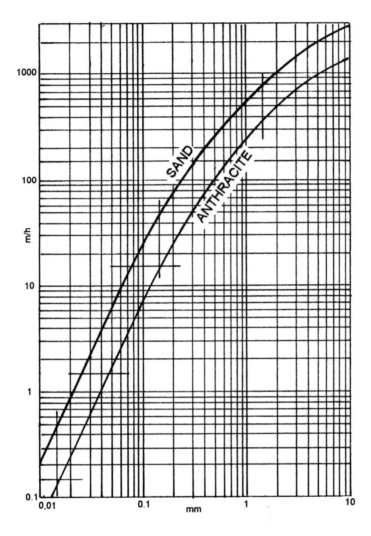

Fig. 3.1 Settlement Rates for Sand and Anthracite in Water at 10°C based on the equation given above.

$$Q = \frac{3\mu + 3\mu_1}{2\mu + 3\mu_1}$$

μ is the viscosity of the continuous phase and μ_1 that of the fluid of the drop or bubble. Thus for small air bubbles the rise rate may be 50% higher than that predicted directly from Stokes Equation. However any adsorption of surface films will tend to produce a rigid surface and encapsulate the fluid. Certainly in dissolved air flotation there will tend to be a shell of floc surrounding the air and the bubble will behave as if it were a solid body.

The above equation only applies where flow is laminar, ie. below Re=0.2. Bubbles large enough to move into the turbulent region are flattened by the impact of the water and eventually break up so that the rise rate never exceeds about 350mm/s. Surfactants which lower the surface tension cause bubbles to break more easily into smaller ones and the rise rate is less.

In the above discussion it is assumed that the fluid remains stationary. Settlement velocities are relative to the fluid. In a large tank even a dilute suspension will have a density different to that of water, and this will cause the fluid to move. For example a suspension of clay at 100mg/l whose particles have a specific gravity of 2.5, (ie. a differential of 1.5) entering a tank of otherwise clear water will produce a fluid with a density difference of $60g/m^3$. This corresponds to a force of 0.59N on a tonne of water. The acceleration will be 5.9 X 10-4 m/s^2, and the velocity after a fall of 3m will be 12.7m/h. Although this may appear slow, it can have marked effect on the performance of a tank operating at only 1-2m/h, and in a poorly designed tank the resultant plunging effect can cause recirculation.

3.3 Hindered Settlement

The above applies only to dilute solutions. As soon as the concentration of the suspension exceeds about 10% by volume the liquid displaced by the suspension starts to interact as it rises past the down going solids. In Chapter 16 the Kozeny equation for flow in granular beds is discussed in some detail. This is equally applicable to suspensions, although the constant changes as the path becomes less tortuous.Restating the Kozeny Equation:-

$$\frac{\Delta p}{h} = \frac{K(1-e)^2}{e^3} . S^2 \mu u$$

For suspensions it is more convenient to use the volume concentration C instead of voidage, although allowance should strictly be made for the residual voidage when the suspension is fully settled. Thus (1-e) =C.

In a fluidised bed or a suspension, the pressure difference caused by the density of the suspension in a depth of h is:-

$$\frac{\Delta p}{h} = (\rho_S - \rho_L) \, g \, . \, C$$

For spherical particles S=6/d.

Substituting and combining the above equations,

$$u = \frac{(\rho_S - \rho_L) \, g d^2}{36 K \mu} \, . \, \frac{(1-C)^3}{C}$$

Part of this equation is identical to the Stokes Equation. Thus:-

$$\frac{u}{u_o} = \frac{(1-C)^3}{2KC}$$

Stevenson (1974) found that floc suspensions followed the relationship:-

$$\frac{u}{u_o} = \exp(-10C)$$

This is easier to use and gives a similar curve to the former.
Richardson and Zaki((1954) found that yet another expression could be used:-

$$\frac{u}{u_o} = e^n = (1-C)^n$$

Richardson and Kahn (1989) have found that n is defined by the equation:-

$$\frac{4.8-n}{n-2.4} = 0.043 Ga^{0.57} \, [-2.4 \, (\frac{D}{D_t})^{0.27}]$$

48

Where Ga, the Galileo Number is defined as:-

$$Ga = d^3 \rho_L (\rho_S - \rho_L) \frac{g}{\mu^2}$$

For sand grains n=2.6. For 3mm floc particles settling at 3m/h, the density difference is 0.67kg/m³. Ga=100 and n=4.5. For practical filter sands with a size range of 2:1 it appears from sand expansion measurements that n is nearer 7.5.

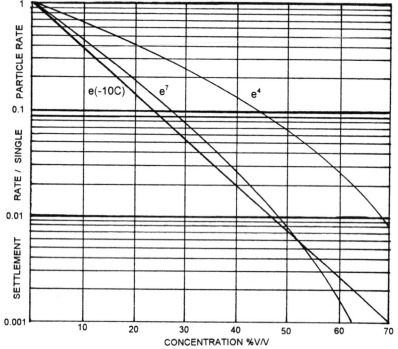

Fig. 3.2 The Ratio of Hindered Settlement Rates to Single Particle Settlement Rates versus Concentration by Volume. Stevenson and Richardson/Kahn Plots.

Thus it seems that Richardson and Zaki's work relates to uniform particles. More work is required to clarify the effect of size range. Stevenson's equation is based on measurements on aluminium floc from a floc blanket clarifier. Exp(-10C) gives very similar results to $(1-C)^7$, as shown in Fig.3.2. This has been confirmed by Warden (1982) in his work on thickening of waterworks sludges.

3.4 Compression and Thickening

As the concentration increases settlement slows down. Kynch (1952) has shown that for any finite element in the bed the rate of settlement is still governed by the permeability of that element and the weight of the material in that element.(In contrast, some other sludges, including sewage sludges, have a structure which requires the dead weight of the sludge above to assist compression.

When a mixed dilute suspension is allowed to settle the particles at the top of a column settle at their own characteristic rate with the larger and denser ones falling fastest. Lower down the particles encounter the bottom of the column and the concentration must therefore increase. This causes a transition to slower hindered settlement. There is a sharp interface that rises up from the bottom. Zones of increasing concentration rise up from the bottom at specific rates. There is a continuous progression with materials such as water works sludge to higher concentrations. On the other hand close cut sands have a relatively high internal voidage and settlement ceases when the incompressible particles have settled to a voidage limit which still leaves a low resistance path for fluid flow.

Kynch's theory provides method of predicting the underflow concentration from a continuous settlement thickener where the underflow is being withdrawn at a constant (average) rate. (Short cycle intermittent withdrawals are equivalent to continuous withdrawal providing that the volume withdrawn is small compared with the sludge inventory within the tank.) Firstly the solids flux resulting from settlement is plotted against the concentration (Fig.3.3). The solids flux is the product of concentration and settlement rate. The latter may be calculated from any of the appropriate equations discussed earlier, or it may be measured directly in the laboratory. Secondly the solids flux produced by withdrawal (ie. the product of the proposed withdrawal rate and concentration) is plotted. These two fluxes are added to produce a third curve which has a maximum and a minimum. The minimum constitutes a limiting flux at the underflow rate chosen. The under flow concentration therefore is this limiting flux divided by the underflow rate used in plotting the second curve. Using a simple programme which calculates the fluxes in small steps the minimum value (having passed the maximum) can be calculated for any underflow rate and hence the underflow concentration for that rate.

3.5 Fluidised Beds

The theoretical basis of fluidised bed is similar to that for settlement. Settlement is a transient state whereas fluidisation is steady state. In a fluidised bed the settlement is

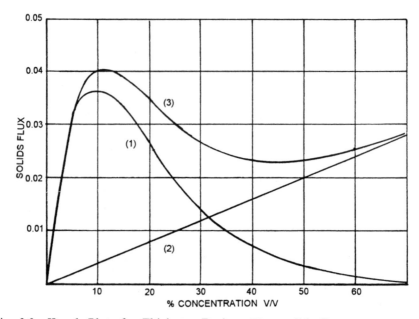

Fig. 3.3 Kynch Plots for Thickener Design. (1) = solids flux vs. concentration m3/m2/h X settlement rate of single particle. (2) Solids flux caused by an arbitrary example underflow . (3) Sum of (1) and (2).

counteracted by the upflow of fluid, in this case water. For a given upflow velocity the solids concentration adjusts itself to match the rise rate of the fluid in the vessel. If the rise rate is too high the particles may be flushed out at the top. If the flow changes then the height of the expanded bed changes in a predictable manner. If solids are added to the bed the interface rises. A settling suspension however varies in concentration with depth whereas a fluidised bed is substantially uniform. One might expect that with a mixed bed that faster settling particles would fall to the bottom and form a sperate layer. This is to some extent true but Al Dibouni and Garside (1979) have found that a difference in fluidisation threshold of at least 2 is required for significant stratification. There is considerable circulation within fluidised bed, which has been discussed already in some detail by Coulson and Richardson (1990) and also Carlos and Richardson (1968). It is perhaps significant that the Carman and Kozeny Equations show that a heterogeneous bed is more permeable than a homogeneous one because the effect of voidage on resistance is extremely non-linear. The implications of the foregoing for the design of treatment units such as clarifiers and thickeners are discussed in the following chapters.

3.6 Flocculation

In Chapter 2 the chemical treatment of water has been discussed. Coagulants are added to precipitate soluble contaminants, to destroy the negative charges and protective layers that stabilise suspensions. Once this has been done the particles start to agglomerate into larger ones which can separate more readily. This agglomeration process is known in the water industry as flocculation.

3.6.1 Precipitation

This is the first stage in the process and one that tends to be overlooked. For good coagulation the chemical added must be mixed fully and rapidly with the main flow. In a diffusion controlled process the rate of precipitation is governed by the following equation:-

$$\frac{dM}{dt} = K_L S (C_{ss} - C_s)$$

dM/dt is the rate of precipitation (moles/s/m^3) in the solution, K_L the mass transfer coefficient, S the available surface area, and Cs and Css are the saturation and supersaturation concentrations. Mass transfer to small particles in a diffusion controlled situation is characterised by the Sherwood Number Sh. where:-

$$Sh = \frac{K_L d}{K_D}$$

For small particles Sh=2.0 (Coulson et al.(1990))
The specific surface area of a suspension of particles of concentration C by volume is 6C/d thus the first equation becomes:-

$$\frac{dM}{dt} = 12 \frac{C K_D}{d^2} \cdot (C_{ss} - C_s)$$

K_D being the diffusion coefficient of the rate limiting species (probably Al^{3+}).

Little data on these parameters is available . However the solubility of aluminium hydroxide at pH 6.5 is 1.1 X 10^{-7} (Anon 1971). The diffusion coefficient of aluminium would appear from comparison with similar ions to be about 0.8X10^{-9} moles/s/m.

Some work on calcium carbonate has shown that supersaturation at nucleation is about 50 times the normal solubility (Al Malawi,1979). No information is available on the figure for aluminium hydroxide, but for the present purpose a similar ratio would not seem unreasonable. Thus a dose of 25mg/l of aluminium sulphate which would hydrolyse to 5mg/l of aluminum hydroxide in a reaction time of say 1 sec in a flash mixer would lead, on this basis, to a particle size of 0.18μm. Certainly the initial crystallites are known to be submicron.

Precipitation will only occur if the solution is supersaturated. For good flocculation particles with a high specific area are required and therefore for a given total quantity of coagulant chemical the mixing must take place in a very small volume at a high specific rate. It also helps if the conditions produce a low solubility so that this may be counteracted by the high surface area.

Stevenson (1964) developed a precipitator which deliberately mixed the reactants very slowly and as a result produced a granular precipitate. This work was extended to ferric sulphate and it was found that a non flocculating precipitate could be formed. Such fine submicron crystallites can approach each other very closely and the van der Waals attraction compared with the particle mass is considerable. They stick once they come into contact.

It also follows from the above that recirculation of solids from subsequent stages is likely to be counterproductive because in this case one does not wish to grow the crystallites further. On the other hand existing particles of clay etc will become coated with hydroxide floc. It is also fortunate the organic impurities present in most raw waters interfere with crystallisation, reduce the mass transfer coefficient and further increase the surface area of the precipitate.

3.6.2 Perikinetic Flocculation

Perikinetic flocculation is the term applied to agglomeration caused by collisions between small particles under the influence of molecular bombardment or Brownian Motion in which the particles move randomly and stand a finite chance of colliding (Fig.3.4). The process is characterised by the following equation:-

$$t = \frac{3\mu}{8K_B T} \left(\frac{1}{N} - \frac{1}{N_o} \right)$$

where N and N_o are the number concentrations initially and at time t. The process is therefore unaffected by external factors apart from temperature. To extend the above example half of the 5ppm of single 0.18μm particles would collide to form binary particles in 153 sec. These binary particles would of course collide

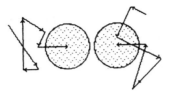

Fig. 3.4 Perikinetic Collisions.

with each other and with the remaining primary particles and so on.

Kruyt (1952) has published a correction to this time when the particles are of unequal size:-

$$\frac{t_1}{t_2} = \frac{1}{4} \left[4 + \left(\sqrt{\frac{d_1}{d_2}} - \sqrt{\frac{d_2}{d_1}} \right)^2 \right]$$

With a size ratio of 40 the time will be reduced 11 fold. Thus perikinetic flocculation is accelerated by the presence of existing solids and it follows that the initial crystallites of hydroxide combine preferentially with other larger suspended solids already present rather than with themselves. This characteristic is further evidence that mixing of the coagulant must be very rapid to ensure that the hydroxide is dispersed with all the other suspended solids to attach to them rather than to other hydroxide particles in a local excess.

In view of the fact that the kinetics are related to a number concentration, which for a given mass concentration is related to the inverse cube of size, perikinetic flocculation slows down rapidly as the mean floc size grows. In the above example 50% removal of 0.36µm particles would take 1200 seconds. Orthokinetic flocculation then overtakes perikinetic flocculation.

3.6.3 Orthokinetic Flocculation

This process involves the collision of particles as the result of the shearing movement (Fig.3.5) of the fluid (water) induced either as a result of flow or from the action of a paddle or turbine.

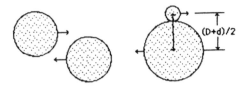

Fig. 3.5 Orthokinetic Collisions.

The rate of disappearance of a particular size of floc is defined by the Schmoluchowski equation:-

$$\frac{dn}{dt} = -\frac{4}{3} GR_c^3 n^2$$

where R_c is the collision radius. ie the distance between the centres of the particles when they come in contact. Hence for particles of equal size:-

$$\frac{dn}{dt} = -\frac{4}{3} Gd^3 n^2$$

However $n\pi d^3/6$ corresponds to the volume concentration C, the particle volume concentration. Thus

$$\frac{dn}{dt} = -\frac{8}{\pi} GCn$$

or

$$\frac{n_t}{n_o} = \exp\left(-8\frac{GCt}{\pi}\right)$$

Tambo (1979) points out that strictly the above applies to laminar flow and that under turbulent conditions the correct equation is:-

$$\frac{dn}{dt} = -\frac{12\pi}{\sqrt{15}} \cdot Gd^3 n^2$$

This indicates a rate some 23 times faster than the above.

The shear rate G is a measure of the energy input and is defined by the Camp Stein equation:-

$$G = \sqrt{\frac{P}{\mu}}$$

where P is the power dissipation per unit volume. For example at 10°C 13w/m³ corresponds to a shear rate of 100s⁻¹.

Jar tests for coagulation with hydroxide floc generally produce a floc volume of about 0.5% after settlement. 70% removal of a given size of floc at this concentration would take 78 sec. using the Schmoluchowski equation or 3.4 sec. with the Tambo Levich equation.

With mixed size floc the collision radius becomes (D+d)/2 and
Equation 24 becomes:-

$$\frac{dn}{dt} = -\frac{4}{3} G \left(\frac{D+d}{2} \right)^3 Nn$$

If D>>d then the concentration C of the larger floc dominates the situation, while dn/dt is still the rate of disappearance of any species of floc much smaller than the large floc.

The situation is complicated by the changing density as the floc grows. Camp (1968) and Lagvankar et al.(1968) suggest that the density varies with the diameter with an exponent of (-0.7). Thus in growing from 1μm to 1mm the specific volume (inverse of density) increases by 125 fold. This makes the larger mature floc very much more effective at capturing virgin floc than the other virgin floc. It also explains why in ajar test no floc may be visible for several minutes. Implications in the design of equipment are discussed elsewhere. It should be noted that the product of shear and time which can be controlled by the designer is known as the Camp Number Gt. It is dimensionless and is a parameter used in the design of many types of clarifier.

3.6.4 Differential Settlement

In addition to the above a third collision process occurs in the absence of any external energy input. In a situation where large and small floc particles coexist in suspension the larger ones are able to fall past the smaller ones (Fig.3.6) and in a situation not unlike filtration through granular media the larger particles may capture the smaller ones.

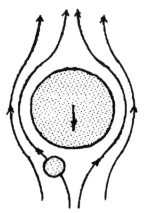

Fig.3.6 Collisions caused by diffferential settlement

Tambo (1979) gives the following equation:-

$$\frac{n}{n_o} = \exp\left(-3\,C_f\frac{dh}{D^2}\right)$$

Tambo also gives an alternative which assumes laminar flow past the large floc particles:-

$$\frac{n}{n_o} = \exp\left(\frac{9}{4}\,C_f\frac{d^2h}{D^3}\right)$$

A 1 meter deep floc blanket (ie, a fluidised bed of floc), with 5mm particles at a concentration of 20% v/v would give 90% capture of 50µm floc on the basis of the former equation and only 2% on the basis of the latter. Experimental data from Tambo appears to support the former.

Overall, flocculation involves three stages ie. perikinetic, and orthokinetic flocculation with in many cases capture by differential settlement as well. The latter only becomes significant with floc more than a few tens of microns in diameter, but it provides an efficienct polishing process.

3.6.5 Floc Strength

The above takes no account of floc strength and breakup. Each type of floc has a characteristic strength which is manifest as a limit to the diameter at any particular rate of shear in the fluid. In filtration for example floc particles break up as they enter the bed and the size involved in filtration may bear little resemblance to the size of visible floc upstream.

It has been shown by Lagvankar and Gemmel (1968) and by Tambo and Watanabe (1979) that hydroxide flocs vary in size with shear:-

$$d = \frac{d_1}{G^x}$$

where x is in the range 0.7 - 1.2. d = equilibrium floc particle size at the prevailing shear rate G, d_1= floc size at unit shear. A typical floc particle might have a diameter of 5mm at unit shear. Experimental data (Stevenson 1972) suggests that there is some variation in the value of x. However the general conclusion is that small floc particles will withstand a high shear, and can therefore be grown rapidly under high shear, but the energy input in flocculation must be tapered downwards as the floc grows. Some floc breakage is inevitable, and a compromise must be reached between growth and breakage. A system in which the local shear rate varies widely, such as with small high speed impellers can give poor results even though the average energy input may be correct.

From the above equation one might expect that the particles would grow to an infinite size at zero shear. However zero shear cannot occur while the particle remains suspended and able to grow. The shear experienced during settlement in a stationary fluid constitutes a minimum value and prevents floc growing to an infinite size. A stationary particle sitting on a horizontal surface cannot of course grow except by diffusion and the above relationship does not apply as there is no shear to convey material to the surface. Centrifuging will reduce the floc diameter. There is probably a limit which will be the size of constituent particles in the original suspension (eg. clay.)

The maximum (equatorial) shear at the surface of a settling particle is $\mathbf{G_{max}}$= **6u/d**, where u is the settling velocity, which may in turn be calculated from the Stokes

Equation. When floc is carried into a filter, probably suffering a fall in a weir, it may experience local shear rates of say 1000 s⁻¹ or more and so will be broken down into fragments of 10 to 30 μm diameter or less.

Within a filter media the floc is subjected to fluid shear which increases as the bed clogs. The average shear rate in the fluid (not shear stress on the grain surfaces) can be calculated by applying the Camp Stein equation again.

$$G=\sqrt{\frac{P}{V\mu}}$$

The power dissipation P in a bed depth h is the product of pressure loss and mass flow and can be calculated from the Kozeny Equation, thus:-

$$P=\Delta pu=\Delta H\rho gu=180.\frac{(1-e_o)^2}{e^3}.\frac{u^2}{D^2}.\mu h$$

The distinction between the actual voidage e and the clean bed voidage e_o is discussed later. D is the grain diameter and μ the absolute viscosity.

The volume of the pores per unit area of bed of depth h is eh. Hence by combining these two equations the shear within the pores is:-

$$G=13.4.\frac{(1-e_o)}{e^2}.\frac{u}{D}$$

The average shear is a widely used parameter which can be calculated from other measurable parameters. There are local variations. For example the shear at the centre of a capillary is zero, and at a maximum at the walls.

As an example, if x=0.7 the following floc sizes might be expected within a filter bed, at the residual voidages indicated, if the above example enters the filter bed. (This is based on 5mm floc particles at unit shear, 0.8mm media size and 2mm/s flow velocity.

e	40%	10%	5%	2%
G (s.₁)	125	2010	8040	50,000
Size (μm)	170	24	9	7

It is not known how far the above equation remains valid as the shear is increased but it is reasonable to expect that the original fragments of algae, clay etc will remain as building blocks and ultimately will persist perhaps as hydroxide coated particles. There is probably a shear rate at which the sizes become relatively independent of shear. It is unlikely that one can extrapolate to µm sizes.

The density of the particles reaching a filter can also vary widely and this will influence the mechanism of the process. Mostly one is concerned with hydroxide floc residues. The density of a large floc particle which settles at a measurable rate can be estimated from the Stokes Equation already discussed. A 3mm particle settling at 3m/hr will have a submerged density of 0.22kg/m3. However Camp (1968) has shown that the volume concentration of floc (ie the inverse of density) is related to the shear by the following equation:-

$$C = C_1 G^{-0.74}$$

where C_1 is the volume concentration at unit shear.

Hence using the above equations it would seem that a 30 µm floc particle would have a submerged density of nearer 7 kg/m3. This is probably more representative of particles reaching the media. It follows that floc which is deposited under higher shear conditions will be more compact and occupy less pore volume than floc deposited at low flows. It is of course stronger and able to withstand such higher shear rates. This feature has not been incorporated into the model discussed in Chapter 17.

3.6.6 Floc Ageing

The age of the floc at the crucial time of separation (ie. the time between coagulation and meeting the media, settling or meeting a floc blanket) can vary from a few seconds in the case of an upflow filter, perhaps a few minutes with a pressure downflow filter in direct filtration, to four hours possibly with horizontal flow clarification and even several days with floc blanket or solids recirculation clarifiers (this being the average age of the floc blanket solids.) Also if floc on a filter is disturbed by a change in flow rate for example, a few days after deposition it may behave very differently from when it was first deposited. It may have lost its adhesion.

Francois (1987) has studied the ageing of aluminium floc with and without added kaolin but unfortunately not with organic matter present. He has shown that floc shrinks with time and its density increases. In his experiments the floc appeared to become stronger initially. The zeta potential (surface charge) changed after several days from positive to negative. Accordingly one would expect any residual coagulating power to disappear.

The author's experience using jar tests on surface water samples (see Stevenson 1986, for details of the procedure) indicates that in practice the deterioration is more rapid. An appropriate dose of aluminium sulphate was added to a sample with rapid mixing. On slow stirring floc was allowed to form in the usual way and the settlement rate determined. On subsequent rapid stirring the floc broke up but was allowed to form again with slow stirring. However the floc became progressively more reluctant to form as the process was repeated and became older. Settlement was progressively slower. There appeared to be both an ageing effect and also possibly an effect caused by the cumulative shear. As is mentioned elsewhere excessive shear in pipe work and channels should be avoided if unfilterable residues are to be minimised. Extending the argument, it is possible that particles that are shed from the pores within the media during filtration may be less adhesive than they were when they first arrived. If this is so the age of each component would have to be recorded, which would make any model almost unmanageable.

The density of floc varies with its age. It appears that the density continues to increase and with it the settlement rate. This is exploited in recirculation and floc blanket clarifiers. The process is known as syneresis and probably involves maturation of the crystallites with a change in their size and packing within the floc aggregate.

Ultimately, hydroxide floc loses its flocculation characteristics and behaves much like the suspended solids in a raw water. If filter washings are recycled they must be recoagulated on the way back into the process.

Weirs must be used with discretion on flocculated water and normally a height of fall greater than 0.5m can cause a deterioration in quality for this reason. Pumping of flocculated water is very risky and often it is necessary to add a polyelectrolyte to repair the damage.

3.6.7 Coagulant Aids (Polyelectrolytes)

These are chemicals that overlap with coagulants but in the present context they are used to increase the strength of floc particles and therefore to improve size, settlement, filtration etc.

The effect is to increase the size of floc produced at a given shear rate. The molecules comprise long chains with hydrophilic groups attached and have been compared with strings having adhesive patches attached. These adhesive patches attach to floc particles to bind then into larger aggregates. However to carry the analogy further the floc particles must be of the right size and number if a good aggregate is to be obtained. If the polyelectrolyte is added with the coagulant the hydroxide crystallites are too small and numerous for the polyelectrolyte to have much effect and the adhesive points are saturated with solids having little substance. On the other hand, remembering that the density and therefore the strength of floc falls as the size increases, if the floc is too large the adhesive patches will pull out pieces from the floc surface, saturating the polyelectrolyte chain but leaving the rest of the floc particle untouched. It is found in practice that there is an optimum contact or flocculation time (not usually involving mechanical flocculation) where the floc matches the polyelectrolyte. A good example of this effect is given in Table 1, in which the contact time between coagulation and the addition of polyelectrolyte in a jar test was varied. The settlement rate shows a dramatic increase under the right conditions, whereas if added incorrectly the polyelectrolyte is wasted. It is wise to provide means of varying the point of addition and the contact time. Generally a few minutes is beneficial. Likewise good mixing is as important as it is for the main coagulant, otherwise excessive irreversible adsorption on some of the floc can occur.

TABLE 1
Example of the effect of delay time between the addition of coagulant and polyelectrolyte in a jar test.

The raw water had a hardness of 240mg/l as $CaCO_3$, and Turbidity 10NTU.
Coagulant - 50mg/l Aluminium Sulphate.

1. Coagulant alone.		Settlement	0.8-1.0m/h
After the addition of 0.2mg/l Magnafloc LT25 polyacrylamide			
2.	Simultaneous addition		0.7m/h
3.	Intermediate contact (2-3 mins)		2.5-3.5m/h
4.	Optimum conditions (~20 mins)		12.6m/h
5.	Excessive delay (> 60 mins)		6.9m/h

The above remarks do not apply to the liquid amine and quaternary polymers which act mainly as coagulants, not as coagulant aids, and can be added simultaneously with aluminium and iron salts.

3.7 Measurement

The measurement of solids in suspension ultimately must be gravimetric. However this is tedious and involves filtering the suspension through a paper or membrane, drying and weighing. Thus turbidimetry is widely used as surrogate. The internationally agreed standard is the formazin suspension which is made up freshly from formaldehyde and hydrazine to a formula given in Standard Methods (Greenberg *et al.*(1985)). Light scattering is measured nephelometrically (measurement of light scattering at right angles), hence the unit NTU which is sometimes referred to as FTU. Formazin has a density close to that of water. Clays which have a particle density of 2.5-2.6 are usually found to give a conversion of 2.5mg/l per NTU.

The scattering and absorption of light varies with particle size. For particles larger than the wavelength of light, which includes most solids in water treatment, the absorption of light is mainly a matter of obscuration. Each particle casts a shadow of area $\pi d^2/4$.

The number concentration is :-

$$N = \frac{6C}{\pi d^3}$$

where C = volume concentration

Combining these equations, with a random dispersion of particles the light intensity will fall exponentially as follows:-

$$I = I_o \exp\left(-\frac{3CL}{2d}\right)$$

where L is the depth.

Thus for example 1 ppm by volume will give the following absorption in 1.0m depth.

Particle size μm	Light penetration
4	68%
2	47%
1	22%
0.5	4.9%
0.25	0.2%

This applies strictly to black absorbing particles. In practice there will be some scattering of light in all directions and the actual absorbency will be slightly less. Scattering to the side will depend on the configuration, ie. the width of the incident beam and the depth of viewing. Hence measurements in arbitrary instruments are calibrated against standard suspensions.

There are many other turbidity standards, some developed for other industries such as brewing. The silica scale was widely used at one time (1 NTU $\equiv$ 2.5 SiO_2)

but the calibration was dependant on the source of kaolin used for calibration. The Jackson Candle in which the depth of water required to obscure a standard candle flame is measured was the basis on which the formazin standard was established. (1 NTU = 1 FTU = 1JTU). In practice the Jackson measurement, being absorptiometric, does not give the same results as the Nephelometric method, particularly at higher turbidities. Fast settling solids will settle to the bottom of the tube and obscure the flame but not scatter light further up. French documents often refer to "goutte de mastic". One instrument manufacturer equates 12.5 goutte to 1 JTU. The same source gives seven other standards but these are not used in the water industry.

Since cryptosporidium oocysts became a concern in treated water particle counting is being introduced as a more sensitive measurement. Laser based instruments can detect and measure the size of particles down to below $1\,\mu m$. On a volumetric basis 1 ppm converts as follows:-

Particle Size μm	Particles / ml
10	1.9×10^3
3	5.2×10^4
1	1.9×10^6

For comparison on a mass basis particle counts should be multiplied by the cube of the particle size. Such a practice has not yet been adopted generally. Also it must be remembered that conversion to mass involves the density of the particles and dried cells of algae which have been counted in the live state will weigh much less than clay particles when dry.

3.8 References

Hadamard, J. (1911) Mouvement permanent lent d'une sphere liquide et visqueuse dans une liquide vuisqueuse", Compte Rendus, **151**,1735

Kruyt, H. R. (1952), Colloid Science. Vol.1. Elsevier. Amsterdam

Kynch, G.J. (1952) "A theory of sedimentation," Trans. Faraday Soc. **48**,166

Stevenson, D. G. (1964) "Design and operation of continuous chemical precipitators", Trans. Inst. Chem. Eng., **42**, T316

Carlos, C.R. and Richardson, J.F. (1968) "Solids movement in fluidised beds. I Particle velocity distribution" Chem. Eng. Science, **23**, 813

Carlos, C.R. and Richardson, J.F. (1968) "Solids movement in fluidised beds. II Measurement of mixing coefficients", Chem. Eng. Science, **23**, 825

Anon. (1971), Committee Report, J. American Waterworks Assn. **63**, 99

Al Dibouni, M.R. and Garside,J. (1979) "Particle mixing in fluidised beds", Trans. Inst. Chem. Eng., **57**, 94

Al Mawlawi, J.H. (1979) Ph.D.Thesis, University of Loughborough, (Chemical Engineering Dept.)

Tambo, N. and Hosumi,H. (1979) "Physical aspects of flocculation.II Contact flocculation." Water Research, **13**,441

Warden, J.H. (1983) Sludge treatment for waterworks, WRc Report TR189. Water Research Centre.

Greenberg, A.E., Trussell, R.R.and Clesceri,L.S. (1985) Standard methods for the examination of water and wastewater. American Public Health Association/American Waterworks Association/Water pollution Control Federation.

Kahn, A.R. and Richardson, J. F. (1987), "The resistance to motion of a solid sphere in a fluid",Chem. Eng. Comm. **62**, 135

Kahn, A.R. and Richardson, J. F. (1989), "Fluid particle interactions and flow characteristics in fluidised beds and settling suspensions of spherical particles", Chem. Eng. Comm. **78**, 111.

Coulson,J. M., Richardson, J. F., Backhurst, J. R. and Harker,J.H. (1990), Chemical Engineering, Vol.2. Pergamon, Oxford, UK.

Further reading.
Bratby,J., (1980), Coagulation and Flocculation, Uplands Press, London.

Amirtharajah, A., Clark, M.M. and Trussel, R.R. (1991), Mixing in Coagulation and Flocculation, American Waterworks Association Research Foundation, Denver.

CHAPTER 4.

EQUIPMENT HYDRAULICS

4.1 Mixing

This often tends to be a neglected subject and all to often a single hose can be seen discharging into a channel without consideration of the consequences. Mixing situations can be separated into critical and routine. In the latter case a homogeneous mix is required before the next stage of the process or before flow splitting, but the time scale is not particularly important. Examples include the addition of pH correcting chemicals, the blending of returned wash water, fluorides, powdered carbon etc. In the critical category is the addition of coagulants, polyelectrolytes and possibly chlorine. In this case the speed of mixing affects the outcome. To illustrate the effect one may carry out laboratory coagulation jar tests and compare a test in which the coagulant is added with rapid mixing to all of the sample aliquot with a test in which the same amount of chemical is added to half of the water and then a short time later the rest of the water added. The quality produced in the second case will usually be found to be significantly worse. This is similar to the situation where a hose is allowed to discharge into a channel and the plume of coagulant slowly spreads laterally across the flow.

Conventional mixing practice is concerned more with the achievement of a high degree of uniformity in a reasonably short time without the expenditure of excessive power. For coagulation rapid mixing throughout the body of the flow with less emphasis on perfect homogeneity is more important because of the high rate of reaction and the ensuing flocculation. This has been discussed in Chapter 3 where it was noted that the coagulant crystallites can flocculate with themselves rather than with the suspended solids in the raw water. The same is true of coagulant aids which can double coat some floc and leave the rest uncoated rather than cover all the floc uniformly.

4.1.1 Hydraulic Mixing

Fortunately some very simple means for rapid mixing are available providing that they are used correctly. Proprietary devices are not always the most cost effective or appropriate. Rapid mixing becomes more difficult as the scale of the operation increases. For example if a chemical is dosed at the centre of a pipe and lateral diffusion allowed to carry it to the walls the angle of the plume tends to remain constant so that the mixing time in such a system increases linearly with the pipe diameter. A common rule of thumb, possibly without much basis, is that mixing occurs in 20 pipe diameters in a straight pipe. At pipe velocity of 1.25m/s mixing will be complete in 0.8s in a 50mm diameter pipe but 9.6s in a 600mm pipe. If this technology is to be retained a two dimensional grid can be made up to distribute the chemical evenly across the cross section. (The design of such grids is discussed later). Such devices have been described in the literature and are appropriate particularly for large open channels.

Any structure which generates a vortex street will assist transverse mixing and the above mentioned grid assists in both by dividing the flow into small sections and also by generating turbulence. Some sources advise the use of forward pointing orifices in such grids on the basis that the jet will spread out more against the oncoming flow. However the volumetric flow of the chemical is usually small. A rearward facing discharge is less likely to produce encrustation by floc over the distribution pipe or grid.

Bends induce a double vortex with transverse circulation across the pipe. A plain orifice also induces centre to edge circulation in the circular vortex street down stream. Thus a chemical added at the centre of the pipe is dispersed rapidly. However if the chemical is added at the side it will tend to remain in the boundary layer on that side and consequently mix slowly. Injection fittings should therefore be designed for the specific pipe diameter. With any chemical capable of forming a precipitate the fitting should of course be removable under pressure to enable blockages to be cleared.

In recent years static mixers have become popular. They are expensive compared with the alternatives. They were originally developed for blending viscous fluids in pipe lines and probably are over designed for water. In any case they do not compensate for poor injection. Experience on a 600mm example showed a significant improvement on transferring from a static mixer to a weir with a distribution sparge pipe.

Dilution with carrier water has repeatedly been shown to improve the rate of mixing with most chemicals, especially the more viscous ones such as concentrated coagulant liquors and polyelectrolytes. Mixing is most rapid when the chemical has a similar viscosity to water as anyone who has tried to mix treacle with water will know. On the other hand over dilution, particularly with chemicals that hydrolyse with the alkalinity present in the water will be counterproductive.

Weirs are probably the most satisfactory method of mixing in open systems. The flow is split into a thin ribbon and the chemical sprayed uniformly over the crest of the flow. Mixing occurs in the turbulence downstream. Janssens has refined the process by adding finger baffles into the descending stream (Fig. 12.2). It follows that the receiving pool should be shallow to confine the energy into a small volume and further reduce the mixing time. The energy input per unit volume remains substantially constant with flow in contrast to orifice systems. It is wise to limit the crest height and upscaling therefore merely involves building longer weirs. The conditions therefore do no change with scale. A 300mm fall has proved satisfactory in many situations. The depth of water over the crest must be limited to produce a free fall at the weir.

Hydraulic jumps have many of the same merits as weirs but are more expensive to construct, and are more difficult to upscale without changing the configuration.

4.1.2 Mechanical Mixing

This has been characterised in detail in the chemical engineering literature and certain standard configurations exist. Many manufacturers of mixing equipment have their own special designs, however not all of them appreciate the special requirements of the water industry. A mechanical mixer implies an impeller and a tank, or possibly two or more of these in series. The implications of tanks in series are discussed in Chapter 5. Impellers may either be radial flow turbines or axial flow propellers. Power dissipation data for some standard designs of mixing impeller shown in Fig.4.1 are quoted by Uhl and Gray(1973). Each design has a characteristic power Number Np.

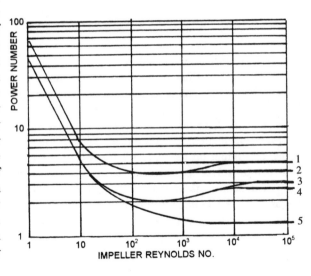

Fig. 4.1 Power Numbers for various 6 bladed Impellers. 1. Flat blades on disc, width D/5. 2. As 1. but full radial blades. 3 & 4. As 1 & 2 but width D/8. 5. 6 full diameter blades angled at 45°, projected width D/8.

The power dissipated is :-

$$P = N_p \rho N^3 D^5$$

The pumping or circulating capacity of the impeller is characterised by the flow number N_Q, where:-

$$N_Q = \frac{Q}{ND^3}$$

Examples of power numbers are:-

6 short blades on disc	W/D =0.2	N_p =5
6 full diameter blades	W/D =0.2	N_p=4
as above	W/D=0.125	N_p=2.7
as above but angled at 45°	W/D=0.125 (W = projected width)	N_p=1.4

These values are valid for an impeller Reynolds No. ($\rho ND^2/\mu$) above 10^3.

Flow numbers range generally between 0.3 - 0.5. For more precise data manufacturers must be consulted.

To avoid swirling of the contents of a tank and a significant reduction in the relative velocity the tank should either be square or if round it should have static radial wall baffles. Normal practice is to provide six at full depth, and project one tenth of the tank diameter. An acceptable alternative in some situations (for example slow speed flocculation) is to provide reversal of rotation at regular intervals of a few minutes, depending on the size. Where two or more impellers serve a single tank contra-rotation solves the problem.

The same authors have collected data on batch mixing times for standard configurations. While there may be minor differences in the value of the constant, the equations are of the general type:-

$$N\theta = 4 \, (T/D)^2$$

N is the speed in revolutions per second, θ the mixing time T the tank diameter and D the impeller diameter. In this equation each half is the number of revolutions to reach complete mixing. The smaller the diameter, the greater will be the number of turns required. Generally the ratio of tank diameter (or size) to the impeller diameter is kept constant as the scale changes to maintain a compromise between cost andmixing time. For a given time the speed must vary inversely with the square of the impeller diameter. However the power will go up inversely with impeller size. Large slow impellers are costly but efficient, and a compromise has to be reached.

Problems arise when the scale of operation changes and this becomes apparent immediately when laboratory scale experiments are upscaled to full size. The power dissipation at an impeller only defines the work put into the system and takes no account of the homogeneity. Much of the energy dissipation is localised at the tips of the blades. The tip velocity is often limited to prevent undue shear. In this case N is proportional to $1/D$.

For a given configuration the power to achieve mixing is proportional to the inverse cube of the required mixing time. The tank diameter for a given height / diameter ratio will be the 2/3 exponent of the throughput. If the tank/impeller diameter ratio is constant the impeller speed for a **given mixing time** must be constant, thus the power dissipation will rise with the 10/3 exponent of the throughput. The shear or G value rises with the 5/3 exponent of throughput and can easily become excessive. Few plants are designed on such a basis. More often a longer mixing time is accepted, and the design based on constant G. This explains why full size plants often do not work in the same manner as pilot plants. One immediate conclusion is that multiple small mixers are more efficient in terms of time and energy than one single large unit. The argument may also be extended to explain why mixing at weirs, where the flow is formed into a thin ribbon, is so effective.

4.1.3 Mixing Techniques

Carrier Water

Mixing is far more rapid if both components are mobile liquids which have low viscosities. Thus better results are usually obtained in a water treatment context where small amounts have to be mixed with large volumes if viscous or semi viscous chemicals are diluted with carrier water. This is particularly true of polyelectrolytes, but the same situation holds for concentrated liquors such as aluminium sulphate and ferric sulphate. Some semi proprietary chemicals such as the complex aluminium polymers however are claimed by the manufactures to suffer if diluted in this manner. Excessive dilution particularly with water of high alkalinity can be counterproductive. Thus the total alkalinity in the carrier water mixed with the chemical should not cause more than say 1% of the chemical to be neutralised or precipitated. Lime may also be diluted but only with water of low carbonate alkalinity or carbon dioxide content otherwise calcium carbonate may be deposited in the dosing lines. Carrier water in this case helps to prevent settling out of lime at low dosing rates. Powdered activated carbon may also settle out. In some installations lime and powdered carbon are metered as dry powders and conveyed hydraulically from a mixing cone or tank to the point of application.

There is no reason why chemicals should not be added via the same dosing line if they are compatible. Coagulants and acids or even chlorine may be added through the same line and precipitation on the walls of the line will be inhibited.

Dilution with carrier water usually occurs immediately downstream of dosing pumps and the dispersion in the flow in the dosing lines provides a degree of smoothing from the pulses of the pump which otherwise alternately overdose and under dose the main flow. The flow of carrier water is not critical providing that an excess is not used.

4.2 Flocculation

The previous chapter discusses flocculation as a combination of three possible processes. Here we are concerned specifically with orthokinetic flocculation in which energy is dissipated within the system in order to promote collisions and the growth of agglomerates. The intensity of energy dissipation is characterised by the shear G which has units of s^{-1}. A process designer will usually define the G value to be attained together with the time which together give the Camp Number Gt.

Energy may be dissipated in a number of ways, for example:-
a. Hydraulically by using a defined hydraulic gradient.
b. Mechanically using an impeller such as a turbine, or "flocculator paddle".
c. Hydraulically by using an external recycled jet or a pulsing mechanism.
d. Mechanically with a reciprocating device at any frequency from < 1Hz to 50Hz

In the previous chapter it was mentioned that floc has a limited strength and can be broken up. This seems to depend on the peak shear that it experiences, like many fracture situations. Thus the ideal flocculation system has a perfectly uniform local value of G. This is not easy to achieve.

It is well known in mixing technology that small high speed impellers are less energy efficient although much cheaper than large low speed mixers. The small ones act much in the same way as pumped recirculation mixers in which the momentum in the jet is transferred to the rest of the contents rather than the contents being driven directly by the impeller. Thus of the above types, c., while used extensively for mixing, is not normally used for flocculation. Examples of Type d. exist in the USA. The higher frequency is used only in a proprietary mixer.

Gt values (a measure of total energy input for a given viscosity) for flocculation in theory should depend on the concentration of suspended solids. Flocculation as previously indicated is a function of GCt ie. shear, concentration and time. Thus more concentrated suspensions flocculate more readily. G values are typically in the range 10-100 s^{-1}, with the higher figure being in the first stage of a 3 stage system. 30s^{-1} is a general purpose figure for settlement. For dissolved air flotation 50s^{-1} is commonly used. Retention times vary from 30-60 min. for one-pass (no recirculation) designs while in dissolved air flotation 20-30 min. is common although some recent papers claim that 5 minutes is sufficient.

The Camp Stein equation includes viscosity. The optimum power input will therefore change with temperature, less power being requited for given G at higher temperatures.

If a true optimum is required pilot tests should be carried out on any particular water. Such tests will need to include the correct number of stages for reasons made clear in the Chapter 5. Most designers use figures similar to the above with margins of safety and a facility to adjust the power input.

4.2.1 Hydraulic Flocculation

The power dissipated by a stream of water is given by:-

$$P = \frac{1}{2}\rho u^2 Q = \frac{1}{2}\rho u^2 \cdot au = \frac{1}{2}\rho au^3$$

where Q= volumetric flow, a= area of flow, u=velocity, ρ= density of fluid, P =power (watts, if the other parameters are in SI units.) This equation will give the overall power input to a tank where the inlet velocity is u.

Alternatively if a head (ΔH) or pressure loss is involved:-

$$P = Q\rho g\Delta H$$

where g is the gravitational constant.

Thus for a pipe or channel one may combine this with the Camp Stein Equation:-

$$G = [\frac{\Delta H}{L} \cdot \frac{\rho g u}{\mu}]^{\frac{1}{2}}$$

where ΔH/L is the hydraulic gradient and μ the absolute viscosity. Pipe or channel flocculators are used by many designers and have the advantage of providing a fairly uniform rate of shear throughout the volume, although with a common serpentine arrangement much of the energy dissipation occurs at the U bends. It is also possible to insert baffles into the pipe or channel (chicanes). Fig.4.2 gives information on the headloss coefficients (multiples of $u^2/2g$, where u is the main full bore channel velocity).

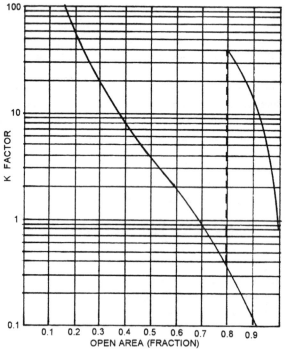

Fig. 4.2 Friction coefficients for orifices and grids in a pipe or channel.

The disadvantage of horizontal hydraulic flocculators is the possibility of settlement of solids and siltation. To avoid this relatively high velocities are required which in turn can produce an excessive value of G and an unacceptable headloss across the unit. Vertical units are better in this respect. General experience suggests that such flocculators are more efficient in terms of required retention time than paddle flocculators, probably because of the reduced floc breakage, but the sedimentation problem would rule out this approach on silty waters unless a presedimentation section is incorporated.

It may be noted that based on the Poiseuille Equation for capillary flow, in laminar flow:-

$$G = \frac{5.6u}{d}$$

or between parallel plates:-

$$G = \frac{3.4u}{d}$$

Stacks of parallel plates have been used in experiments with dissolved air flotation at WRc. Also advertisements for devices using corrugated sheet flocculators have appeared. These have the adjacent plates arranged in parallel but "out of phase" so that there is a series of chambers with a constriction between them. The constriction can of course be adjusted and some control over the energy dissipation can be exercised. However in a commercial situation on a plant of any size the cost of any such device will tend to rule it out compared with the simplicity of an open tank with a paddle.

It is a debatable point as to whether the pulsing system used in some types of settlement tank is a means of flocculation or not. The primary purpose would seem to be one of maintaining suspension of the floc blanket. However pulsing to provide reciprocation between two chambers is a viable method of flocculation which would have many of the merits of hydraulic flocculation. A perforated wall between the two chambers provides a means of controlling the energy dissipation. The vacuum blower arrangement used on the Pulsator clarifier could be used to drive several chambers either operating in series or parallel.

4.2.2 Mechanical Flocculation

The general equation for the dissipation of power from a blade moving through water (or any fluid) is:-

$$P = \frac{1}{2} \cdot \rho u^3 S C_n$$

where S is the surface area presented and C_n the force coefficient of the blade. u in this case is the relative velocity between the blade and the water. In a cylindrical tank without baffling the water can move round almost as fast as the impeller and power dissipation will be much less then expected.

Converting the above equation to a narrow paddle rotating about an axis parallel to its length:-

$$P=\frac{1}{2}\rho . L . W . C_n (\pi DN)^3$$

where L and W are the length and width respectively, D the diameter of rotation of the centre of the blade and N the speed in revolutions per second.

These equations can be converted into one for a radial blade :-

$$P=\frac{1}{8}\rho WD^4 C_n (\pi N)^3$$

In this case D is the diameter to the blade tip. Note this is for a single arm only.

The force coefficient is defined by the following equation valid for Re between 10^3 and 10^6.

$$C_n=1.10+0.02\left(\frac{W}{L}+\frac{L}{W}\right)$$

Where the blades are angled the power dissipation is reduced by the factor:-

$$F=\left[\frac{2\sin^2\alpha}{1+\sin^2\alpha}\right]$$

There seems to have been some disregard for the underlying theory in the design of many flocculators. A common traditional design is the picket gate type. Bearing in mind that the power dissipation falls off with the cube of diameter of rotation the inner blades do little good. Also the local shear at the edges of the blades is probably detrimental and one would expect a single blade located at the full diameter to be the most efficient. To go further one would also expect that apart from the lower force coefficient a tubular blade would produce better results than a sharp edged blade. Little work seems to have been done on this aspect.

Because of floc shear limits are placed on the tip speed, normally about 0.6m/s with 1.0m/s for the first stages of a set..

Mixer manufacturers regard flocculators as their territory but there are additional considerations that may not always be appreciated. Angled propeller types are becoming more popular. Oldshue and Trussel (1991) found a significant difference in the residual turbidity using different impellers. It would appear that larger diameter propeller blades of the aircraft type with an angle that varies inversely with diameter (hydrofoil blades) can give results similar to the traditional slow speed radial (gate type) flocculators. Axial flow impellers with a fixed angle and marine propellers are less satisfactory. One might conclude therefore that the ideal impeller should act as a pump to propel the water without dissipating energy at the impeller itself.

Griffiths (1996) has studied hydrofoil blades in some detail and concludes that they have a considerable advantage over conventional picket flocculators in that they induce flow in the tank rather than merely moving through the water. Thus they are able to are able to provide a flocculating action in a larger and more irregular tank than the latter. Fewer units are required and the cost is likely to be less. Griffith's objective was to dissipate a given amount of energy more efficiently and with a lower maximum shear stress leading to less floc breakage. Unfortunately he gave no design procedure other than to create some doubt about the conventional approach and one is therefore left in the hands of the suppliers with no means of independently checking their offers. The valid point is made that one cannot merely measure the overall electrical power input. Flocculators tend to be heavy although they dissipate relatively little power. The motor and gearbox may absorb a significant proportion of the total input.

All of the types of paddle must sweep a large diameter, usually at least half of the minor tank dimension and tanks with an aspect ratio greater the 1.5:1 are normally avoided. For settlement there is often an advantage in having the impeller lower down in the tank so that any settlement is counteracted. American practice has often been to use low level horizontal units but the cost and additional maintenance have discouraged their use elsewhere.

Flocculation is only important at low concentrations in horizontal flow clarifiers, in radial flow units without recirculation, in dissolved air flotation and occasionally before filtration. There tends to be confusion in other types of clarifier but mostly the the energy input is needed to maintain suspension or pumping,, not to provide flocculation as such. G values are not relevant. However floc can still be damaged and limits may be placed on tip speeds. It would seem that flocculation prior to filtration is often a means of controlling floc size to assist penetration into the media rather than primarily for aggregation. Most filters do not have mechanical flocculation although a contact time to achieve complete reaction and perikinetic flocculation may be incorporated.

It should be noted that conditions in a continuously flowing system are very different to those in a batch system such as the laboratory jar test. This aspect is discussed in Chapter 5. It should be noted that where two or more stages of flocculation are placed in series steps should be taken to avoid back flow between the stages. Thus the transfer velocity based on the open area in the transfer port should be at least as high as any residual wash from the impeller. A figure corresponding to the tip speed would be safe but probably a little excessive.

Also the inlet and outlet must be placed so that the flow does not bypass. In most designs the flow pattern will be governed by the paddles with induction at the shaft and radial flow from the centre, tending to form twin toroidal vortices. Good practice with vertical impellers is to have alterative top and bottom ports.

4.3 Manifolds and Flow Distribution

In most water treatment plant the flow is divided both between process units and within process units. Also flow may have to be collected uniformly from the surface of a tank. There are basic principles that must be observed but also relaxations may be possible if the characteristics are understood.

4.3.1 Feed Manifolds

The flow from a closed manifold discharging into a tank via orifices or fittings (Fig. 4.3) will vary along the length as the result of frictional effects with the end nearest the inlet passing the highest flow and also as the result of the Bernoulli effect (velocity head) in which case the furthest end gives the highest flow.

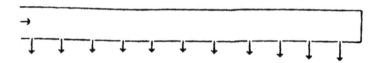

Fig. 4.3 Pattern of velocities in a discharge manifold.

The overall behaviour may be calculated using a computer, starting at the dead end of the manifold with a nominal flow and working back. At each upstream branch the effective head (pressure) providing the flow through that branch will be the head at the end branch plus the accumulated friction losses downstream of the branch under consideration less the velocity head of the flow past this branch. The flow through all the branches is summed and the data corrected to bring the sum into line with the intended total. (It is assumed in this case that the branches are in the walls of the manifold and do not project into the flow. Where this is the case the model must be modified. Stems, as in many filter floors, add to the friction component.

A simple rule of thumb applicable to situations where the frictional element is small is:-

$$u_{max} = 1.4\sqrt{\Delta H}$$

This provides a flow range not greater than 5%. Frictional effects initially improve the situation. Where uniform flow distribution is required from a simple perforated pipe and the spacing of the orifices is not important it is possible to adapt the programmes written for the above to specify a variable spacing for the orifices so the distance is proportional to the flow. This can give a uniform flow per unit length even with high manifold velocities. However it will be necessary to specify a maximum orifice spacing. If the velocity is too high water may be sucked in and entrained at the early orifices.

The same argument may be applied to open channels. In this case the varying head is reflected in changing water levels, which in turn affect the velocity past each branch. The reference branch at the end will have a water level related to that of the tank into which the flow is being discharged, which will in turn depend on the method of decanting. With a closed system the relative variation in the flow through the branches is usually constant as the throughput changes, providing flow is fully turbulent throughout. In open systems the changing flow cross section caused by the level changes will affect the distribution and calculations must be made for both the maximum and minimum flows. It is of course possible to have a dry floor at one end of a channel if the branches lead out from floor level into a tank at lower level..

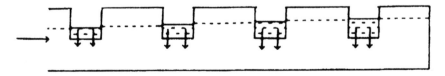

Fig. 4.4 Varying flow rates from weirs in distribution channels.

Side splitting weirs are used extensively, Fig.4.4. The flow over a rectangular weir is given by the general equation:-

$$Q = \frac{2}{3} \cdot BC_d\sqrt{2g} \cdot H^{1.5}$$

where B is the width of the weir and C_d the coefficient of discharge.
 If this equation is differentiated and divided by itself:-

$$\frac{dQ}{Q} = 1.5\frac{dH}{H}$$

In a low friction situation the term dH corresponds to the Bernoulli velocity head $u^2/2g$. dQ/Q is the fractional flow variation.
 Thus for example for 5% flow variation:-

$$u_{max} = 0.654\sqrt{H}$$

where H in this case is the crest height over the weir. Thus for a velocity of 0.6m/s the crest height must be 0.55m for 5% variation. It is possible to vary the height of the weir to correct errors but such a correction will only be valid for one flow.
 The errors in a channel system may be reduced if the channel is tapered to maintain a constant velocity. Low velocities may lead to fallout of floc and siltation. Tapering of channels to avoid low velocities at the end is doubly beneficial.
 Weir splitting boxes are also commonly used but here again surges and waves produced by the incoming flow can disturb the surface and cause uneven flow. If a high weir crest cannot be tolerated a compromise must be sought. Submerged orifices in which the flow is proportional to the square root of the head instead of the 1.5 exponent provide better distribution but with a large flow range adjustable orifices (valves or penstocks) are needed.

4.3.2 Collection Manifolds

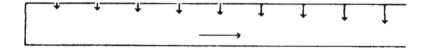

Fig. 4.5 Varying flow in collection manifolds.

The design of an enclosed collection pipe (Fig. 4.5) is somewhat similar to that for a feed manifold. The water at the dead end has to accelerate to water entering down stream and the resultant velocity head acts as a back pressure. Frictional effects also provide a back pressure. On the other hand the Bernoulli effect increases the effective headloss on the down stream ports. Thus in contrast to feed manifolds collecting manifolds produce a higher flow at the ports near the outlet end. Computer calculations again start at the dead end with a nominal headloss and work out the frictional loss between ports, the acceleration head and the Bernoulli head. The latter two are the same so that a single u^2/g may be used instead of $u^2/2g$. Again even decanting may be obtained by varying the spacing of the ports to match the flow through each port. This enables smaller bore manifolds to be used.

4.4 Decanting Systems

Even flow is usually obtained in plant units such as settlement tanks by using open channels or tubes operating in channel mode so that decanting orifices or notches are above the water line on the down stream side, and the varying water level in the channel merely leads to differences in the free fall.

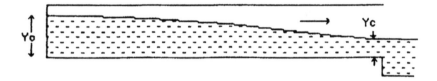

Fig. 4.6 Flow in decanting troughs.

The channel must be capable of handling the maximum flow and calculations of the critical depth are needed. This is the depth which leads to a minimum energy situation, the energy components being the kinetic energy and the potential energy. The kinetic energy is reflected in the difference in depth between the dead end of the channel and the discharge end (after taking account of friction losses), and the potential energy in the depth of water at the point of the critical depth. For rectangular channels the critical depth is :-

$$Y_c = [\frac{Q^2}{gb^2}]^{\frac{1}{3}}$$

where Yc is the critical depth, Q the flow and b the channel width. If the water level in the receiving channel is higher than the critical depth the level in the decanting channel will back up. The dead end of the channel will have an additional depth corresponding to the velocity head at the critical depth. Decanting notches and orifices must be above this level.

The theory may be applied to other shapes of channel, eg trapezoid, U shaped or circular by calculating the minimum energy situation using a sum of thin depth elements with widths to simulate the non-rectangular profile. Some examples are given below.

Some times decanting orifices are located at the bottom of channels. The design is then similar to that of a pipe collecting manifold, although allowance must be made for any variations in level. Such an arrangement is useful where tanks are liable to freeze over.

Decanting orifices permit the following flows:-

$$Q = \frac{\pi D^2}{4} \cdot C_d \cdot \sqrt{2gH}$$

Values of C_d vary with the length to diameter ratio of the orifice and data is given in Fig.4.7. (Also it varies with any rounding of the mouth).

The flow over rectangular notches is:-

$$Q = \frac{2}{3} \cdot Cd \cdot B \cdot \sqrt{2g} \cdot H^{1.5}$$

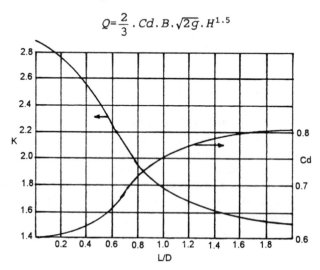

Fig. 4.7 Discharge coefficients for orifices of different length to diameter ratio.

For triangular notches the equation is:-

$$Q = \frac{8}{15} \cdot C_d \cdot \tan\theta \cdot \sqrt{2g} \cdot H^{\frac{5}{2}}$$

where θ is the half angle of the notch.

On large plant built in concrete it is difficult to level all the notches or orifices accurately. Adjustable weir plates are often used but are still an added expense. Orifices provide the least variation in flow for an given error, rectangular notches next and V notches most. On the other hand the level variation with flow (not normally something that matters much) is least in the reverse order.

4.5 Decanting from Clarifiers

Traditional spacings of channels and notches have been based more on past practice rather than any analysis of the situation. With one pass clarifiers which operate with a very dilute floc the density of the fluid and the suspension are virtually identical. The decanting system must therefore draw uniformly from the surface otherwise settling particles may be drawn up. However as will be discussed in a later chapter residual currents from the inlet are probably more important. French (1981) has analyzed the situation in a free volume and has concluded that troughs should be spaced at a distance of no more than twice the reference depth at which separation is expected to occur.

As soon as hindered settlement is involved or even at lower concentrations the density of the floc blanket mass has a stabilising influence. The height of a plume caused by a surge in the rise rate across a fluidised blanket can be estimated from the equation:-

$$h = \frac{\rho}{\Delta \rho_b} \cdot \frac{u^2}{2g}$$

where $\Delta \rho_b$ is the density difference of the floc blanket. As an example, a typical floc particles might have a diameter of 3mm and a settlement rate of 10m/h. The density will be 0.75 kg/m^3. If the blanket concentration is 15% by volume the blanket density will be 0.11 kg/m^3 and a settlement rate of 2.2m/h. A surge of 83m/h would on this basis produce a hump in the blanket surface of only 230mm which would be negligible.

The same argument may be applied to the flow approaching a decanting trough or orifice. Geometrical considerations show that the velocity approaching a single orifice at the surface is:-

$$u = \frac{Q}{2\pi r^2}$$

or for a trough:-

$$u = \frac{Q}{L\pi r}$$

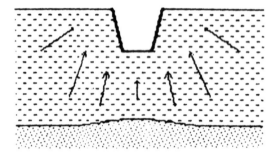

Fig. 4.8   A rising blanket caused by flow converging to a trough or orifice.

A hump in the surface of a given height (Fig. 4.8) would be maintained if the velocity of water flowing radially towards the point of discharge was similar to the surge velocity discussed in the previous paragraph. Thus a single orifice or bell mouth at the surface serving an area of say 4.5 by 2.0m above the above blanket would balance a hump 60mm high if the outlet is only 84mm above the blanket.

The above argument may seem extreme but it provides an approximate basis to validate much wider trough and orifice spacings than have been advocated in the literature for horizontal flow clarifiers without solids recirculation. Direct observation has shown an absence of floc loss even with a blanket level with the base of decanting troughs. In further tests decanting troughs were blanked off and the flow decanted through the side walls of a clarifier without loss of floc. However it will be appreciated that the floc concentration in a "once through" settlement tank may only be about 0.5%, not 10-15% and there is little stabilisation. The design of the decanting system for a blanket or solids recirculation tank differs for good reasons. To use the horizontal flow tank rules in the latter case involves needless expense.

In practice surges usually occur at walls or in corners. For the best results decanting troughs and orifices are sometimes placed to draw off from the areas where the quality of the water at the surface is best.

4.6 References

Uhl, V.H. and Gray,J.H. (1966/1973) Mixing, Theory and Practice, Vol.I, Academic Press, N.Y.

French, J.A. (1981) "Flow Approaching Wash Water Troughs", J. Environmetal Eng. **107**, 359

Oldshue, J.Y. and Trussell, R.R. (1991) "Design of Impellers for Mixing", Mixing for Coagulation and Flocculation, American Waterworks Association Research Foundation, Denver.

Griffiths, S. (1996) "The Effect of Impeller Design on Maximum Shear Stress and the Resulting Impact on the Flocculation Process", J. Chartered Inst. Water Environmental Management, **10**, 324.

CHAPTER 5.

CHEMICAL REACTION ENGINEERING
CONTINUOUS FLOW PROCESSES

5.1 Introduction

This is a branch of chemical engineering which covers the design of equipment for carrying out chemical, biological and physical processes. Laboratory tests such as the coagulation jar test are usually batch tests in which the components are mixed and the entire volume remains in the vessel for the same period. Data on the kill rate for bacteria when exposed to disinfectants for example are usually based on batchwise bench top measurements. Water treatment plants operate on a continuous basis and there may be a considerable element of by passing which may have a dramatic effect on the outcome. For disinfection in particular very high efficiencies are required and the distribution of the residence time for the elements of the flow is particularly critical.

Levenspiel (1972) has written a comprehensive reference work on this subject. Unfortunately the relevant concepts are not appreciated as widely as they should. McNaughton and Gregory (1977) referred to Levenspiel's book but did not develop the concepts given in it. Indeed there has been very little to guide designers of contact tanks and even Warden (1985), when contributing to the BEWA Guidelines on Disinfection could only offer some undimensioned sketches of baffled serpentine flow tanks with baffles in the longitudinal and transverse direction labelled 'good' and 'bad' respectively.

5.2 Flow in Tanks and Pipes

In a flowing system the fluid seldom moves in a piston or plug flow manner. There is usually a degree of longitudinal overtaking and mixing with part of the fluid passing through more quickly and part lagging behind. Two models are used to describe the hydraulic dispersion of such a system, the **dispersion** model (Fig.5.1) and the **tanks-in-series** model (Fig.5.2). The former is more applicable to flow in a pipe or channel where one can envisage a sharp front or band of dye being injected. As the fluid passes along the pipe the sharp band disperses into a fuzzy smear. Ideally the concentration of the dye along the pipe follows a statistically normal distribution with the median ideally being at the mean residence time point defined by the volume divided by the flow.

Levenspiel gives the following equation for the concentration versus time:-

$$C = \frac{1}{2\sqrt{\pi \, (D/uL)}} \cdot \exp\left[-\frac{(1-\Theta)^2}{4 D/uL}\right]$$

where Θ is the dimensionless time (ie the ratio of the actual time to the mean residence time based on the volume ÷ flow). C is the concentration of the injected tracer as a ratio of the concentration that would have occurred if the contents of the tank or pipe had been fully mixed throughout the tank immediately after injection. The group D/uL is known as the `dispersion number' (sometimes the inverse is called the Bodenstein number) The diffusivity D is a measure of the rate of mixing,. The velocity u and the length of the path L assume a pipe or channel configuration.

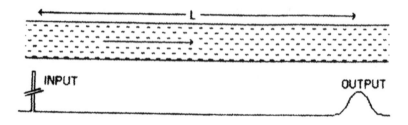

Fig. 5.1 Flow dispersion model.

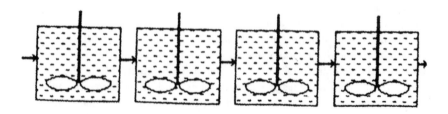

Fig. 5.2 Tanks-in-series model.

These parameters have little significant meaning in a homogeneous fully mixed (or back mixed) tank and the tanks-in-series model is then more useful. This model actually assumes a set of perfectly mixed tanks in series flowing into each other.

If the number of tanks is J then the following equation may be applied in place of the above:-

$$C = \frac{\Theta^{(J-1)} \cdot J^J}{(J-1)!} \cdot \exp(-J\Theta)$$

For values of J between about 5 and 20, D/uL is approximately equivalent to 1/(2J), but above 20 tanks-in-series factorial (J-1) becomes unmanageably large for computation. If J=1 then D/uL is infinite and the relationship no longer holds. Thus for computing purposes the equations used are changed at an intermediate point, say between J=10 and 20, without any apparent discontinuity. It may be noted that the statistical variance σ^2 of the distribution curve is simply 1/J, and the standard deviation σ is the square root of the variance.

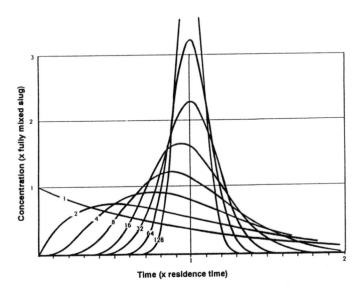

Fig. 5.3 Dispersion curves for a slug injected into various numbers of tanks in series.

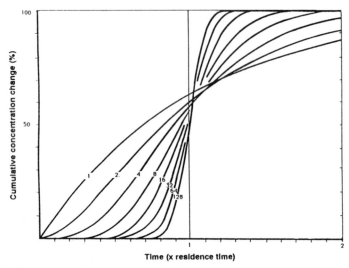

Fig. 5.4 Cumulative change curves following a step change in various numbers of tanks in series.

The above equations may be summed to give a cumulative curve which represents the output from a step change in the feed composition. Levenspiel notes that the 16th and 84th percentiles in these cumulative curves (ie. the times at which the change is 16 and 84% complete) are 2 standard deviations (2σ) apart. This provides a simple way of analysing tracer experiments used to calibrate tanks. In this case a suitable tracer is dosed or dosing is stopped, the concentration versus time in the outgoing stream is then analyzed and the result plotted on probability paper to enable the above percentiles to be read off from the resultant straight line. Lithium salts are often used for this purpose because lithium is easily measured by atomic absorption spectrometry at very low concentrations and also it is not usually present at significant levels in the water, neither is it toxic.

Figs. 5.3 and 5.4 are computer prints of the dispersion and cumulative curves for a logarithmic series of values of J from 1 to 128. It is suggested that the J value presents a convenient measure of the tank performance, although chemical engineers tend to used the dispersion number, D/uL, more frequently.In the water industry some other more ad hoc parameters have been coined. Like many such parameters they have no sound basis and in this case they are not easily used in mathematical calculations. The 10 percentile time has been advocated by some. This, expressed as a fraction (t_{10})

of the mean residence time, would be an equivalent parameter. Also some American literature refers to the Morril Index, which is the ratio of the 90 percentile to the 10 percentile times. The programme used to plot the curves in Figs. 5.3 and 5.4 has been used to calculate conversion curves to relate t_{10} and Morril Indices to the number of tanks in series or J number (Fig.5.5).

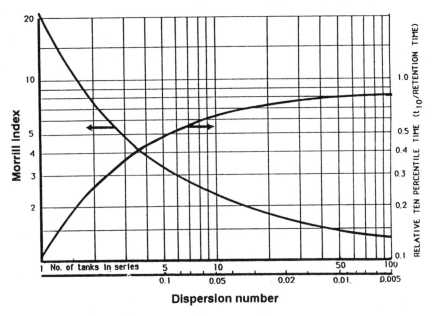

Fig. 5.5 Conversion curves for dispersion numbers to Morril indices and ten percentiles.

Like most ideal models both of the above are approximations of reality. It is possible to have hybrid versions. One immediate failing of the tanks-in-series version is the assumption of an infinite rate of mixing in each stage. A tracer will take a finite time to reach the outlet of a practical tank even if in the long term the tank does become fairly homogeneous. The practical model may correspond to a sequence of plug flow plus mixed flow. Simple square or circular tanks may develop a rotary motion with a tracer pulse producing a cyclic set of pulses which decay with time. However such tanks will have a poor performance as a reactor. Also one can have dead zones which reduce the effective capacity of the tank. It is possible from the shape of the output curves to analyze the situation but this is only an interesting academic exercise if the disinfection efficiency can be measured directly by other means. Computational fluid dynamics computer programmes, for example CONTANK™ which is aimed at disinfection, analyses the system as a set of small cells and calculates the progress of disinfection. Such programmes are not affected by these approximations.

5.3 Effect of Dispersion on Reactor Efficiency

For a <u>batch</u> process behaving as a first order reaction (for example, obeying the Chick Watson Law for disinfection) the time t_b for the reaction to achieve a given efficiency, which may be expressed as a conversion or disinfection factor F, is defined by:-

$$t_b = \frac{1}{k} \cdot \ln\frac{1}{F}$$

where F =(concentration in)÷(concentration out) ie. N_o/N, k is the first order reaction constant ($N=N_o \exp(-kt)$ for the batch situation). N_o/N may also be quoted as a disinfection or reaction efficiency index, $Log_{10}F$.

Levenspiel gives the following equation for the time t_J required to achieve the same efficiency in a dispersing system with J effective tanks.

$$t_J = \frac{J}{k}[F^{1/J} - 1]$$

Thus the ratio t_b/t_J may be regarded as an <u>efficiency factor</u>. The reciprocal, a **compensation factor** `CF' is more useful.

$$F_I = J\frac{[F^{1/J} - 1]}{\ln(1/F)}$$

`CF' is therefore the extra time (or the extra disinfectant) required to achieve the same efficiency in a given flowing system as in the ideal batch system, expressed as a ratio.

Table 1. which is based on the above equations gives some idea of the extent of the risk inherent in tanks with various dispersion numbers and the benefit of a little extra time.

Table 1.

Actual Efficiencies and Compensation Factors for various Nos. of Tanks in Series (or Dispersion Numbers) and Target Efficiencies

No. of tanks in series	Target disinfection index	Actual disinfection index	Compensation factor
1	100	5.6	21
2	100	11	3.9
4	100	21	1.9
8	100	38	1.4
16	100	57	1.2
32	100	74	1.1
64	100	85	1.05
1	10,000	10	1100
2	10,000	31	21
4	10,000	120	3.9
8	10,000	460	1.9
16	10,000	1400	1.4
32	10,000	3300	1.2
64	10,000	5500	1.1
1	1,000,000	15	72,000
2	1,000,000	63	140
4	1,000,000	390	8.9
8	1,000,000	3,100	2.7
16	1,000,000	21,000	1.6
32	1,000,000	97,000	1.3
64	1,000,000	270,000	1.1

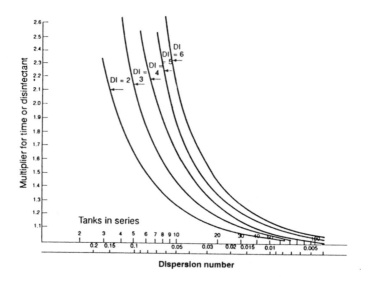

Fig. 5.6 Compensation Factors for first order reactions.

The magnitude of the Compensation Factor is greatly influenced by the disinfection factor. There is a considerable penalty if the required disinfection factor is overestimated. Polluted river water may contain 10^4 bacterial counts/100ml. After coagulation and filtration it may contain only 100 counts/100ml. Zero values are of little use in design calculations. Thus if one were to aim for say 99% compliance with zero in the disinfected water one would expect no more than 1 in 100 samples. The value for N would then be 0.01 and the Disinfection Index would be 4 or 6 for the filtered and raw water respectively.

Fig.5.6 presents the results of further calculations and shows the relationship between the Compensation and Conversion/Disinfection Factors. One can see how the dispersion number has to fall as the disinfection factor or index rises if the Compensation Factor is to be maintained at a given value. From the above arguments therefore a practical tank for a doubtful water should have a value for J of 25 or 45-50 respectively for a treated or untreated water respectively, ie. a Dispersion Number of 0.02 or 0.01 respectively.

At this point the rigorous theory starts to desert the engineer. However if one reverts to the dispersion model some help is available. For turbulent flow at Re>5x10⁴ the Dispersion Number (D/ud_h) for a length of pipe equal to one diameter (or one hydraulic diameter in the case of a noncircular pipe or channel) is given by Levenspiel as 0.2. (d_h = the hydraulic diameter ie. 4 times the cross section ÷ wetted perimeter). Thus for J=50, Dispersion Number=.01 a channel or pipe 20 hydraulic diameters long would be required. Such a configuration would be an inconvenient way of holding the necessary volume and hence the usual serpentine flow pattern within baffled tanks has been adopted.

Computational evaluation of such tanks shows immediately that some of the efficiency of such tanks is lost in the U bends because the flow tends to stream down the outer wall for a few diameters. (Miller (1990) provides data on hydraulic losses that show such streaming to affect the headloss at a subsequent bend up to 30 diameters downstream.) This can be rectified with small ancillary vanes or baffles (not the expensive curved vanes as used in oxidation ditches). This is where CONTANK™ comes into its own.

Plain rectangular or square tanks have been at a considerable disadvantage in that the efficiency is intrinsically low, furthermore, no means have existed to enable even an approximate assessment to be obtained. Fortunately programmes like CONTANK™, for example, are now available to enable such tanks to be appraised and the location of baffles etc. to be investigated fairly quickly.

5.4 Flocculation

This is more complex than disinfection because in the latter case the used chlorine or dead bacteria have no effect on the process. In flocculation the initial floc particles collide and disappear exponentially but they produce a new generation of larger less dense particles which accelerate the process. The components are discussed in Chapter 3. The collision rate varies with the cube of the diameter of the larger particle of a pair in collision so that some back mixing is very beneficial. Also in contrast to disinfection the efficiency required in flocculation is much less. In any clarification process only 90% removal or less may be required and the characteristics are such that the dirtier the raw water, the faster and more efficient the process becomes.

No model of flocculation in flowing system based on the available theory appears to have been published. However there are indications that the design of the flocculation system is important and for flotation for example two stages rather than one are generally used. Sinuous pipe flocculation which would give a sharper residence time distribution appears to offer reduced flocculation times. However it is possible in this case that true plug flow may be detrimental because the larger floc particles would not encounter the smaller ones and accelerate the process.

5.5 Oxidations and Adsorption

Levenspiel gives data on dispersion through packed beds but not through fluidised beds. One may assume however that the latter are close to the former. The author's experience in fluidised bed ion exchange suggests that the resin beads are fully mixed in each bed and have a uniform composition but the water passes through with a relatively low dispersion and certainly with no element of flow reversal. One may therefore compare such a system with the blanket in a sludge blanket clarifier. The sludge will have a uniform composition. Thus if powdered activated carbon is dosed the carbon at the top of the blanket will be as exhausted as at the bottom and the equilibrium concentration which defines the maximum removal will be relatively high unless a large excess of carbon is added.

In contrast granular activated carbon in a filter does not mix much and such filters are often deliberately managed to minimise mixing so that a sharp absorption front advances through the bed. The water passes out through virtually uncontaminated carbon.

Flow in granular beds is not of a perfectly plug flow nature. Levenspiel (1972) gives the Dispersion Number for packed beds as De/ud_p as 2.0. A bed 1000 particles deep will correspond to about 100 tanks in series.

5.6 Disinfection

5.6.1 Kinetics

Many, if not most, chemical reactions in solution tend to be 'first order' and proceed at a rate proportional to the concentration of one of the unreacted species present. This is indeed the case with disinfection in water treatment. The residual concentration of bacteria is then defined by the exponential equation:-

$$N=N_o \exp(-kCt)$$

where N_o and N are the initial and final concentrations or counts,and t is the time (in seconds if SI units are used). C in this equation is the concentration of the disinfectant (usually expressed in mg/l), which is assumed to be constant for the purposes of this equation and process. k is the reaction constant for the particular species of bacteria or other organism and for the particular conditions of temperature and pH.

This basic behaviour was discovered early in the 20th century by Watson and Chick using disinfectants such as calomel (mercurous chloride) and phenol and has become known as the Chick-Watson Law. The chemistry of the disinfection process has been reviewed relatively recently by Haas (1990) for the American Waterworks Association. Copious data on Giardia has been provided by the A.W.W.A. (1991), but the situation regarding E.Coli is less well defined.

In the literature the product of concentration and time (in minutes) has become an accepted parameter for defining disinfection conditions, and values for 99% kill under batch conditions, the so called 'CT' numbers, are listed. The reaction constant k is related to the CT number as follows:-

$$k = 0.076/'CT'$$

As an example Haas gives a CT value for free chlorine on E, Coli at 5°C and pH 8.5 of 0.4 mg-min/l. Hence k would be 0.19mg-s/l.

Unfortunately the American Guidelines for the Surface Water Treatment Rule, von Huben (1991), have abandoned the above definition of CT and have introduced 99.9% and 99.99% versions, ie. '3 log' and '4 log' versions. In this case:-

$$k = 0.076/'CT_{99}'$$
$$= 0.115/'CT_{99.9}'$$
$$= 0.154/'CT_{99.99}'$$

As indicated earlier the CT number is meaningless in practice without data on the hydraulic behaviour of the system. With this exponential relationship a small percentage of the flow passing though more quickly than intended will cause the disinfection efficiency to fall very significantly.

5.6.2 Existing Standards

The current WHO guidelines recommend 0.5mg/l of free chlorine at a pH <8 for 30 min. Published data for the indicator organism E.Coli at 2-5°C at pH8.5 show the 99% CT value to be 0.4mg-min/l. At lower pH values where free hypochlorous acid is present the 99% CT value is about 10 times lower. Taking the conservative former figure, 96s are required under batch conditions to achieve a disinfection index of 4 or 144s for an index of 6. The WHO guidelines would seem to provide a considerable margin of safety (12.5 fold) such that efficient contact tanks may not be necessary. It is not known whether any such considerations went into these guidelines. In theory it should be possible to reduce the contact time with a more efficient tank but the basic data is imprecise. However it might be dangerous to design on the specific indicator organism E. Coli. A good tank will also improve the efficiency with respect to more refractory organisms. Conversely it seems that from a regulatory point of view, if the above conditions have to be maintained then an efficient tank design is unnecessary.

If one were not constrained by regulatory conditions then it would follow that the size of contact tanks could be reduced if the dispersion number is low. It seems on examining the literature that at the present time the basic data is insufficiently definitive to provide a sound basis for the strict chemical reaction engineering design of disinfection tanks. The available data on disinfection kinetics is not entirely consistent and they need to be confirmed and gaps filled in.

The chlorine demand of a raw or treated water is variable and in any case cannot be measured with great precision. ±20% is probably a pragmatic figure at low levels. The disinfection reaction depends on the residue after the demand has been satisfied. One may argue whether at, say, 0.5 mg/l residual an extra 0.1 mg/l is significant. The extra 0.1mg/l is indeed less significant if the chlorine demand is, say, 1.0mg/l or more. Thus an Compensation Factor of 1.2 probably constitutes a practical ideal situation which is close enough to the batchwise condition not to warrant further action. At such a value the penalty is submerged in the uncertainty of the basic data. In any case one would have to consider the cost penalty of extra disinfectant or the cost of a slightly larger tank over the amortised life of the tank against the capital cost of extra baffling.

There may be a temptation to go for tanks with low dispersion but the optimum solution may well be a compromise with a simpler tank of larger volume (particularly depth) or to accept the higher running cost of extra disinfectant.

E. Coli bacteria have been taken as an indicator of bacterial pollution, and as far as the U.K. is concerned and indeed most other parts of the world this is the case. The WHO recommendations also have this basis. There are however very many organisms present, some of which are more resistant. The Surface Water Treatment Rule, Anon (1991), in the USA is based on the removal of Giardia Lamblia. Confusion can occur unless this is appreciated.

Deficiencies in tank design can be counteracted by increasing the concentration of the disinfectant or the contact time. The present practice of purchasers who demand that the contractor shall guarantee the bacteriological quality of the final treated water is hardly sound or fair to either party. The contractor has no control over the bacteriological quality of the raw water or of the disinfectant demand (the amount absorbed before any residue for disinfection is left). However he can increase the dose to compensate for deficiencies in the design. The purchaser will want evidence that he has a cost effective process unit. In theory this may be the case but hitherto it has not been easy without expensive testing to arrive at a good design.

The American Waterworks Association documents on the Surface Water Treatment Rule, Anon (1991), are some of the first in the water industry to highlight the importance of tank design. Examples of poor, average and superior baffling are given but none of these compare well on the basis discussed below. Computational fluid dynamics as a science has now reached the point where dedicated user friendly programmes such as CONTANK™ * are available. This programme is aimed at the design engineer rather than the computer specialist.

5.6.3 Modelling (CONTANK™*)

This has been mentioned above. It is a fairly straightforward finite difference analysis program intended for the practising engineer which allows the plan of an existing or proposed tank to be drawn on the screen on a defined grid, inlets and outlets added and any baffles drawn in, all with the aid of the 'mouse'. Various input options such as weirs, submerged pipes, mass vertical flow, are available. The program accepts data on disinfection reaction constants and other process parameters. The output, which can be dumped to a printer, includes a dispersion number, t_{10} and t_{90}, a disinfection index contour map overlaid on the plan, flow vectors, level variations (which can be used to calculate loads on light baffle walls) and the residence time distribution curve. To keep the program within the capability of a desk top PC it uses a two dimensional simulation, which is valid for tanks of uniform depth. There are limits to the number of computational cells.

It does not require great computing expertise. Experience has shown that the computer literate engineer can master the programme in an hour or two. Each simulation run takes only about an hour to run. It has been found to provide most useful quantitative information on most contact tank configurations and constitutes a major step forward where the designer has previously been operating blind. It can be used for other applications such as flocculation tanks but is limited to homogeneous single phase systems and therefore is not applicable to an ozone contact tank with gas sparging.

Figs. 5.7 and 5.8. illustrate the velocity vector and disinfection index screen dumps of an example. A detailed understanding on the part of the user of hydraulic theory is not required. However the user must define the disinfection rate constant k as the programme may be used to simulate many disinfectants and conditions.

Programmes of this type will enable designers to develop standard designs for their own use as well as being able to assess existing tanks. It is however beyond the scope of this paper to explore the characteristics of specific designs and to recommend standard configurations.

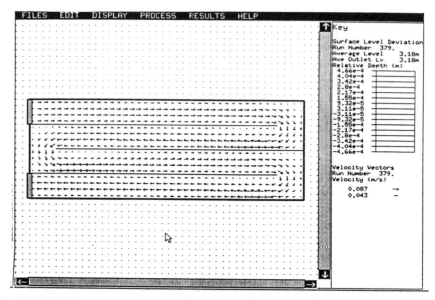

Fig 5.7 Velocity vector diagram produced by CONTANK.

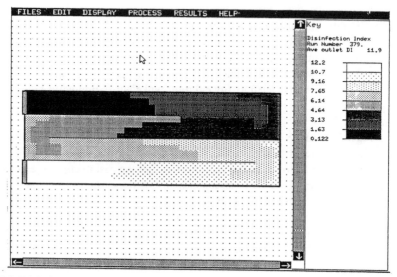

Fig. 5.8 Disinfection Index Diagram produced by CONTANK.

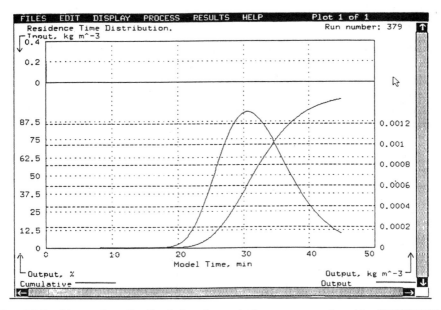

Fig. 5.9 Residence time distribution and cumulative curves produced by CONTANK.

5.7 References

Anon. (1991) Guidance Manual for Compliance with the Filtration and Disinfection Requirements for Public Water Supplies using Surface Water Sources. American Waterworks Association, Denver, ISBN 0-89867-558-8

Anon.(1993) Guidelines for Drinking Water Quality, WHO, Geneva, 1993

Anon. (1993) Process Guidelines for Drinking Water Disinfection, BEWA, (now British Water, 1, Queen Anne's Gate, London SW1H 9BT

Haas, C.N., (1990) Water Quality and Treatment, Ch. 14, American Waterworks Association, Denver.

von Huben, H. (1991) Surface Water Treatment :The New Rules. American Waterworks Association, Denver, 1991, ISBN 0-89867-536-7

Levenspiel, O. (1972) Chemical Reaction Engineering, Wiley, Earlier editions 1966,1969.

McNaughton, J.G. and Gregory, R. (1977) WRc TR60.

Miller, D.S. (1990) Internal Flow Systems, BHR, Cranfield, England, 1990.

Further Reading.
Lawler, D.E. and Singer, P.C. "Analysing Disinfection Kinetics and Reactor Design: A Conceptual Approach Versus the SWTR", J. American Waterworks Assn., **85**, 67, (1993)

CHAPTER 6.

PRETREATMENTS

6.1 Screens

A water plant treating a river water must be protected against large solids that would block or damage the process units. floating tree trunks and carcases of dead cattle present extremes that may be encountered. Fish may be a nuisance in some circumstances but even grass cuttings, especially from the removal or management of riverside vegetation can be troublesome. The possibility of oil floating down rivers must also be considered and the intake arrangements may have to be designed to prevent ingress. Leaves in the autumn, especially large ones such as horse chestnut are a hazard.

The first requirement is to prevent damage to the low lift or primary pumps. The size of screen will of course depend on the size and type of pump. The degree of protection needed by the process units depends very much on their design and method of operation. Thus for example a hopper tank clarifier, a Reactivator type solids recirculation clarifier, a direct filtration plant or a dissolved air flotation plant normally need little protection. No small orifices are involved and a floating cabbage for example would not interfere with operation although eventually it would have to be skimmed off. However if any of these units were proceeded by a corrugated plate type static mixer then fairly fine screening would be necessary (orifice mixers or weirs are often as effective and avoid the need for such ancillary equipment.) Pulsators and other flat bottomed clarifiers, Accelators and many other clarifiers often have 25-50mm inlet orifices. Even many filter designs have surface flush toughs with similar orifices. In some instances such small orifices could easily be eliminated and with it the need for a stage of screening. The most severe situation is probably presented by upflow filtration of the type where the flow enters through a nozzle floor with apertures of only 6-8mm, or where the wash water must pass through a retaining screen finer than the media size. Screening in such cases will be an added cost which must be set against any savings involved in such equipment.

6.1.1 Coarse Protection

Where oil or floating debris is anticipated **floating booms** and static screens are normally fitted to the works inlet. Generally, the larger the works the more extensive the screening barriers installed. Where bankside storage is used it may often be wise to incorporate a second line of defence against oil, such as an interceptor tank at the pump discharge into the reservoir. It is easier to control oil with a constant level than with a varying one. Means of skimming the oil must of course be included. Standby screens will normally be provided for all moving versions.

Coarse bar screens normally have a bar spacing ranging between 50-100 mm. Installed at 60-80° to the horizontal. They may be hand or mechanically raked and between 1-4m wide and 1-15m deep depending on the size of works.

6.1.2 Medium Screens

Medium screens normally have a bar spacing ranging between 20-50 mm. They are installed at 60-80° to the horizontal and also may be cleaned by hand or mechanically.

6.1.3 Fine Static Screens

Fine screens normally have a bar spacing ranging between 5-20mm. A minimum of two (duty/standby) screens is usual with an angle of 80° to the horizontal. Debris is removed either by mechanical rakes or air backwashing. The approach velocity is less than 1.4 m/s to avoid drawing in fish.

6.1.4 Band Screens (Fig. 6.1)

These normally have panels with a 2-10 mm mesh spacing, linked to form a continuous band like the track of a tracked vehicle but arranged vertically. They are used for screening of river intakes with a high flood level. Debris is removed by backwashing the mesh panels with high pressure jets set above the deck level. The immersion depth should be specified by the supplier and is dependant on mesh size and throughput capacity. They are usually installed downstream of primary bar screens.

6.1.5 Drum Screens (Fig.6.2)

These are available as single or double entry drum/cup screens. They occupy more space than the band screen but need less maintenance. Drum screens normally have 0.5-10 mm mesh. Like the band screen debris is removed by high pressure water backwashing of the screen above the deck level. The drum diameter should be sufficient to extend 1.25 m above the deck and downwards to a sufficient depth to have enough surface area submerged to pass the required throughput. These are supplied in sizes ranging from 1.5-10 m in diameter and 0.2-3 m effective width. A minimum of two are usually required (duty/standby).

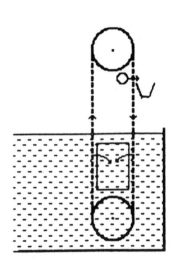

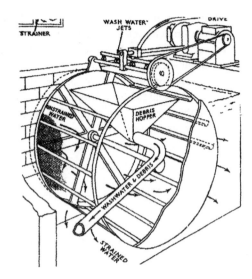

Fig.6.1 Band screen.   Fig. 6.2 Drum screen.

6.1.6 Microstrainers

Microstainers are similar to drum screens and are used to remove fine solids from raw water by passage through a fine mesh fabric. Usually they comprise aslowly rotating drum sealed at one end, and fitted with panels of fabric over the cylindrical surface. The top of the drum remains out of the water. The drum is generally 1.5 - 3m diameter and up to 3 m wide and has a variable speed drive.

The filtering fabric is made of either stainless steel mesh or nylon cloth with aperture sizes of 15, 23, 35 or 64 µm for filtration or larger for fine screening. The design allows sections of fabric to be replaced easily.

The backwash water jet system includes a trough for collecting the screenings. Growth of algae and blockage by zooplankton can be a problem and ultraviolet lamps or washing with sodium hypochlorite are often employed to control this. Acid washing may also be required to remove calcium carbonate scale if the water is supersaturated (algal blooms often raise the pH above pHs.)

As solids loading increases, the strainer will block more rapidly and cleaning frequency must be increased by increasing the speed of rotation. The maximum is normally 5 rpm. The maximum headloss is typically 0.3m. The washwater consumption is normally between 1 and 3% of volume treated. Disposal of this water can present a problem because of regulatory restrictions. The microstrainer in contrast to the other mechanical screens may be regarded as more of a treatment stage rather than a protector of treatment stages. It is often used to remove filamentous algae which would tend to block filters yet penetrate settlement tanks. Dissolved air flotation has probably reduced the popularity of microscreens.

Screens generally are mechanical devices which are manufactured by specialist companies which are able to offer a proprietary range and can provide advice.

6.2 Reservoir Mixing

While some might be surprised to see this listed as a treatment process it is sometimes as important and effective as many others. Algae tend to proliferate at specific depths where the light intensity is optimal for their growth. While they rely on photosynthesis for growth they often decline if there is too much light. Vertical mixing moves them from their preferred level and hinders development. Also deeper reservoirs often form an anoxic layer at the bottom where manganese and iron present in the bottom soil and mud dissolve under bacterial action. Mixing prevents such stratification and prevents this action. Without it wind and inversions caused by cooling of the surface water, particularly in the autumn can cause a sudden rise in the manganese level that may catch operators unaware.

Mixing of reservoirs was reviewed by Tolland (1977). It may be achieved either by high pressure jets suitable arranged or by air bubbling from sparge pipes laid on the bottom of the reservoir. Each installation is a special case but computational methods are available.

Mixing should be considered if treatment plants for reservoir or lake waters are being planned. It may not be practicable or economical on very large reservoirs whose main function is for hydroelectricity. Again the advent of dissolved air flotation has removed much of the case for mixing, in that algae are now less of a problem in treatment than was the case previously.

6.3 Aeration

It is necessary to distinguish between ground and surface water as the requirements are very different. Surface water seldom needs aeration unless the water is drawn from lower anoxic layers of reservoirs. The EC directive on Water Quality (1980) gives no guide levels for dissolved oxygen but gives a comment that it should not be less than 75% saturation. British regulations and WHO recommendations give no figures. Thus aeration may be selected or disregarded on the basis of being a means for meeting other requirements. It is also possible to have excess dissolved oxygen arising from algal activity or a rise in temperature and this can cause air blinding in filters even though the supernatant water depth is correct. Aeration will strip off the excess gas as well as correcting a deficiency . The advent of dissolved air flotation has made aeration largely redundant because the bubble blanket will only form if the water is already saturated and conversely this blanket presents a large surface area close to the surface and strips off excess dissolved gas at a pressure close to atmospheric.

There may however be situations where aeration or air injection may be beneficial even with dissolved air flotation. As discussed in Chapter 12 an anoxic water will have about double the air demand of saturated water for flotation. The deficit can be made up using less power by aeration techniques than by increasing the recycle.

Biological pretreatment has been used to oxidise ammonia, manganese, tastes and odours. One example is the fluidised microsand bed of which there are three examples in Britain. The amount of oxidation is limited by the dissolved oxygen. For higher levels multiple passes may be necessary with intermediate re-aeration.

Ground water may have to be aerated to strip carbon dioxide and raise the pH to precipitate iron and manganese as well as adding oxygen for the same purpose. The percentage efficiency with respect to the approach to equilibrium required for removing carbon dioxide is often much higher than for the solution of oxygen. However such stripping may cause calcium carbonate to precipitate, which will encrust filter media and pipework. Carbonate equilibrium calculations are essential. In such cases it may be necessary to inject air into a closed system that does not allow carbon dioxide to escape.

Aeration is a mass transfer process that is limited by the equilibrium situation. Efficiency is measured in terms of the fractional approach to this limit. Thus:-

$$Efficiency = \frac{C^* - C_{out}}{C^* - C_{init}}$$

where c^* is the equilibrium value, eg. the solubility of oxygen in air at 1 bar.

The Water Pollution Note referred to below introduces the concept of Deficit Ratio (R) where:-

$$R = \frac{1}{(1-E)} = \frac{(C_* - C_{init})}{(C^* - C_{out})}$$

This is used as a measure of efficiency in equations given later.

6.3.1 Cascade Aeration

This is the simplest technique and is suitable for up to about 50% approach to equilibrium both for stripping and solution. A Note on Water Pollution (1973) gives some design data for calculating the efficiency. (See also Tebbutt (1972) and Downing(1958).

$$R = 1 + 0.38ah(1 - 0.11h)(1 + 0.046T)$$

a is 1.8 for clean water and 1.0 for moderately polluted water, h is the height of fall (m) and T the temperature (°C).

Cascades more than about 2m in height offer little further benefit. The receiving pool is an integral part of the process and entrained air bubbles contribute to the mass transfer. Cascade aeration has the merit of being resistant to fouling or blockage with debris or calcium carbonate. It is particularly suitable for surface water and needs no fine screening beforehand. The retention time is short.

It is possible to build cascades within pressure vessels such as filters (Fig.6.3). If these operate above atmospheric pressure the efficiency is increased considerably although account must be taken of the blanketing effect of nitrogen (See

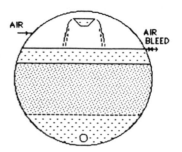

Fig. 6.3 Enclosed (submerged) weir over a horizontal filter .

air absorbers in Chapter 12). Thus the addition of 3mg/l of oxygen from air would require the same height as 1.5 mg/l from an 8:1 nitrogen/oxygen mixture in a closed vessel at 1 bar absolute or 15mg/l at 9 bar gauge. (The oxidation of 1mg/l Fe requires only 0.71mg/l O_2.)

6.3.2 Bubble Aeration

This is less applicable to water treatment plants. Its main application is where a continuous supply of oxygen is needed over a longer contact period. Thus it is appropriate for ozonation where a few minutes reaction time is needed and the ozone must be fed in continuously. A similar situation exists in the activated sludge process. Bubbles can useful for mixing both in reservoirs as already mentioned and also for chemical tanks.

6.3.3 Turbine Aeration (Surface aeration)

The applications are very similar to bubble aeration, with the added limitation of the size of tank needed to accommodate such devices.

6.3.4 Spray Aeration (Fig.6.4)

This is less popular than it used to be because of the advent of cheap plastic tower packings. Fine spray nozzles are used to produce small droplets. The process cannot readily handle surface water without fine screening and is therefore applicable to ground water. The efficiency is moderate, not as good as blown packed towers, but

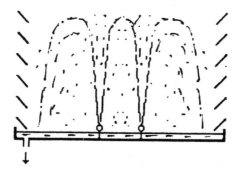

Fig. 6.4 Spray aeration.

it is not troubled by precipitation of calcium carbonate or fouling with iron bacteria. The efficiency depends on the precise nozzles used and these must be calibrated. Efficiencies up to 80% are possible. However the spray chamber must be well ventilated to allow the carbon dioxide to escape and a major problem is caused by wind drift, which apart from loss of water can cause iron staining on surrounding structures.

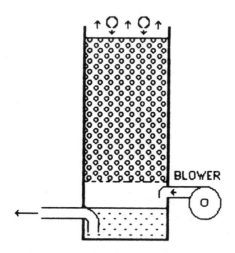

Fig. 6.5 Packed tower (forced draught) aeration.

6.3.5 Packed Towers (Fig.6.5)

Modern towers use lightweight plastic rings between 25mm and 50mm in diameter or saddles of proprietary design. The height depending on the efficiency required may be up to 3m or more. Efficiencies can easily be 90% or greater. They are also suitable for stripping volatile organic compounds such as chlorinated solvents, THMs etc. (Bilello and Singley 1986, Mumford and Schnoor 1985, Adams and Clark 1991,) as well as for oxygenation, stripping of carbon dioxide and even hydrogen sulphide. Surface rates based on the plan area are usually in the region of 50-100m/h. The retention time is again very short and significant reactions are not expected to occur within the unit.

For high efficiency a countercurrent forced air flow is essential. The volumetric flow may be about 10 times the water flow, but specialist books should be consulted for precise designs. For low efficiencies such as for plain oxygenation unblown packings may be used, in which case the air is entrained with the water. The efficiency is still higher than a cascade, and the configuration converted from linear strip to a plan area.

Packed towers avoid the wind drift problems of spray systems and are far more compact. However they are prone to fouling with iron (iron bacteria) at levels above about 1mg/l Fe. This high efficiency can lead to encrustation with calcium carbonate. These problems may be avoided by limiting the air flow and controlling the efficiency or by the use of sodium hexametaphosphate or other sequestrants which prevent deposition and crystallisation. However these additives will interfere with any downstream solid/liquid separation processes. By analogy with softening processes recirculation of calcium carbonate sludge or microsand should reduce deposition on packings to the point where occasional acid washing would suffice but this requires verification.

The efficiency of a packed tower varies with temperature and the size of the packing but as a guide a 2.0m tower will give about 90% efficiency at 10°C. Packing suppliers should be consulted for more precise data.

6.3.6 Air Injection

This is frequently used on ground water treatment plants where oxygen is required without loss of carbon dioxide. Few designers wish to provide a separate process unit and a compressed air supply is connected to existing vessels or to the incoming pipes. Little design data is available. There is a danger of air flowing along the crown of the pipe and not distributing into pressure filters uniformly. Static mixers may assist but the most positive arrangement is to feed individual filters with the air and to provide an internal cascade to assist dissolution (Fig.6.3).

Another method of dissolving air is the vertical shaft (cf. The Deep Shaft activated sludge process). The so called microflotation process uses this principle in which air is entrained in a down going shaft flowing at a velocity in excess of the bubble rise rate (>0.35m/s). Air either partially or fully dissolves under the increased hydrostatic pressure. Presaturation in such a manner can reduce the air demand for flotation and reduce power costs.

6.4 References

Adams, J.Q. and Clark, R.M. (1991), "Evaluating the Costs of Packed Tower Aeration and GAC for Controlling Organics", J. American Waterworks Assn. **83**,49

Anon. (1973) "Aeration at Weirs" Notes on Water Pollution No. 16, Dept. of Environment. London.

Anon. (1980) "Council Directive relating to the Quality of Water intended for Human Consumption", Official Journal of the European Communities. 80/778/EEC

Bilello, L.J. and Singley,J.E. (1986) "Removing Trihalomethanes by Packed Column and Diffused Aeration", J.American Waterworks Assn. **78**,62

Downing, A.L. (1958) "Aeration in Relation to Water Treatment"Proc. Soc. Water Treatment & Examination, 7,66

Mumford, R.L. and Schnoor, J.L. (1985) "Mass Transfer of Volatile Organics in a Packed Bed Stripper" Proceedings of American Waterworks Assn. Congress, p 485

Tebbutt, T.H.Y. (1972) "Some studies on Re-aeration at Cascades", Water Research, **6**,297

Tolland, H.G. (1977) "Destratification/Aeration in Reservoirs", WRc Technical Report TR50

CHAPTER 7.

NON-FLOCCULATING SETTLEMENT UNITS

7.1 Introduction

The design of most settlement systems used in water treatment exploits the particular properties of hydroxide floc and it is difficult to separate the features that relate purely to the settlement of inert particles that are not modified or flocculated in the process. In the context of water treatment non-flocculating processes may include presettlement to remove coarse material that will settle at an acceptable rate without treatment, or to remove high concentration of suspended solids (say >1-2000 mg/l). These would interfere with the operation of separation units designed for lower concentrations, eg. flat bottomed floc blanket tanks. Wash water recovery, traps for washed out filter media and thickeners are other examples. The basic principles also apply in waste water treatment in storm water tanks, primary and secondary settlement tanks.

The basic design parameter is the surface area, which is selected on the basis of the largest of either the maximum over flow rate to avoid entrainment of solids of the intended cut off size, or the underflow rate required to give the desired concentration of sludge.

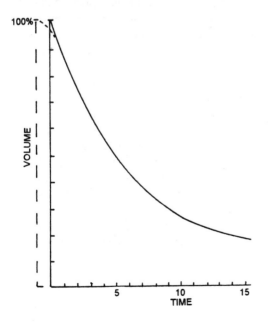

To determine the overflow rate samples of the suspension are allowed to settle and the maximum settlement rate of the suspension measured (always expressed in linear terms, not percentage or fractional volume) , after allowing for the initial turbulence to cease. (The rate falls as the concentration of the suspension increases). See Fig.7.1.

Fig. 7.1 Example of a settlement curve. (Based on a concentration of 5% and a settlment rate of 0.2 height units per time unit).

Alternatively it may be decided that the unit is to remove particles above a certain size in which case the settlement rate may be calculated. The area of the tank regardless of whether it has a horizontal or vertical flow will be:-

$$A = u_s \cdot Q$$

where Q is the volumetric throughput. The underflow rate may be determined using a pilot plant or by an extension of the laboratory test. In the latter case it must be remembered that the deposit in a measuring cylinder is not uniform and that the material at the bottom will be more concentrated than that at the top. Thus it is necessary to measure initial settlement rates at various concentrations by allowing the suspension to settle further after measurement of the rate and then to decant off the supernatant, to remix and settle again, thus obtaining a set of data for higher concentrations perhaps twice, four times and eight times, until little measurable settlement is obtained. The data may be plotted or a mathematical relation obtained. If the rate is multiplied by the concentration the mass flux is obtained. This may be plotted as a Kynch diagram as discussed in Chapter 3. From this the underflow rate may be derived. For presettlement tanks handling raw water and where concentrated sludges are not required or even not desirable the overflow rate is usually the limiting factor. In thickening the underflow will usually govern the size, particularly if the main clarification unit already produces a reasonably concentrated sludge (>1% w/w). It must be remembered that there must be a mass balance. Thus if a presettlement tank has to handle a 10,000 mg/l (1%) and the underflow is not to exceed 10% to avoid possible blockage then 10% of the incoming flow will emerge as sludge. The incoming flow must therefore be increased to maintain the output in such circumstances. Concentrations of the above level are not uncommon at times in rivers which flow through arid areas after storms or at times of glacial melt.

Chemical treatment may be used to assist settlement in many applications where non flocculating tanks are used but because flocculation is so fast at high concentrations no account has to be taken of this aspect in the design.

Experience has shown that conventional coagulation may be relatively ineffective on waters with high levels of suspended solids. Inorganic coagulants merely add to the solids and a very slow settlement rate is obtained. Pre-settlement to remove the bulk of the suspended solids, possibly with added polyelectrolyte to improve the settlement rate, will allow the residual solids, possibly 50-100mg/l, to respond to conventional coagulation at low doses in a subsequent stage and to yield a good quality settled water.

7.2 Square Pyramidal or Conical Tanks

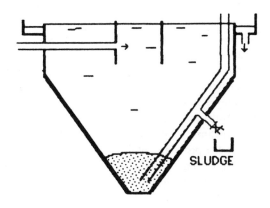

Fig. 7.2 Pyramidal settlement tank.

These are the simplest and are suitable for small installations and industrial units. Conical ones are the most common when constructed in steel and pyramidal in concrete. A 60° slope is used almost universally but there is evidence that the slope may be as low as 52°. Lower slopes increase the likelihood of sludge hanging up on the walls. The upper part above the nominal separation level (normally about 1.5m below water level) usually has a vertical wall.

Conical or pyramidal tanks usually have a centre upturned inlet surrounded by a stilling ring (Fig.7.2) which diverts the radial flow at the surface downwards. The stilling ring is normally about 15% of the tank diameter and may project about 1m below the surface. Decanting is usually from the periphery, through orifices or over weir notches (the latter in the case of sewage treatment, the former often in water treatment.) Floating scum where likely, is retained by scum boards in front of the weirs and a scum scraper must then be provided.

The sludge outlet at the bottom must be designed so that it is not likely to choke. A downward going outlet at the bottom is prone to blockage and a rising outlet from the bottom to a bend through the tank wall is safer as the sludge cannot settle in the outlet. A widely used arrangement is to have a rising pipe to the surface and a branch to the sludge bleed valve part way up. The pipe may then be rodded from above while the tank is full and even while attempts are being made to bleed sludge. A bottom drain will still be required for emptying and means for rodding must be included.

As a general rule in all cases bleeding sludge intermittently through a full bore valve is more reliable than attempting to bleed continuously with a throttled valve. The latter can easily choke and as sludge tends to thicken the flow may fall off, which allows the sludge to thicken further, leading ultimately to cessation of flow. Adjustable bell mouths have proved moderately successful but are still less reliable than a full bore intermittent flow.

Such tanks are not self regulating. Discharge of sludge must be monitored either by checking the concentration of the discharge or the sludge level in tank. If sludge is allowed to remain for some time it may consolidate and "rat hole", leaving a thick sludge on the tank walls and a core of thin sludge or water going down to the outlet. If this is a problem means of moving the hanging sludge must be provided. A motor driven scraper solves the problem but destroys the simplicity of the tank. Such sludge usually slumps when the tank is drained down. Air lancing has also been used for this purpose, this being applied while there is no overflow.

Square tanks often follow the design of conical tanks. However there is no reason why the inlet should not be along one wall with means for directing the incoming flow downwards. Like wise the decanting weir in this case is on the opposite side. This arrangement is usually cheaper and as effective. It is possible to have a conical tank with crescent inlet and outlets each covering a 90° sector, but extra cost would be involved. Perforated pipes may also be used for decanting if these enable the cost to be reduced.

7.3 Rectangular Hopper Tanks (Fig. 7.3)

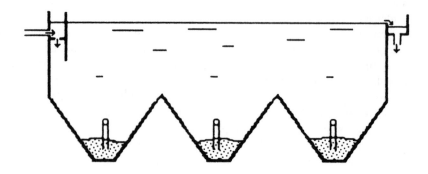

Fig. 7.3 Rectangular multihopper settlement tank.

For larger installations in situations where civil engineering construction is cheap compared with steel fabrication or where maintenance skills are limited, it is possible to have sets of hopper tanks arranged in line with a common vertical boundary wall above the sloping section. Such an arrangement corresponds to a horizontal flow tank as far as the upper section is concerned but the sludge can be removed by hydraulic means from each individual hopper. The early hoppers will tend to collect the most sludge in this case.

7.4 Flat Bottomed Horizontal Flow Tanks (Fig. 7.4)

These come in various types ranging from purpose made settlement tanks to sludge lagoons and even reservoirs. In the simplest versions sludge may be allowed to accumulate over a period of possibly months or years and eventually drained down in rotation and excavated. In developing countries where labour is cheap such settlement tanks are very appropriate. Even where front end loading machinery is available such tanks may still be an effective means of dealing with sludges. In mineral recovery settlement lagoons are widely use and to avoid instability on sloping sites the slurry is applied at the dam side so that the coarser particles settle out on parts that may be required to withstand a thrust. Finer more colloidal particles settle nearer the decanting section often in an excavated area at the inner edge of the lagoon.

The size of the tank or lagoon will in such cases be determined by the quantity of solids to be stored. A useful rule of thumb is that solids will settle as long

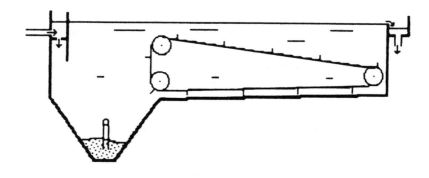

Fig. 7.4 Scraped horizontal flow flat bottomed settlement tank.

as the surface velocity is less than 17 times the settlement rate. Thus if particles settling at 3m/h or more are to be removed the velocity should not be greater than 50m/h. For example, to settle a water with 100mg/l suspended solids from a flow of 1000 m3/h in a tank with a 4:1 length/ breadth ratio, where the solids settle to a density of 0.5% and 3 months storage is required, 43,200 m3 would be required, ie a lagoon 60m X 120m X 3m deep. The sludge would collect to a depth 333mm below the surface, ie. the total depth would be 3.33m. Such tanks are however designed on average loadings, not peaks.

Scraped tanks with continuous removal of solids are designed on the basis of a surface rating as with the hopper shaped tanks already described. Such tanks provide the ideal model on which Hazen based his equation. However in practice they do no perform entirely in an ideal manner. There is an element of bypassing which is difficult to eliminate. To achieve ideal plug flow the length to breadth ratio should be as large as possible. A figure of a least 4:1 is commonly quoted. True ideal ratios calculated from chemical engineering principles would not give a practical tank. Baffles as used in good contact tank designs would no doubt be beneficial but would interfere with sludge removal. In contrast to contact tanks the density of the suspension causes the incoming flow to spread over the floor so that separation becomes vertical. The residual velocity of the incoming water also produces a residual surge or jet that clings to the tank floor and walls and persists possibly to the other end of the tank. Inlets on such tanks may consist of an array of downward facing orifices. The resultant jet may be defected by an upstand wall, or diffused by finger baffles. In some designs perforated walls are used as diffusers. However in non flocculating tanks the final quality demanded is not usually as high as after coagulation and a compromise may be reached.

Temperature however is a greater problem. A 5° rise or fall (10-15°C)will change the density of water relatively by 6 X 10^{-4}. This is similar to the submerged density of a floc particle or the density of a 40mg/l suspension of clay. Temperature changes of a fraction of a degree can cause stratification and short circuiting. Higher rates with shorter retention times reduce the effect of temperature. Shallow rivers produce more difficulty than reservoirs but the worst situation occurs with long feed mains exposed to the sun.

A common hybrid is the horizontal flow tank with a hopper at the inlet end. Coarse solids fall direct into the hopper, which is not subject to the conveying limitations of the scraper. The hopper also serves as a drain for the scraper, as the solids conveyed by this are pushed into the hopper.

For continuous sludge removal full width chain and flight scrapers are used. These use a sprocket chain and are expensive both to install and maintain. Speeds of 3-4 m/min are normally used.

For low loadings cable operated scrapers may be employed. These use a single bogie which is pulled across the floor. In one design the scraper blade is lifted on the return run by the reverse pull of the cable. Reciprocating scrapers, now used in sewage treatment works, do not appear to have been used for scrapers for non flocculating water treatment purposes.

7.5 Circular Scraped Tanks

Rectangular horizontal flow tanks offer a convenient solution for fill and draw sludge removal. For continuous removal chain and flight scrapers are expensive and a circular tank with a rotating scraper is usually a more economic proposition. There is only one moving part and apart possibly from a steady bearing under water all the parts requiring maintenance are above water.

Tanks used for presettlement are very similar to sewage primary settlement tanks, and have a central inlet, which may be combined with the centre post supporting a half bridge scraperwith a peripheral drive. On smaller or heavier duty tanks the scraper and inlet pipe may be suspended from a full bridge. The principles are the same as those of a conical hopper tank.

Sludges from surface sources are denser than sewage sludges and peripheral friction drives may be inadequate. Centre driven scrapers are normally used both for this purpose and for thickening.

For presettlement peripheral decanting is usually adequate. Depths of tanks vary from about 3-4 metres. This is governed by the water velocity particularly at the centre. The floor slope to the centre is purely for drainage on emptying. The conveying of sludge depends entirely on the scraper.

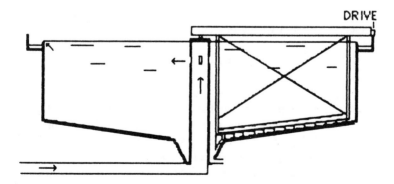

Fig. 7.5 Half bridge circular scraped settlement tank.

7.6 Scraper Design

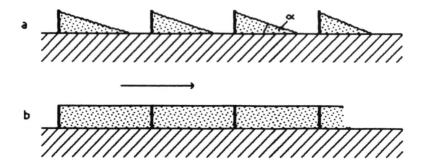

Fig. 7.6 Chain and flight scraper - conveyance of sludges.
a) individual heaps. b) full flow.

7.6.1 Chain and Flight

Two approaches may be taken depending on the nature of the sludge and the slope of the floor. In one case a mound of sludge will build up in front of the blade in a similar manner to the spoil in front of a bulldoser. In the other case with a more cohesive sludge and close blade spacing the sludge may flow as a carpet of thickness equal to the blade depth. In the latter case the volumetric flow rate will be :-

$$Q_s = u_{blade} h . W$$

If the spoil merely forms a wedge in front of the blades the flow will be:-

$$Q_s = W . \frac{h^2}{2\tan\alpha} . \frac{u}{L}$$

where α is the angle of repose of the sludge in front of the blade. Q_s will be the product of the volumetric underflow surface rate and the area of the floor to be scraped. The spacing between the flights may be quite small and the possible conveying rate is high compared with rotating scrapers.

The side of horizontal tanks may be benched (ie to make them slope in) to reduce the width of the scraper. This puts a heavier load on the scraper and does not affect the duty. The edge of the scrapers however is likely to overflow before the centre has reached full capacity.

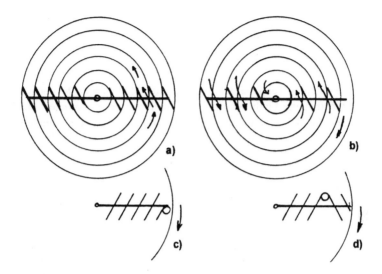

Fig. 7.7 Circular Scraping. a) Transfer of windrows by the scraper blade. b) blades set for inward scraping to a centre well. c) blades set for outward scraping to a suction pipe. d) blades set for two way scraping to an intermediate suction pipe.

7.6.2 Rotating Scrapers

These may convey inwards or outwards or both ways, Fig. 7.7

Inward Scraping

A scraper blade located on any pitch diameter must convey all the sludge falling outside this diameter and pass it on to the next blade. The conveying capacity of that blade will be its frontal projected area multiplied by its forward velocity. (it is assumed that like a plough blade the scraper pushes the windrow left by the upstream blade over to the position of the down stream blade.)

Thus if u_p and u_z are the peripheral speed and the underflow surface rate respectively:-

$$u_z \cdot \frac{\pi}{4} \cdot (D^2 - d^2) = u_p \cdot \frac{d}{D} \cdot \sin\theta \cdot h \cdot L$$

The blade height may therefore be calculated:-

$$h = \frac{\pi}{4} \cdot \frac{u_z}{u_p} \cdot \frac{1}{\sin\theta L} \cdot (D^2 - d^2) \frac{D}{d}$$

D and d are the outside sweep diameter and the pitch diameter of the particular blade. It will be noted that the blade height varies with the pitch diameter and becomes infinite at the centre. To solve this the inner most blade must be a complete spiral to take the sludge into the centre or the centre well must be large enough eg 15% of the tank diameter to keep the blade down to a reasonable height. The equation assumes that there is a single arm with a full set of blades which convey the sludge forward at each turn. In practice it is more likely that there will be two arms with alternate blades missing (a hit and miss arrangement). The conveying capacity may be increased by adding additional arms, or providing a full set of blades on each arm. Stub arms are sometimes used to compensate for the reduce conveying capacity near the centre to keep the blade height to a reasonable figure.

Some tanks have a conical outer section and scrape an inner flat floor only. In this case the D in the D^2 term should be the outer tank diameter while D in D/d is the sweep diameter of the scraper.

Outward Scraping

In this case the left hand side of the first equation above will be:-

$$\frac{\pi}{4} \cdot u_z d^2 = u_p \cdot \frac{d}{D} \sin\theta . hL$$

$$h = \frac{\pi}{4} \frac{u_z}{u_p} \cdot \frac{1}{\sin\theta . L} . d . D$$

The blades increase in height from the centre to the outside but do not reach the size of the inward version. The optimum situation is a hybrid one where sludge is conveyed from the centre to 70% of the diameter and from the outside to the same point. The centre well outlet provides a convenient way of evacuating the sludge from the tank but on large tanks transfer pumps mounted on the scraper can be used to collect the accumulated sludge and transfer to a centre sump.

There appears to be some lack of understanding among suppliers about the angle of the scraper blades. Many are set at 45° to the direction of travel. Some sludges do not slip at this angle and tend too build up in front of the arm. 30° to the direction of travel (60° to the radius) is safer and corresponds the angle of conical settlement tanks.

Example. Consider a 30m tank with an underflow of 0.05m/h. The peripheral speed is normally about 4m/min (240m/h) and the blade is 2m long set at 60° to the arm. Scraping inwards the blade height at the centre well at 4.5m diameter will be 960mm. Scraping outwards it will be 147mm. The two will be 107mm at the 70% diameter with two way scraping.

Data collected from various sources suggests that many designers use an arbitrary 150mm blade height in all situations.

7.6.3 Blade Loading

Little independent information is available. Some information is given in lbf/ft (N/m) whereas the loading should be proportional to area. It would be wise for a designer to determine his own values for blade loading for various sludges. Sanks (1975) quotes figures which have been converted on the basis that the data relates to 150mm high blades.

Alum sludges in clarifiers	$600N/m^2$
Lime sludges in clarifiers	$1800N/m^2$
River silt in clarifiers	$4000N/m^2$
Alum sludge in thickeners	$2500N/m^2$.

If scrapers stall and sludges and silts consolidate an extra load must be applied to move the mass on restarting. Some designs provide a retraction mechanism to lift the blades in such circumstances.

Reciprocating Scrapers

These are a variant on the chain and flight design. One is the cable operated design in which a transverse blade mounted on a bogie is pulled along the tank by a cable system wound on a twin drum to maintain parallel movement (the same principle being used on some drawing boards). In one design the bogie carries a linkage that raises the blade on the return run. The solids carrying capacity is limited as there is only one blade for the full length of the tank. The same limitations apply to a design in which a scraper blade is propelled by a reciprocating ratchet bar. In this the pawl flips over at the end of the travel, to return.

A design which is closer to the chain and flight is the Swedish Zickert scraper which has a grid of wedge shaped flights covering the full area to be scraped. These reciprocate at an amplitude equal to the flight spacing. The sludge runs up the sloping face but is propelled by the vertical face (Fig. 7.8). It is driven by a hydraulic cylinder above the tank through a bell crank. A similar design is offered by a German company for discharging rectangular silos.

122

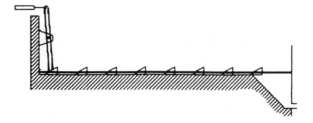

Fig. 7.8 Reciprocating wedge grid scraper.

7.7 Control

Settlement tanks of the type discussed above need little control unless the concentration of sludge to be discharged is to be kept at a maximum and water wastage minimised. Scrapers convey all that falls in the area scraped up to the maximum discussed above. They can usually idle without harm. The sludge discharge system also can be left to bleed at constant rate and the concentration allowed to vary. However if there is an incentive to minimise waste either the concentration of the discharge or the level of sludge in the final hopper must be measured. A pair of sampling pipes which collect from a high and a low level will enable the operator to initiate or stop the discharge. Sensors may be used on more advanced plants.

It is wise to run all sludge bleeds for a short while at a minimum frequency to ensure that such pipes are not blocked. It is also a wise precaution to provide means of backflushing sludge bleeds with high pressure water, eg. from a fire hose. In some situations with long discharge lines or fast settling solids flushing with clean water after a discharge is a regular practice. This avoids settlement and ensures that flow starts in clear line.

7.8 References

Sanks, R.L. (ca 1975), Water Treatment Plant Design, Ann Arbour Science.

Further Reading
Chelminski, R. (1952) How to Calculate Power Loads for Thickeners. Eng. and Mining J. **153**, (4) 89

Willus, C.A. and Fitch, B. (1980) Determining Thickener Torque Requirements, Water 1979, American Inst. Chem. Eng. Symposium Series 00656-8812-2910-0197 p352

CHAPTER 8.

SINGLE PASS FLOCCULATING SETTLEMENT TANKS

8.1 Introduction

These process units, some times called Clarifloccculators, produce a continuous version of the conditions in a laboratory jar test. The incoming flow is mixed with coagulation chemicals and flocculated usually by low speed paddles to form a settleable floc which is allowed to pass on to a settlement basin. The floc in suspension is only as old as the retention time in the flocculation zone and settlement rates are low by comparison with more recent devices. The settling floc is dilute and differential settlement plays little part in the capture process.

8.2 Horizontal Flow Clarifier (Fig.8.1)

8.2.1 Design
The most basic version is the Horizontal Flow clarifier, which is rectangular, with flocculation at one end and a decanting system at the other. Most examples are found in the USA where they have persisted as the result of conservatism and in some cases as a result of regulations defining treatment conditions. In Europe with some exceptions they have been superseded by designs that offer higher surface rates, less maintenance and better quality.
Much basic information on process conditions will be found in the American literature. Unfortunately such data is sometimes applied to other designs where the same conditions and principles do not necessarily hold.

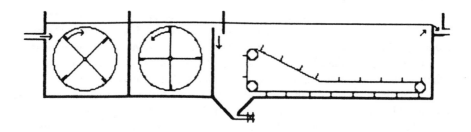

Fig. 8.1 Horizontal flow clarifier with scraper.

Mechanical mixers may be used for flash mixing, typically with G values of 500s^{-1} for 1 minute. Flocculation involves mechanical shear and exploits the orthokinetic process at the concentration produced by the combined solids resulting from the added coagulant and the solids present in the raw water (typically 0.25-0.5% v/v as floc). The required shear (G value) is achieved by the correct design of paddle. (see Chapter 4.). G values may range from 50s^{-1} down to 10s^{-1} for multistage flocculation. The lower limit is that at which the floc tends to fall out in the flocculation compartment. A typical flocculation time would be 45 min, giving a Gt (Camp Number) of 50,000 to 100,000. The exact values required depend on the concentration of solids and such tanks perform best at high concentrations. For this reason horizontal flow clarifiers tend to demand higher doses of coagulant than the designs discussed in the next two chapters, although corresponding to the doses found necessary in jar tests. The local shear rate is highest at the tips of the blades and the tip velocity is normally limited to about 0.5m/s or less.

To control short circuiting in such a mixing system (see Chapter 5) at least two stages of flocculation are normally used, some times three. The dividing walls must have ports small enough to prevent back mixing between the stages. However the flow patterns within each stage are determined by the paddles and providing the inlets and outlets are not adjacent the configuration is not particularly critical.

The transfer from the final flocculation stage into the settlement basin however is more critical. Significant jets can persist in the settlement basin and interfere with the ideal plug flow which is being sought. In particular the residual tangential circulation within the flocculation chamber with vertical flocculators will cause the water to move transversely as it emerges into the settlement basin and possibly continue in a helical pattern towards the far end. In spite of the higher cost and maintenance horizontal flocculators are use extensively. These eliminate the tangential component and provide a smoother flow pattern in the settlement basin. Such paddles are also able to maintain suspension at low G values over the full floor area of the flocculation tanks.

The settlement section as in all cases is designed on a surface area basis. Surface rates without polyelectrolyte tend to be low, eg. 1-1.5m/h. (m^3/m^2) The settled sludge may be allowed to accumulate on the floor, to be removed on draining down the tank or a mechanical scraper installed. Chain and flight, cable operated, or reciprocating scrapers as discussed in the previous chapter may be used.

Settlement tank depths are usually between 2.5 and 4m, shallow ones being more suitable for continuous sludge extraction. The velocity along the tank above any accumulated sludge must always be well below the conveyance limit of 17 times the settlement rate. The flocculation section design is based on volume (retention time) and the depth need not be the same as the settlement section. Each compartment should ideally be square when viewed parallel with the flocculator shaft.

8.2.2 Control

    This is much the same as with non-flocculating tanks except that the chemical treatment must be matched to the requirements of the water at the time. The equilibrium treated water quality is attained after one or two units of the retention time in the system, when starting from an empty tank, and adjustments in chemical dose are reflected in a similar time. American regulations at one time demanded a 4 hour retention time in such tanks (a requirement that bears no relation to the design basis).

8.3 Corridor tanks (Fig.8.2)

    A development which reduces the surface area of the settlement tank is the corridor tank, which has been built at several French plants. This comprises up to 4 horizontal basins built above each other, only the upper one having an exposed surface. These stacked settlement basins are served by a common flocculation system but the settled water is brought up through shafts to separate decanting weirs so that the flow through the individual basins is controlled. The structure is much deeper and the flocculation section likewise matches the rest of the structure.

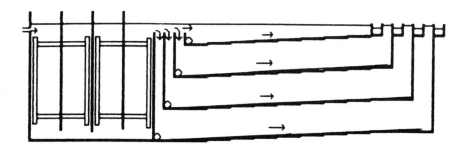

Fig. 8.2 A 4-deck corridor settlement tank

8.4 Over and Under Tanks (Fig. 8.3)

This is another horizontal design which was constructed in some numbers in the UK and elsewhere. It has a doubtful scientific basis and examples do not work particularly well. It comprises a rectangular tank with transverse baffle walls allowing an over and under flow pattern. No flocculation tank was included. It would appear that some of the principles of the vertical flow tank contribute to its operation. Sludge was normally drained out from a ridge and furrow floor with perforated drainage pipes lying in the furrows. These themselves were not designed on the principles which are now well understood and tended to block up except at the discharge end. Such tanks had therefore to be emptied regularly and cleaned out. Often they provide a structure which can be adapted to other processes.

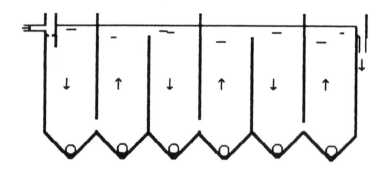

Fig. 8.3 Under and over settlement tank.

8.5 Circular Clariflocculators.

While there has been a number of innovative designs three will be discussed.

8.5.1 Centrifloc. (Fig. 8.4)

This consists of a circular tank with a central ring curtain wall suspended either from a bridge or incorporated in a raked beam structure rising from the side of the tank. This ring wall divides the flocculation section from the settlement section in the outer annulus. The water to be treated is dosed upstream is separate mixers, possibly at splitter weirs and fed through a radial launder to a distribution ring discharging into the centre zone. Two or four gate type flocculator paddles are

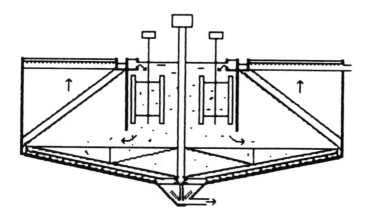

Fig. 8.4 Centrifloc full bridge clarifier

installed in the centre zone, the number depending on the size. The flocculated water passes under the ring wall to the settlement zone while the clarified water is decanted off through radial troughs or on "cheap" versions through a peripheral trough. The settled sludge is scraped to a central well by a centre driven twin arm scraper and discharged through a variable level bell mouth or a timer controlled valve.

The flocculation zone is effectively a single stirred tank and considerable by passing may be expected. However it would appear that in practice this design is not a true single pass unit. The flocculators re-entrain some of the sludge being scraped along the floor to produce an element of solids recirculation.

The treated water quality is not normally up to the standard of a solids recirculation tank (typically 1-5 NTU) and like the horizontal flow tank it is more suited to water with a moderate to high turbidity whether municipal or industrial. The surface rate is normally no more than 1.7m/h. Being scraped it can handle high concentrations and heavy solids. The depth is similar to that of other tanks at 3-4m.

The main criticism of the design apart from it not having universal application is the cost of the structure. Examples have been built in diameters up to 47m.

8.5.2 Centre Post Clariflocculator (Fig.8.5).

This operates on an identical basis to the above. The dosed water is fed in through a hollow centre column, with ports near the top. The column acts as the centre pivot for a rotating half or "three quarters" bridge which is normally peripherally driven.

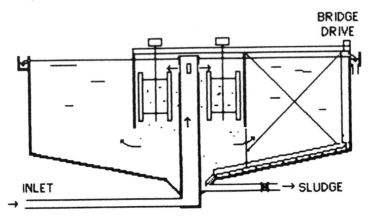

Fig. 8.5 Centre post , half bridge flocculating clarifier.

There are two versions. In one a steel or plastic sheet ring curtain baffle defines the flocculation zone. This curtain is mounted on the scraper bridge and rotates with it. The bridge also carries two flocculators which progress round and are better able to pick up sludge from the floor to assist in flocculation. Scraper blades are suspended from the bridge but these will tend to have a low capacity otherwise the bridge will twist or lose traction.

In the second version the curtain wall is carried on columns rising from the floor. The inlet arrangements are as before but the scraper bridge carries a divided scraper system, the outer one moving the sludge to a ring of columns and the inner one continuing to the centre well. The section of floor at the ring of columns slopes at 60° to allow the sludge to fall freely from the outer annulus to the centre. In another variation a ring of sludge hoppers is provided outside the support columns and the flocculators relied on the keep the inner floor clear.

There are other possible variations in design but the operating principle remains the same. This family of clarifiers is able to accept both inward and outward scraping. A transfer pump on the bridge can move sludge to a centre drain within the centre post structure, or to an outer peripheral gutter.

8.5.3 Picket Clariflocculator (Fig.8.6)

This differs from the others in that instead of having flocculators driven by independent motors the scraper bridge carries an organ pipe array of blades while the base of the flocculation zone also carries similar upstanding picket blades that intermesh with the hanging blades. It would appear that the energy dissipation for flocculation is achieved by the continuous extrusion of the water through the spaces between the blades. The speed of the scraper will be highest with high concentrations of solids in the feed but this will run counter to the flocculation requirements. Such units appear to have been used more in industrial applications.

8.6 Control and operation.

This is the same as for the horizontal flow clarifiers. Start up is rapid. There is no discharge of previously generated sludge should the chemical treatment fail. On shut down the scrapers can be run on and sludge discharged to clear the floor after flow has ceased.

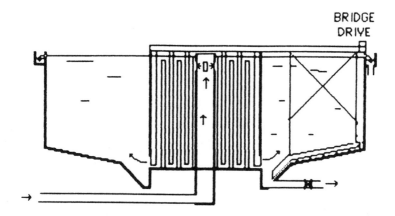

Fig. 8.6 Picket clariflocculator.

The sludge bleed is normally controlled to obtain a reasonable concentration (4-5% solids should be obtainable). If the bleed is insufficient the scraper torque will increase. A common procedure is to increase the bleed until thin sludge is obtained and throttle back again. Excessive bleeding will give a dilute sludge.

CHAPTER 9.

RECIRCULATING CLARIFIERS

9.1 Introduction

The settlement tanks described in the previous chapter flocculate the water being treated at the solids concentration produced by the suspended solids already present plus the solids from the coagulant. Schmoluchowski's equation for flocculation (see Chapter 4) shows that the rate of flocculation is directly dependent on the solids concentration (by volume). Time and concentration are interchangeable. So are shear rate and time but the shear is limited by floc breakage and already conditions are close to the boundary in this respect.

The concentration in the flocculation zone may be increased by recirculating sludge back from the settlement zone to the flocculation zone or earlier. This principle is exploited in two generic designs which are long past their patent expiry dates and have therefore been adopted by more than one supplier.

9.2 Reactivator™ Solids Recirculation and Related Types (Fig.9.1)

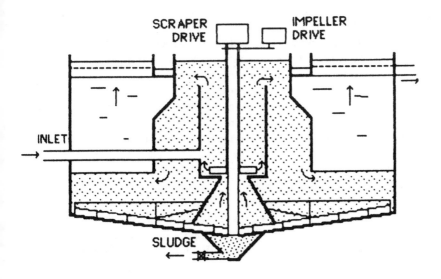

Fig. 9.1 Reactivator type solids recirculation clarifier.

9.2.1 Design

This design is usually built in a cylindrical tank with a bridge carrying the internals. These consist of a curtain baffle to separate the settlement section from the "flocculation" section, a centre cylinder, with a turbine mixer/pump in the base and a scraper system. The latter is driven at low speed while the turbine impeller is mounted on a tubular shaft concentric with and outside of the scraper shaft. These have separate drives.

Dosed water enters the centre cylinder where it meets recirculated sludge which is drawn up from the floor region above a centre sludge well. The impeller acts as a slow speed centrifugal pump and the resultant mix flows up to the surface and down the annulus of the centre zone. Flocculation takes place rapidly at the prevailing concentration of 5-10% V/V by a mixture of orthokinetic flocculation and differential settlement.

The flocculated mixture flows out under the curtain wall into the outer annular settlement zone. Although the flocculation zone is a single stage fully mixed vessel any by passing in the reaction engineering sense will be compensated by the high rate of flocculation. The concentration is some 10 times the figure of non - recirculating systems. Also the presence of large developed floc greatly accelerates the early stages of the process as discussed in Chapter 4. Differential settlement in the settlement zone also acts a filter towards residual small floc particles.

It will be noted that the impeller in this design only pumps the recirculated sludge. The recirculation rate may be 100-200 % of the net flow. The scraper conveys settled sludge towards the centre sludge well and for good measure this is provided with pickets to give a degree of thickening. 5% w/w sludges or more can be obtained from river sources. However calculations made from drawings of such units have shown that the scraper in typical units is considerably undersize if it is to convey the entire recirculation flow. It appears therefore that the main function of the scraper is to convey heavy silty material and also to prevent consolidation of the sludge on the floor. The main flow to the centre will therefore be in the form of a concentrated but mobile slurry that has not had time to reach a consolidated condition.

This design is unique in that it will handle silty river waters as well as thin coloured upland water and also precipitation softening, for which it was first developed. Surface ratings are similar to those of floc blanket clarifiers and rates of 3-4.5m/h are common. The concentrated sludge is also a unique feature for this type of clarification. One 45m square unit in Australia has a pantograph extension on the scrapers to reach into the corners. Corrugated GRP was used for the curtain divider to keep the weight down.

9.2.2 Control

This type of clarifier is not self controlling. The floc separation level is normally at the bottom of the curtain divider, as the floc flows out to the tank wall. There should be no change with flow. If it rises then the flow must be reduced and the sludge bleed increased. The concentration in the centre flocculation zone has to be monitored and the sludge bleed adjusted up or down to meet the target figure. This procedure may of course be automated. If too much sludge is bled off the concentration will fall and with it the performance. As discussed in Chapter 3, the settlement rate of the slurry falls as the concentration increases and hence there is an upper limit for any throughput. The basic free fall velocity of the floc can of course be modified by the use of polyelectrolytes and this provides a further degree of freedom, but the concentration must still be monitored.

Some types of clarifier may be uprated within the hydraulic capacity merely by adding polyelectrolytes. In this design the recirculation system will limit the throughput so that even with fast settling floc the concentration will fall. If the impeller is speeded up floc breakage may set in. A larger impeller and transfer port will be needed.

From the above the scraper speed would seem not to be too important for clarification duty. However with silty water and for precipitation softening the scraper must convey the heavy solids. The scraper torque is a useful parameter. Both light and heavy sludges are discharged from the centre sludge outlet. Again a fully open or shut valve will be found most reliable and for softening flushing of the pipe after each discharge has been found advisable.

As mentioned elsewhere some designers use a blade angle of 45° which can prove marginal in some situations.

9.3 Accentrifloc/Accelator

9.3.1 Design

While this may have a superficial resemblance to the above there are profound differences. The tank no longer has a substantially flat bottom but a truncated conical one. The flocculation zone is divided into a lower conical zone and an upper one, these being separated by a low speed turbine, again acting as a pump.The incoming flow enters to bottom zone through a set of orifices from a ring shaped duct at the top. Recirculated sludge enters at the bottom through an annular slot at the base of the outer settling zone. The impeller in this case has to pump both the net flow and the recirculated flow. The top zone acts as a feed chamber to a set of ports which discharge the floc into the outer settlement zone. Clarified water is decanted off through radial troughs, and most of the settled sludge returns to the flocculation zone.

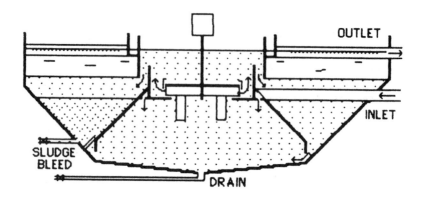

Fig. 9.2 Accelator solids recirculation clarifier.

The impeller carries hanging blades which act as a turbine to maintain full mixing in the lower zone. The number of blades can be varied according to the duty. The mixing can also be controlled by varying the speed of the impeller but this may cause the recirculation rate to be too high. To offset this a shroud ring is fitted round the impeller to throttle the discharge.

The discharge of excess sludge is achieved by partitioning off part of the settlement annulus to form sludge hoppers. These are located below the level of the incoming floc so that they fill up preferentially. The hoppers may be taken out of service by opening flap valves at the bottom to allow flow back into the lower flocculation zone. Sludge is bled from the working hoppers by timer operated valves.

From the foregoing it will be apparent that some skill is required to operate such clarifiers and although large numbers have been built these were mainly before 1970. Some utilities have found difficulty in maintaining good performance as result of dilution of skills with time.

This type of clarifier has also been used extensively for all applications including precipitation softening. The unscraped floor in the centre may lead to settlement if neglected or too low a mixing intensity used. Nevertheless it is capable of an excellent performance within its hydraulic capability up to about 4.5m/h. Like the above design there are intrinsic limitations in the recirculation system that prevent very high rates from being obtained with polyelectrolyte. If the feed rate is increased too far the flow through recirculation slot will reverse when the impeller flow is overtaken and correct operation will cease abruptly.

9.3.2 Control

Like the earlier design sludge bleeding must be controlled by monitoring the volumetric solids concentration in the flocculation zone for which sampling lines are usually provided. The duration of the intermittent bleed is advanced or reduced as necessary. If dilute sludge is discharged and the concentration still rises extra concentrators must be brought into service.

A typical volumetric concentration is 5% in the flocculation zone and 10% in the settlement zone. This difference will fall if the recirculation rate is increased. As before there will be a maximum sludge concentration above which the settlement rate cannot be maintained.

The floc interface in the settlement zone again remains steady. If it rises the floc concentration is probably too high. The impeller speed is often left fairly constant but it should in theory be increased with the throughput to maintain a constant recirculation ratio, which is indicated by the concentration ratio between the settlement and flocculation zones.

9.4 Other Designs

There are other variations that fall mainly within the above category, including a scraped tank which is otherwise similar to the Accentrifloc. As a class these clarifiers are probably obsolescent, having been overtaken by the cheaper floc blanket tanks except for precipitation softening duty.

It is possible to upgrade the single pass designs discussed in Chapter 8 either by providing for greater internal recirculation, by adapting the flocculator impeller or by splitting the sludge bleed and recirculating a large part (cf. Activated Sludge). If this is attempted a low shear recirculation pump will be needed. A simple weir type splitter box by the inlet channel would divide the flow. Some recirculation designs follow this route.

9.5 Modelling

These units may be easily characterised by writing a programme which uses the Kynch approach to obtain the underflow concentration emerging from the settlement section at various underflow rates. The concentration after this has been blended with the incoming flow may then be calculated and from that the maximum rise rate that can be sustained. If required the scraper design may also be calculated. The following is an example based on a rise rate of 3m/h with a floc that settles at 10m/h on its own.

Recirculation ratio	0.1	0.2	0.3	0.4
Max. rise rate m/h	5.5	4.0	3.3	2.9
Floc Concn. %	6.0	9.1	11.1	12.3
Underflow Concn. %	61	52	46	42

If the free fall velocity of the floc is increased (eg. with the addition of polyelectrolyte) to 20m/h higher recirculation rates may be used:-

Recirculation ratio	0.1	0.2	0.3	0.4
Max. rise rate m/h	10.1	6.9	4.3	3.0
Floc Concn. %	6.8	10.6	15.2	19.0
Underflow Concn. %	70	61	52	42

If the scraper or hopper system is unable to convey the thickened sludge thin material will be recirculated. The floc concentration and rise rate will be higher but the quality less. From reported recirculation rates which are higher it would seem that the scrapers may be under designed for ideal performance. An inadequate scraper will lower the concentration and increase the required recirculation rate.

The impellers in such clarifiers have probably been designed empirically. Some rough idea may be gained from an examination of turbine mixer data. Uhl and Gray (1966) (see Ch.4.) discuss the subject and give the following equation for flow from a turbine.

$$\frac{Q}{ND^3} = 0.6$$

There will be a limitation to the peripheral speed of the impellor because of floc breakage in this application of about 0.6m/s.

9.6 Reference

Uhl, V.W. and Gray, J.B. (1966) Mixing, Theory and Practice, Academic Press, New York & London

CHAPTER 10

FLUIDISED FLOC BLANKET SETTLEMENT TANKS

10.1 Introduction

Considering the flood of information published on filtration and dissolved air flotation it is surprising how little is available on floc blanket clarification. These differ fundamentally from the tanks already described in very many respects. The basic concept is believed to have been discovered by a British contractor operating in India in the mid 1930's. while modifying hopper bottomed tanks. The principle is that of the fluidised bed which also is encountered in the back washing of filters. In this case the particles are floc particles but the behaviour is similar and it is possible to model some aspects of the process using fine sand. In contrast to the designs discussed in Chapter 9 no mechanical elements are involved and the structure is far simpler with the result that such clarifiers have largely displaced the latter for settlement applications other than for precipitation softening.

The first examples were pyramidal hopper tanks with a central downward pointing inlet. These are difficult to upscale beyond a certain size and multiple units are required. On larger plants multiple hoppers were constructed without dividing walls above the top of the conical base. The upper 1.5m depth was common to the entire set, ie. there were no dividing walls above the top of the sloping section. The design progressed to a trough configuration with a row of inlets at the bottom. One example was built with a set of small 0.6m miniature hoppers each with a 20mm inlet to form an almost flat bottomed tank. Finally during the 1950's and 60's true flat bottomed tanks started to emerge. One contractor at an early stage introduced a pulsing arrangement to control the energy input. All these designs operate on the same principle although advertising literature sometimes clouds the issue.

10.2 Hopper Clarifiers (Fig. 10.1)

The basic concept evolved empirically originally and for 30 years the limiting features were not fully understood. They are however extremely simple and robust. Because of the similarity with the Dortmund tank there has been some confusion and some utilities have regarded them as obsolete technology. However in contrast to the Dortmund tank which has a top inlet and stilling ring the hopper clarifier (sometimes known as the Candy Tank) has a bottom inlet pointing into the bottom of the hopper which may be conical or pyramidal depending on whether it is to be constructed in steel or concrete.

139

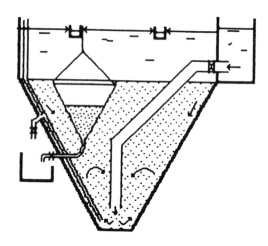

Fig. 10.1 Hopper bottomed clarifier.

The inlet jet is diverted by the structure into an upward wall jet which finally disperses part way up the side of the hopper and returns in a toroidal circulation pattern in the bottom half of the tank. The turbulence further up decays so that when solids are present flow in the upper region of the hopper is fairly quiescent. This system keeps solids present in suspension as a fluidised bed. The solids inventory is fairly well mixed and the concentration in the fluidised bed is also fairly constant apart from near the walls. The fluidised solids form a fairly sharp interface which is dependant on the solids inventory and the expansion. The latter like any fluidised bed depends on the free settling rate of the particles and the surface rate at the bed (or blanket as it is known in the industry) interface.

In the late 1960's misleading publications suggested that the mechanism involved a plug flow pattern with a decreasing velocity as the water rose up in the tank. This concept would suggest that there is a zone of clear water (or low concentration) below the base of a blanket as well as clear water above. Such a system would be unstable with a denser zone being balanced above a less dense zone. If the floc blanket in a hopper tank is sampled it will be found to be fairly uniform. A transparent model of a half hopper was constructed and operated with fine sand. The visible flow patterns confirmed the above conclusions.

The inlet velocity is fairly important. Although some flexibility is available it must be in the right region. A low velocity can lead to boiling of the blanket surface at the centre of the tank. An excessively high velocity will cause the wall jet to continue up the whole side and cause boiling at the edges, particularly in the corners. Thus if a tank is to be uprated it is often necessary to change the diameter of the inlet to maintain the correct velocity which for clarification will usually be between 0.5 and 0.7m/s. Ives developed a type of diffuser to control boiling of the blankets in hopper tanks. This redirected part of the flow up into the body of the blanket but extensive trials on full size plants have shown that changing the diameter of the tail pipe is simpler and more cost effective. Such trials have also shown that larger inlets on tanks not being uprated can be counterproductive. The Candy Filter company used to instal a variable velocity nozzle in tanks intended for softening so that adjustment could be made while in operation.

The hopper tank is unique in that the surface area of the floc blanket varies with the blanket level. As a result there is a degree of compensation when the flow changes. It is possible to calculate the blanket level and actual surface rate for various throughputs and apparent surface rates by calculating the volume of the floc in the tank at the operating rate and concentration, then constructing a curve based on different concentrations and surface rates and hence a revised surface rate and blanket area from the following equations. The free settlement rate of floc is:-

$$U_o = \frac{U}{e^{-10C}}$$

The floc volume inventory for a 60° pyramidal hopper will be :-

$$V_f = 0.289L^3 . C$$

For a new concentration and the same inventory the linear dimensions will be:-

$$L_{new} = (\frac{V_f}{C_{new}})^{1/3}$$

The new surface rate will be:-

$$U_{new} = U_o \exp(-10 C_{new})$$

From the new surface rate and the size of the blanket surface the new throughput can be calculated.

This procedure provides a means for deriving the change in blanket level with any given flow change. The only measurement needed is the blanket concentration at a defined rise rate.

The main control requirement with floc blanket tanks of this type is regulation of the blanket level. If too much is bled off the efficiency of the tank will fail. On the other hand it is essential to have a floc concentrating hopper of sufficient size to handle the peak solids flux. Many earlier tanks failed in this respect as they were only provided with minuscule hoppers. A typical size is 3% of the tank area per m/h rise rate. It is possible to design precisely using the Kynch calculation but the maximum solids load is difficult to forecast. Some tanks are provided with side concentrating hoppers but the most popular are lightweight suspended ones, up to 2.7m in diameter. If a constant bleed is provided the concentrator will act as weir towards the floc blanket. When the blanket level is low no solids will be lost.

Various refinements have been added to minimise water wastage such as operation of the sludge bleed triggered by the weight of the concentrator, or photoelectric level measurement.

One disadvantage of this type of settlement tank, which is common to all the floc blanket and recirculation tanks is the slow start. It can take several days for the best quality to be achieved while the solids inventory is accumulated. The process can be accelerated by overdosing (with pH correction if needed), or by the addition of fullers earth etc. During this phase the unit must be operated at reduced rate. Normally once the tank is in operation the floc inventory is cherished. With several tanks floc may be pumped across to restart a tank that has been serviced. For short periods such tanks may be shut down and restarted without difficulty at full rate. This limitation has not prevented such tanks from being very widely used. The other disadvantage is that in the event of failure of the chemical treatment the accumulated solids may rise and spill over onto the filters. Vigilance on the part of the operators was once necessary but with modern monitoring such failures can be detected and rectified at an early stage.

Flow rate changes cause changes in blanket level as discussed above. The sludge concentrators must be set at a level that provides a margin of safety for expansion when the maximum upward change in rate is applied. It is possible to have concentrators at different levels for this purpose.

In a few examples the inlet has been shortened and fitted with a perforated diffuser the orifices pointing sideways and upward. The intention was to leave the base of the hopper undisturbed as a sludge concentrator. This can lead to consolidation of sludge and septicity. In any case the self regulating aspect of a high level concentrator will be lost. It is also possible to fit a diffuser at the bottom of the tank and still use a high level concentrator to regulate the blanket level. Diffusers with small holes can block with debris but are claimed to give a somewhat more stable blanket.

The inlet system on a hopper tank installation usually consists of large pipes 150-300mm in diameter and fine screening is only needed for the protection of pumps. Such tanks are usually provided with bottom sludge (silt) removal pipes similar to those on non flocculating tanks. These are used only occasionally.

The single inlet does not give such a uniform distribution of water into the blanket as in a flat bottomed clarifier but nevertheless it is a very effective design capable of providing a product equal to that from other flocculating settlement tanks at surface rates up to about 4.5m/h. on lowland river sources with appropriate chemical treatment. Many older tanks have been uprated by modifying the inlet and sludge control arrangements.

Such a system cannot be down scaled with out introducing some unrepresentative features. The depth of the floc blanket needs to be at least 1.5 -2.0m, usually about 3.0m to obtain good results. If the horizontal dimensions are less than the depth wall effects can stabilise the bed and rise rates higher than those in a full size tank may be achievable. Gregory and Hyde (1975) have discussed this and report that at WRc it was found possible to run 0.3m diameter units without polyelectrolyte at 5.4m/h, whereas on full size plants 4.0m/h is rarely achieved and 1-1.8m/h is more common. This is because the inlet energy causes some reduction in floc size which in turn produces a less dense blanket near the bottom. In a large tank this would be unstable and the lower density zone blends or breaks through the denser zone above. Walls prevent this inversion. In experimental columns with windows the blanket at the bottom is clearly less dense whereas sampling of a full size unit shows that the floc concentration is uniform. The above authors give details of experiments in which vertical "egg box" baffles were installed. Improvements in quality were reported, but the limitations of the plants prevented a full exploration of potential improvements in throughput. Pilot clarifiers need to be at least 2.5m in diameter to give representative results. Full size hopper tanks up to 12m square have been built. Trough tanks with multiple hoppers have reached 68m X 12m.

A variant on the hopper theme is the so called swirl tank discussed by Mashauri and Viitasaari (1993). It is known that rotation of the contents of a conical hopper tank imparts extra stability because the rotational momentum about a vertical axis deters horizontal rotation involved in the boiling of floc blankets. However there are also swirl concentrators in which solids migrate to the centre of the tank. These are used particularly in the primary treatment of sewage were the solids are more shear resistant. (The design of these is discussed by Smisson (1967)). The above swirl clarifier appears to be a conventional hopper tank with a tangential inlet at the bottom, and combines the features of the hopper tank and the Spiractor pellet softening precipitator. The sludge concentrator must be located centrally. It must of course be circular and therefore best suited to smaller plants in steel.

10.3 Flat Bottomed Clarifiers (Fig. 10.2)

These, as already mentioned, are a development of the above. Early examples appeared in the late 1950's and 1960's in Singapore, Northern England and France. The absence of the hopper at one time caused some scepticism owing to the misunderstanding of the operating mechanism. In this case a matrix of jets is directed at the floor. Each jet is diverted into a radial floor jet which eventually meets the wall of the tank or the neighbouring jet and then curves up into a toroidal vortex. Above a level corresponding to about half of the jet spacing the turbulence fades into a fairly steady vertical flow. The design is merely an upscaled version of the chemical engineers fluidised bed reactor. Indeed it would be possible to operate with a perforated floor.

As in the hopper tank the microfloc particles produced by coagulation are captured by the combined effect of orthokinetic flocculation and differential settlement. Shear (Camp or G numbers) is not relevant. The main requirement is for an inlet velocity high enough to maintain suspension without undue floc breakage.

The water enters, is clarified and leaves within the same module in contrast to sludge recirculation tanks. Even the excess sludge is collected in the vicinity. Typical units may be from 6m X 6m up to 24m X 50m. The depth remains constant. Beyond a certain size the feed and settled water channels become so large the subdivision is preferable. Because the design is modular there is in fact no limit to the size. By contrast the flow pattern in a recirculation tank takes the water repeatedly out to the edge and back to the centre, and to maintain velocities the depth has to increase with diameter.

In most other respects the flat bottomed tanks is identical to the hopper bottomed tank. Great care is necessary in the design to achieve the necessary uniformity of flow, but this is a straight forward exercise in manifold design. The inlets may be arranged on a square grid or from linear manifolds. Designs have included clusters of small pipes on 900mm centres, trident inlets on 1500mm centres, (Fig. 10.2), stub pipes fitted to low level manifolds and even plain perforated manifolds. The latter tend to give an oblique jet because of the forward momentum in the pipe. Large diameters must be used to control this. The inlet velocity is usually about 1-1.5m/s on steady flow units.

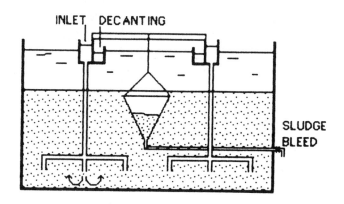

Fig. 10.2 Trident type flat bottomed, floc blanket clarifier.

Pulsed units (Fig.10.3) are widely used. They usually draw some of the feed flow into a vacuum column and then discharge it over 10 seconds or so in a 60 second cycle. The peak inlet velocity appears to be similar to that of a steady flow unit. The pulsing provides a means of maintaining the energy input at low flows but at high flows many users have found that the pulsing is not needed. The disadvantage of the pulse is that closed pipes and ducts are required between the pulse chamber and the inlet jets. These can allow fallout of solids which can only be removed on shut down, whereas many steady flow units have open channels all the way and can be cleaned without shut down. The pulsing involves blowers which consume energy (examples have blower motor ratings of 12kw/m³/s, equivalent to 1.2m headloss). Otherwise the only energy loss is involved in a fall of 300-500mm though the units.

The sludge level is controlled either by built in concentrators or suspended cones as with hopper tanks. In contrast to the hopper tanks there is a constant area for the floc blanket regardless of depth and the latter will vary inversely with the concentration as the flow changes. Again the change in depth on flow change can be

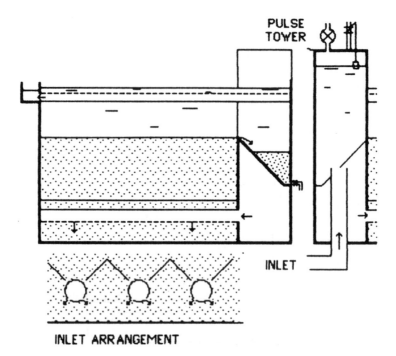

PULSE
TOWER

INLET

INLET ARRANGEMENT

Fig. 10.3 Pulsed flat bottomed clarifier, showing the vacuum tower.

predicted and the appropriate distance left between the rim of the sludge concentrators and the decanting troughs to avoid loss of sludge.

Calculations show that the sludge yield produces a negligible cross flow velocity even if the concentrators were located at the end of the tank. The limiting parameter is the crest height of the sludge flowing over into the concentrator and the length of weir for a given duty. Thus the weir equation may be modified for this application:-

$$Q = \frac{2}{3} Cd\sqrt{2g.\frac{\rho_b}{\rho}}.h^{1.5}$$

Thus taking the blanket densities referred to in earlier chapters (say 0.1kg.m3), a 200m2 clarifier operating at 3m/h on a water generating 1.0% of floc, ie. 6m3/h of floc, would require 650mm of weir at 150mm crest height. In practice much more is required to deal with surges when the flow changes.

While the inlet jets are capable of maintaining suspension of floc their is usually some fallout on the floor when river waters are being treated. Beyond a certain point pulsating clarifiers are provided with presettlement tanks. Steady flow flat bottomed tanks with open channel inlet systems are sometimes fitted with floor scrapers which can be operated while the unit is in operation. Cable operated scrapers as used in some types of wastewater storm tank have been used and reciprocating rakes would also be suitable.

The flat bottomed design whether pulsed or unpulsed has largely replaced the recirculating type clarifiers for clarification duty, because of its greater simplicity and lower cost. It is also a good vehicle for exploiting polyelectrolytes because no limitations from recirculation capacity exist. Experimentally, rates up to 15-20m/h have been achieved. The limit is dictated by hydraulic capacity and the correct combination of coagulant and polyelectrolyte with contact/mixing conditions as discussed under flocculation The contact time between coagulation and the addition of polyelectrolyte is particularly important.

Tank depths are usually in the region of 3.6-4.5m, but if the hydraulic gradient makes a deeper tank more convenient this will have no detrimental effect. Several specialist contractors build tanks to this general design and it is no longer a particularly proprietary one.

Temperature gradients greater than 1°C per hour can be troublesome but normally less so than with single pass tanks. In some cases diurnal problems have been found to be caused by a change in pH in the raw water and have been corrected by modifying the coagulant dose. Higher rates of operation and the consequent reduction in retention time (and the higher blanket density produced by the polyelectrolyte) help to reduce the problem. Steel tanks (both for hoppers and flat bottomed tanks) cause more difficulty than concrete and white tanks are better than dark colours in this case.) Boiling caused by positive temperature gradients usually occurs near the edges of tanks and in corners. The decanting system can be adapted to draw from the better regions of the surface. Often the appearance is worse than reality as quantities of "light floc" may be observed in the supernatant water while the turbidity may be almost the same. Most treatment plants operate filters on a 24 hour cycle and their behaviour is more dependant on the average load over this period.

10.4 Other Versions

The main remaining example of a floc blanket tank is the Precipitator, which has an annular trough with a central contact zone. The particular version built in the UK has a bottom inlet slot and inevitably a low inlet velocity which was prone to blockage. Several examples are still in operation but the design has been discontinued as far as new works are concerned. There are also examples of trough tanks with rows of bottom inlets comparable with the hopper tank.

10.5 Control

It is perhaps unfortunate that floc blanket clarifiers became well established just before dissolved air flotation and before modern control techniques became generally used in the water industry. As a class they are remarkably simple and effective, and if polyelectrolytes are used they are competitive with dissolved air flotation, having in many cases a similar overall footprint (for clarification only) and a much lower power consumption. In all cases proper control of chemical treatment is essential.

Recent advances in control of the sludge bleed have overcome some of the earlier problems but there are still improvements in flexibility that could be made if proper monitoring and on line modelling were included to control blanket levels according to the throughput. Nevertheless the high level sludge concentrator is fairly fool proof providing that the suspended version is not called on to hold its volume full of water when the tank is drained down.

The hopper and flat bottomed clarifiers have no recirculation systems to restrict the upper limit of performance. Providing that the inlets and decanting system will handle the flow the only other limits are the settlement rate of the floc and the capacity of the sludge removal system. Although, as dicussed elsewhere, inclined plates can improve the rise rate polyelectrolytes on their own, when correectly dosed can give some very high surface rates. Delay times between dosing and entering the blanket and and also between dosing of the coagulant and the polyelectrolyte are very important and many older clarifiers were deficient in this respect. Stevenson (1981) has given jar test figures showing how settlement can vary from 1.7 to 12m/h merely by varying the delay. Yadav and West (1975) have confirmed this sensitivity in studies with pilot clarifiers. The author's experiments have shown that a limit is reached where the floc becomes so heavy that it cannot be suspended reliably by the incoming water. The selection of polyelectrolyte and delay time is important as is the control of the dose.

At the time of writing there is no proven on line monitor to regulate the blanket density and hence the polyelectrolyte addition. A 3:1 turndown ratio is not a problem but with polyelectrolyte a 6:1 ratio can be achieved. With the flat bottomed tanks it is possible to overdo the latter and also to run too slowly with the result that the blanket can fall out onto the floor. This can usually be resuspended on turning up to high rate.

The performance of all clarifiers suffers when the tempearature of the water rises. A floc particle with a free settlement rate of 10m/h and diameter 3mm will have a density of 7×10^{-4} relative to water. A blanket with a volume concentration of 15% will therefore have a density ratio of 1×10^{-4}. This corresponds to the density change which occurs with a 4°C change in temperature. When the temperature rises the mixed blanket at the bottom of the tank will therefore tend to boil up through the cooler blanket above. Gould (1982) published data suggesting that even a 0.25°C per hour rise could cause up welling. Fortunately the situation usually appears worse than it is. The supernatant water may be filled with stray floc particles that obscure the blanket surface but the measured turbidity is not normally excessive. At night the opposite effect produces a very stable blanket and excellent turbidities. The total load passing over onto the filters is of course a mixture of the two.

The higher the rate of a clarifier the less will be the temperature difference between the water entering and the water already in the tank. This, coupled with the denser blanket used (with polyelectrolyte) in high rate tanks, tends to counteract the effect. Steel tanks are more prone to boiling in direct sunlight than concrete tanks. Selective decanting, ie. placing the decanting troughs away from the walls, is effective.

10.6 References

Gould, B.W. (1982), Thermistor Thermometer for Thermal Convection Studies in Clarifiers, Effluent and Water Treatment J. April, p157.

Mashauri, D.A. and Viitasaari, M. (1993), Potable Water by use of Swirl Concentrators, Water Sci. Tech. **27**, (9), 7

Smisson, B. (1967), Design, Construction and Performance of Vortex Overflows, Symp. Storm Sewage Overflows, Inst. Civil Eng. p99.

Stevenson. D. G. (1981) High Rate Sludge Blanket and Solids Contact Clarification, Australian Water and Wastewater Assn. Conv. p8.1

Yadav, N.P. and West, J.T. (1975) The Effect of Delay Time on Floc Blanket Effiiciency, WRc Report TR9.

CHAPTER 11.

LAMELLA CLARIFIERS

11.1 Introduction

There is some dispute about the origins of multiple layer or lamella settlement. Culp et al. (1967) attributes the origin to Hazen in 1904 following his basic work on the theory of settlement tanks. However Boycott in 1920 observed that blood corpuscles settled more rapidly in inclined tubes than in vertical ones. Pearce (1962) published a detailed paper on multiple inclined plate packs with theoretical back up in a paper to the Institution of Chemical Engineers. Hansen and Culp (1967) described horizontal tube settlement and with others proceeded to inclined plate systems in 1968. Yao (1973) among several other authors has discussed the design of high rate settlers. It appears that the concept of inclined plate settlers was first conceived outside the water industry. Since then there has been a plethora of devices and many specialist suppliers have sought to include at least one version in their product range.

Lamella clarifiers provide a means for greatly reducing the size of the separation unit. They may be compared with horizontal flow clarifiers in which the settlement basin has been compressed. Unfortunately they have no effect on the flocculation process and this remains as large as before. If the solids concentrations are high and floculation rapid, as in many industrial situations, or if the solids settle without coagulation eg. primary sewage settlement, then lamellas provide a useful and competitive choice. Caution must be exercised if there is any tendency for deposition on surfaces, and precipitation softening for example would not be a good application.

11.2 Horizontal Settlers (Fig.11.1)

In the Hazen Equation (A= Q/u) the depth of the tank does not appear. Thus as far as solids separation is concerned, one may in theory achieve a very much more compact settling system by using a set of thin tanks stacked above each other. The solids retention is of course limited and the problem of supporting the layers must be solved. Settlement of solids in horizontal flow was discussed in detail by Sinclair (1962). Various equations were reviewed by him, the simplest being that settlement occurs if the forward velocity is below 17 times the settlement rate. Thus if a settleable suspension is fed into a horizontal or almost horizontal tube the solids fall out in the length predicted by the Hazen Equation. The solids accumulate on the base of the tube until the velocity over the top reaches the limiting value for deposition. After this point the solids entering the tube are carried forward, many by rolling to the front

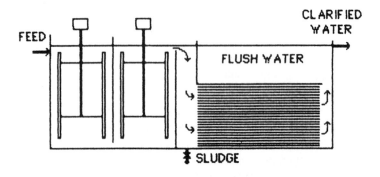

Fig. 11.1 Horizontal type of lamella separator with flocculation.

edge of an advancing sand bank. Banks of tubular ducts may easily be formed by stacking corrugated plastic sheets, particularly the trapezoid variety, and large compact settlers can be assembled.

For the best performance flow should be laminar, otherwise eddies will cause transverse mixing and reduce the efficiency. The Reynolds Number must be less than 2000, where:-

$$Re=\frac{uD\rho}{\mu}=<2000$$

For non circular tubes the hydraulic diameter should be used.

Some claim that the Reynolds number should be less than 500. Laminar flow develops progressively at the entry end of any lamella system and a significant proportion of the length may still contain an expanding boundary layer and therefore not perform as predicted by the Hazen Equation. Coulson et al. (1991) give the following experimental equation for the entry length:-

$$L_x=0.0288.Re.D$$

Thus at Re=500, the velocity in a 50mm diameter tube at 10°C will be 0.013m/s. The entry length will be 0.72m. This length varies linearly with velocity. Thus with practical lamella systems a major part of the length may be in a transitional situation, part laminar and part turbulent. Short units are thus less effective than long ones for a given surface loading.

If the flow is turbulent but the shear at the walls is low enough to permit settlement the concentration of solids will decay exponentially with length instead of linearly over the Hazen length (the length to give the required area from the Hazen Equation). Thus for turbulent flow:-

$$C = C_o \exp\left(-\frac{L}{L_{Hazen}}\right)$$

The efficiency in this case for one Hazen length is only 63% instead of 100% for laminar flow. There are thus several reasons why the projected area for a lamella system needs to be rather larger than that predicted by the Hazen Equation, and most new designs are calibrated experimentally.

The tubes must be cleaned after breakthrough has occurred and in one design the tubes slope upward slightly so that by draining down the feed tank the solids may be backflushed. In another design the forward end upstream of a weir was drained rapidly to flush the solids forward rapidly. A traversing backflush nozzle has also been used.

Horizontal lamella settlers have not proved as popular as inclined plates. Horizontal gravel filters may be regarded as a form of lamella settlement. These are use for preteatment in remote rural situations.

11.3 Inclined Settlers (Fig.11.2)

The Hazen Equation also holds true for inclined plates but in this case the **horizontally** projected area is the important parameter. The effective area is then less than the area of the plates by the cosine of the angle of inclination. If the angle is greater than 55-60° the solids settling on the plates will slide rendering the system self cleaning. In this case the water may flow upwards (counter current), downwards (co-current) and across the sheets (transverse). All three have been the subject of commercial versions of settler. In typical designs plates are separated by 50-75mm and dramatic reductions in the size of a settler can be achieved. Thus with 2m long sheets inclined at 60° and separated by 50mm 26 sheets overlap in any vertical section, thus the surface rate should theoretically be 26 times higher.

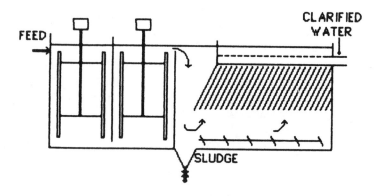

Fig. 11.2 Inclined plate lamella clarifier with flocculation.

In this simple design the sludge settles out into the incoming flow and tends to be carried back into the unit. This greatly reduces the efficiency and most of the more developed designs have attempted to overcome this limitation. Unfortunately another problem is the fact that for the majority of water treatment applications flocculation is required and the compact lamella clarifier still needs a full size flocculator (45 mins.) even though the lamella section may be reduced to a few minutes.

As discussed below lamella separators do not work so well with concentrated suspensions or floc blankets. The simple floc blanket clarifiers using simple rectangular tanks with little internal furniture are cheap, robust and last a long time, whereas thin plastic sheets have a limited life.

11.4 Concentrated Suspensions

It is fairly easy to construct an inclined plate separator and to instal it over a settlement tank. It would be attractive therefore if plate packs could be installed over simple floc blanket tanks. To achieve the much higher separation rates theoretically possible together with the accelerated and simple flocculation given by such a tank. At low concentrations single particles follow the vector resulting from the inclined upward flow of the water and downward settlement (see Fig. 11.3). Unfortunately as soon as the concentration of the suspension increases so does the difference in density between the separated water and the settling floc. The clear water that separates under

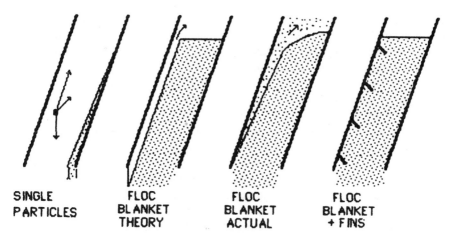

SINGLE	FLOC	FLOC	FLOC
PARTICLES	BLANKET	BLANKET	BLANKET
	THEORY	ACTUAL	+ FINS

Fig.11.3 Behaviour of dilute and concentrated suspensions in inclined tubes and plates.

the downward facing surface is ejected at a significant velocity and re-entrains the suspension, as illustrated. This effect is recorded in Pearce's sketches of 1962. A floc blanket with a density difference of 0.1 kg/l produces a headloss of 0.2mm in 2m depth. This improve head is that of a velocity of 65mm/s (220m/h) and many times the settling rate. Floc is carried up in this rising stream and forms a plume above the settling floc mass. This dilute floc does not settle as well (presumably having suffered a higher shear rate than that in the floc blanket). The problem was overcome by Degremont in their Super Pulsator by adding fins to the underside of the inclined plates to mix the separated water back into the main body of floc. This however would appear to counteract the inclined plate principle so that the plates in that device (separated by about 300mm) would appear to act as stabilisers and improve the rise rate by the same mechanism as found in the thin (300mm) vertical pilot plant floc blanket tanks operated by WRc which were capable of significantly higher rise rates than full size tanks. In this case the plates might serve the same purpose if hung vertically. Casey et al (1984) claim an uprating factor of 1.7 to 3 in flat bottomed floc blanket clarifiers with the use of inclined plates. Corrugated PVC sheets were used, the corrugations being horizontal. The effect of the corrugations is assumed to match that of the fins. Inclined tube packs are now being promoted in high rate blanket tanks (Pulsatube Clarifier). When used above the blanket rather than in it they can act a a polishing device to capture stray floc which increases at high rates and return it to the blanket. Short plates and tubes are less likely to procude the plume effect referred to above. Much data on the performance of lamella systems in clarifiers is confounded by improved mixing and the introduction of polyelectrolyte at the same time. Reports have appeared in which two variables have been changed at the same time and improvements not necessarily correctly attributed. Likewise much descriptive material is promotional and the basic principles involved in high concentration applications need further elucidation.

11.5 Types of Lamella Clarifier

The patent literature contains a wide variety of devices, too numerous to review here. The text book situation is the horizontal flow tank complete with flocculators in which simple countercurrent inclined plates of tubes are installed in the settlement zone (Fig.11.2). This is subject to the limitations already discussed. There is no control over the flow through the many tubes in parallel as with the low resistance of the tube packs eddies and residual jets from the flocculation compartment and the differential pull of the decanting system can cause uneven flow.

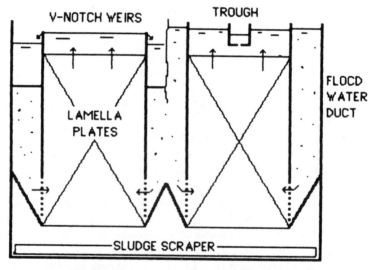

Fig. 11.4 Side entry lamella plate system. (Flocculated water flows from the side ducts into the lamellas. Top decanting trough and pannier trough arrangements shown.)

Temperature changes have been found to cause short circuiting. With a rising temperature the warmer less dense water can stratify and the whole flow may pass up the tubes nearest the inlet. More advanced designs provide decanting launders evenly over the surface of the packs.

A major step forward came with the development of side entry and cross flow inclined plate systems. In the countercurrent version, Fig. 11.4, flocculated water if conveyed in ducts between the plate packs and the flow enters the latter at the side. Water is decanted from the top either over invididual V notches or into a trough running the full length of the plate array. The sludge is able to slide downwards into a collecting hopper or onto the scraped base of the structure, without being swept back by the oncoming flow. In this design the plates are assembled into a space frame at the works and these are installed directly into the chambers at the plant.

In a parallel development the flocculated water was fed to the top of the plates and flow of water and sludge was co-current downwards, Fig.11.5. This had the advantage that the angle of inclination could be reduced to 35°, thereby making better use of the sheet material. Cos 35° is 0.82 whereas cos 60° is 0.5. Thus 42% less material is required. The stratified water was scooped off by a full width slot across the flow channel and ducted to a launder above the plate pack. The design was more expensive and this together with problems of fouling of the duct, particularly when polyelectrolyte was used led to the design being abandoned.

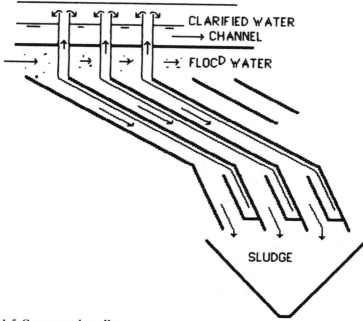

Fig. 11.5 Cocurrent lamella.

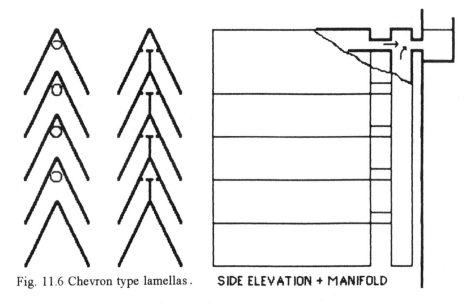

Fig. 11.6 Chevron type lamellas.   SIDE ELEVATION + MANIFOLD

Another design offered by Eimco uses short chevron shaped collectors as shown in Fig.11.6. The size of plate served by each decanting orifice is small.

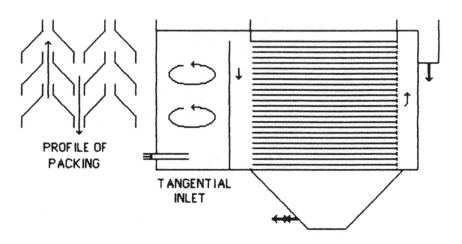

PROFILE OF
PACKING

TANGENTIAL
INLET

Fig. 11.7 Serpac Lamella clarifier.

A cross flow pattern has been used in a number of designs. The Serpac (Fig, 11.7) which has a venetian blind arrangement to separate the falling floc from the water, and a perforated end wall to extract the clarified water.

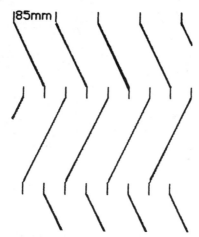

Fig. 11.8 Configuration of Japanese horizontal flow inclined plates (Kawamura et al.)

In Japan the cross flow arrangement has been used in several large plants which were reviewed by Kawamura and Trussel (1991). In these there is a cascade of staggered plates of alternating slope in an arrangement that passes the sludge from plate to plate (Fig.11.8). In both arrangements the lost tank area which occurs at the end of each stack is avoided.The plate separation is 85mm. A further version which appears in manufacturers literature has deep fins on the upper surface of the plates, supposedly to generate vortices.

Other cross flow designs are similar to the counter current ones, having plain inclined plates 1.5-2m long , arranged with side entry and a decanting wall on the opposite side. This design introduces another problem, namely short circuiting through the sludge collecting space below. This has to be baffled in some way.

A further space saving version is the combined lamella/thickener which was first introduced by Passavant (1972) in Germany as an addition to the Axel Johnson cocurrent lamella but later used with the counter current version (Fig. 11.9). This has been copied by other manufacturers. There are constraints in the configuration because the design must be square at the level of the lamella plate packs. Corner benching converts the square to a circle at scraper level. The overall system is very compact. It has been found that the sludge concentration produced from the thickener is remarkably high, (19% on a plant operating on the Rhine in Germany), presumably

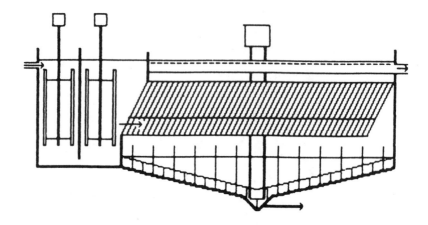

Fig.11.9 Passavant LME combined lamella and thickener (used with the lamellas shown in Figs. 11.4 and 11.5.)

because the floc is not damaged in any way as it falls from the lamella plates into the thickening zone. A similar configuration is found in the Densadeg, in which concentrated sludge with added polyelectrolyte is recycled to form a hybrid device. occupying a small area.

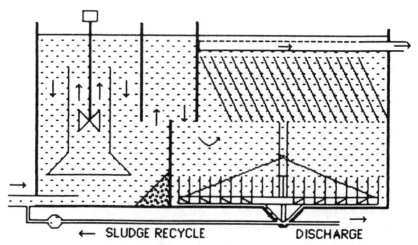

Fig. 11.10 Schematic of Densadeg clarifier.

Undoubtedly lamella systems offer some reduction in the size of the installation for a given throughput, but they have their limitations, particularly in flocculation. Where polyelectrolytes are accepted and also where floc blanket technology has become established such devices have made little headway. The main market seems to have been in uprating horizonal flow clarifiers, where being an add-on their adoption has not been the subject of extensive trials to conservative regulators. Their use in Japan appears to have been influenced partly by the attitude of the regulators to polyelectrolyte.

Advantages include immunity form diurnal temperature variations, because the retention time is very short and the depth for convection trivial. The performance is not affected by wind and rain and being easy to cover they can be housed indoors. They could be particularly appropriate for presettlement where flocculation is necessary or very quick. Otherwise the operation is similar to that of horizontal flow tanks. Start up and shut down are quick.

Disadvantages include the cost of the sheeting coupled with the limited life expectancy compared with concrete. Where plastic sheets are used protection from the sun may be necessary when the tank is empty (even in temperate zones, although white sheets are more resistant to damage in sunlight). The installations are in any case more fragile than traditional ones. Growth of algae and weed, particularly where chlorination is not favoured leads to increased maintenance costs and polyelectrolytes have also caused problems with adherent floc. On large plants the sludge scraping system will be cumbersome, although the recent Swedish reciprocating scraper is particularly appropriate for such systems.

As for the future, it is likely that the growing popularity of dissolved air flotation which has swept many parts of the world and is now becoming accepted in the USA will act as a brake on further progress with lamellas except perhaps for presettlement.

11.6 References

Boycott, A. E. (1920) Title not available, Nature, **104**,532.

Casey, T.J., O'Donnell, K. and Purcell, P.J. (1984) Uprating of Sludge Blanket Calrifiers using Inclined Plates, Aqua, **2**, 91

Coulson,J.L., Richardson,J.F., Backhurst,J.R. and Harker, J.H., Chemical Engineering. Vol.I, 1991.

Culp, G.L., Hansen,S.P. and Richardson, G. (1968), "High rate Sedimentation in Water Treatment Works", J.American Waterworks Assn. **58**,681.

Hansen,S.P. and Culp G.L. (1967) "Applying Shallow Depth Sedimentation Theory", J. American Waterworks Assn. **57**, 1134.

Kawamura, S. and Trussell,R.R. (1991), Main Features of Large water Treatment Plant in Japan", J. American Waterworks Assn.,**81**,56.

Passavant A.G. "Method and Apparatus for the Treatment of Liquids", British Patent 1,373,189

Pearce, K.W. (1962),"Settling in the Presence of Downward Facing Surfaces", Proc. Symp. Interaction between Fluids and Particles, Inst. Chem. Eng. London, p30.

Sinclair, C.G. (1962) ""The Limit Deposit Velocity of Heterogeneous Suspensions", previous reference p 78.

Yao, K.M. (1973), "Design of High Rate Settlers", A.S.C.E. J. Environmental Eng. Div. **99**,621

CHAPTER 12.

DISSOLVED AIR FLOTATION

12.1 Introduction

This involves the use of air dissolved usually at high pressure in a side stream which is released into the coagulated main flow to form micro bubbles which become attached to the solids to be separated and carry the solids to the surface as a scum. The process is an inversion of settlement and there is a similarity between flotation and the processes already discussed. Together with the widespread introduction of ozone and granular carbon it constitutes one of the most dramatic changes in water treatment practice of the late 20th century. Probably the first attempt to publicise the process generally was the Felixstowe meeting (Melbourne and Zabel 1976). An update on the considerable progress is given in the 1997 conference proceedings (Anon. (1997). A typical arrangement is shown in Fig.12.1. A closely related process some times called microflotation supersaturates the whole flow at an increased hydrostatic pressure in a U shaped shaft for example, so that the air comes out of solution as the water rises. The air bubbles are not so well controlled and the process has not proved successful or commercially attractive. One paper on this (Hemming and Cottrell) was included in the above conference. Another variant is electroflotation, described by Barrett (1975). In this fine bubbles are generated at electrodes. The equipment is extremely simple but the power cost on typical water

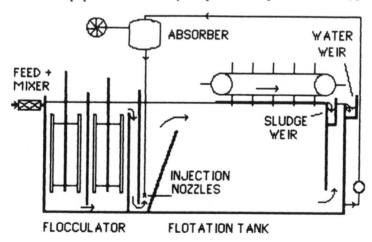

Fig. 12.1 A common configuration for a dissolved air flotation plant.

sources of only moderate conductivity is high. Also the hazards from the hydrogen and oxygen generated would make the process unpopular with safety officers if used indoors.

Dispersed air flotation as used in the mineral industry is a different process and uses coarse air bubbles. It relies on partial wetting of mineral particles at the air/water interface rather than capture by fine bubbles in a manner similar to that of floc in a filter or floc blanket clarifier.

Dissolved air flotation has been used industrially for several decades in industries such a paper and oil refining. Interest in water treatment use in the UK started in early 1970s after staff from WRc visited South Africa. The technology received a considerable boost at the 1976 conference. At that time there were 15 plants were already in use in Scandinavia and three under construction in Britain, and experimental work was already proceeding in South Africa. Melbourne and Zabel reviewed the process in some detail in 1980, and quoted Van Vuuren (1965) as the earliest paper. However two patents (Purac 1966 and 1967) of similar date (the latter for a combined flotation/filtration system) indicated that practical design details were already established at that date. Longhurst and Graham (1987) reported on the performance of about 24 in Britain and missed at least one further one. Since then the process has swept the United Kingdom and virtually displaced settlement for new plants to the extent that over 80 are reported to be in operation in 1996 (Anon, 1997). The same is true to a fair extent in Australia and many examples also exist in South Africa. The USA and countries more directly influenced by its technology has resisted it until recently, partly because of the influence of the regulatory system that discourages the use of new technology until proven. However the process is now becoming accepted there as well. Being a fairly new process it has attracted very considerable attention in recent years with several conferences and detailed papers, much repeating earlier information.

Flotation is used both for sludge thickening and for clarification. In the latter case coagulation and flocculation are required. Chemical treatment for this purpose is basically the same as for settlement or direct filtration and no short cuts are possible. Being a type of contact process the chemical doses are very similar to solids recirculation or floc blanket clarification. Claims that the process reduces coagulation doses appear to be based on comparisons with non contact clarifiers.

Dissolved air flotation, like other processes, can be analyzed and the design put on a rational basis. Again there are examples where historical precedents have influenced designs. The common element in current practice is the dissolution of air in a side stream, usually recycled, at an elevated pressure, and release of this pressure through a nozzle or valve under conditions where the microbubbles that form are mixed with the flocculated water. The process is not particularly critical and a number of variations are in use. Valade et al. (1996) conclude that the process is robust and gives a consistent performance despite variations in the raw water. The main disadvantage is the power consumption, typically equivalent to the loss of 6m head for the main low or twice the loss through a typical filter installation. The energy loss is 10 times more than in a floc blanket clarifier but only about twice that of a recirculation clarifier and trivial compared with the energy involved in distribution.

12.2 Mechanism

The capture process would appear to be primarily interception by differential motion. The equations will be similar to that for interception in filtration. Air bubbles generated by the release of pressure form the recycled water have a size reported to be about 50-70 μm. The volumetric concentration is usually about 0.5% by volume. Thus there is a voidage of 99.5%. The bubbles rise under their own buoyancy following Stokes Law as discussed in Chapter 3. At 10°C the rise rate for 50-70μm bubbles will be 1-2mm/s (3.6-7.2m/h). In practice higher rates are attained (up to 15m/h). In a disperse mass the larger bubbles will grow at the expense of the smaller ones. Thus it is likely that bubbles continue to grow after intercepting the floc. Also one would expect the bubble/floc aggregates to be larger than the original bubbles. The bubbles sweep through the floc for a short period between the points of mixing and separation, which is usually about 60 seconds in most designs.

In Chapter 17 equations for interception and diffusion are derived. These may be adapted to the present situation, assuming the air bubbles to act in the same was as sand grains. (This approach is favoured by Edzwald, (Anon. 1997). However it will be found that at the concentrations and rise rates involved the efficiency predicted is very low for interception and low but plausible for diffusion, particularly for particles of 1μm and less. However if the flocculation equations for collision with large floc based on Schmoluchovski, see Chapter 4, are applied assuming modest shear values ($G=10s^{-1}$) very rapid capture is indicated.

$$N = N_o \exp\left(\frac{-GC_b t}{\pi}\right)$$

where C_b is the bubble concentration.

The contact zone of the plant in which the recycled water and air bubbles mixes with the raw flocculated water is thus a form of flocculator. The energy is partly intrinsic as the flow passes round the bends and partly derives from the residual energy from the aeration nozzles, which although having diffusers to shield the floc from the high velocity in the orifices nevertheless produces an emerging velocity of 0.5-1.0m/s. Shear rates are likely to be about 15-20s^{-1}. This is the approach favoured by Fukushi and Tambo (1994).

Interestingly the capture efficiency predicted using an orthokinetic flocculation approach is independent of floc particle size and only dependant on the bubble concentration. In contrast to floc blanket clarification, to which the same equation may be applied, the air bubbles will withstand a considerable shear rate without breaking down. This provides theoretical support for short flocculation times and the observed absence of large floc.

The floc is hydrophilic and one would not expect adhesion by lack of wetting and a finite contact angle between the solid and air water interface, as exploited in mineral separation by dispersed air flotation. It would appear that the bubbles rapidly accumulate surface films of surface active material which make them behave as flexible solids. For this reason also small bubbles behave hydraulically like solids and obey Stokes Law as they rise.

The bubble/floc aggregates rise to the surface of the tank and form a thin film. If the scum is not continuously removed this grows and accumulates to form a thick foamy mass as the bubbles coalesce. Foam drainage follows and if left the upper part of the scum (the tip of the iceberg) becomes denser than the lower buoyant submerged material which does not drain. This situation tends to cause inversion of segments of the floating carpet and eventually fallout may occur.

The bubble size is clearly important as the shear stress the surface increases with bubble size. Beyond a certain size floc may be sheared off, a phenomena encountered also in filtration. The mean shear stress will be:-

$$R' = \frac{\rho g D}{6}$$

Thus a 100µm bubble will have a mean shear stress of 0.16N/m^2 or a surface shear rate of 123 s^{-1} at 10°C.

It has been claimed that bubble nucleation in this context requires small particles. The internal pressure in a bubble is:-

$$\Delta p = 4\frac{\gamma}{D}$$

The surface tension of (clean) water is 0.071N/m. The pressure in a 50µm bubbles is therefore 5.7 kPa. (570mm water gauge) However such bubbles grow from small ones and at 1µm the pressure will be 2.8 bar gauge, 3.8 bar absolute. This pressure increases the solubility pro rata and unless nuclei are available and the excess saturation is sufficient bubbles will not form. Cavitation or pre existing particles are therefore necessary to provide such nuclei and the excess saturation must be sufficient to ensure that the cavities grow and do not redissolve under the excess pressure.

The subsequent growth of bubbles after nucleation is a mass transfer process similar to the growth of crystals. The relevant equation is given in Chapter 3. In common with crystal growth the size of the bubbles formed will depend on the solubility of the gas because the concentration gradient driving growth increases with the solubility. Thus smaller bubbles would be expected from hydrogen and larger ones from carbon dioxide.

12.3 Process Design

12.3.1 Mixing

This is as important as with settlement, in fact Janssens (1989) considers it more so and has developed a weir with finger baffles in the cascade (Fig.12.2) to mix the coagulant with the flow more rapidly rather than to rely on the plunging of the stream into the receiving pool. The contact time between the point of mixing and the process, which is important with floc blanket clarifiers, is provided by the flocculation stage.

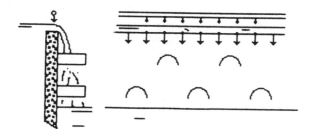

Fig. 12.2 Janssens finger weir for rapid mixing.

The chemical treatment used with dissolved air flotation is the same as with other separation processes except for one or two obvious points. One would not prechlorinate otherwise residual chlorine would tend to evaporate into the air above the flotation unit. Likewise if preozonation is used residual ozone must be absent.

12.3.2 Flocculation

The technology established by WRc calls for flocculation times of about 25 mins and G values of about 50 s^{-1}. Two stages of flocculation is most common. These standards are now being challenged. Bunker and Edzwald (1994) claim that 2.5-5 mins flocculation is adequate and no further improvement is gained by increasing the time. The shear rate was apparently affected to some extent by the type of coagulant. It was also reported that polyelectrolytes were beneficial with low turbidity water at low temperatures with aluminium sulphate but they were unnecessary with polyaluminium chloride. Dahlquist et al (1996) suggest that three stages of flocculation may be preferable with short flocculation times, which would be logical from chemical reaction engineering considerations as discussed in Chapter 5. These authors also concluded that there was little difference between shear rates of 30 and 70 s^{-1}. Discussion at the 1997 conference suggested that many had found that shear was less important than time and some had even shut off flocculators without loss of performance. The general consensus was that 15-20 mins was necessary for cold climates but 10 mins was adequate where the water temperature was above 15°C.

Most plants in the UK use vertical shaft paddles, whereas in Australia horizontal paddles have often been used. The tangential velocity component from vertical units can cause uneven flow through the riser section, a problem common to horizontal flow clarifiers. A tortuous path with some headloss is needed so that the tangential flow in the flocculator is not passed forward. In dissolved air flotation such uneven flow will lead to part of the flow being overdosed with air and part underdosed. Horizontal paddles give a uniform flow when viewed from above.

Some recent plants have been built with propeller and so called hydrofoil type flocculators, which are presumably cheaper. The maximum local shear for a given duty has traditionally been regarded as higher in this case and Dahlquist et a. (1996) have found the propellers gave higher turbidities and particles numbers in a comparative evaluation of these and gate type designs. This is contrary to the point made by Griffiths (1996). However no details were given and as discussed in Chapter 4. subtle differences in the design may have a significant effect on the flocculation efficiency. The hydrofoil according to Griffiths has a lower maximum shear stress for a given energy input and works more by causing flow rather than shear. Such flow will be in an axial direction and should cause less difficulty with tangential movement in the riser than conventional gate paddles.

Hydraulic flocculation has been used successfully in at least two plants using conventional flotation. The performance of one was included in the survey by Longhurst and Graham (1987) and was within the normal range for a clarified water and capable of being filtered to the usual standards. The disadvantage with hydraulic flocculation is that the energy must be obtained from the inlet velocity between stages which requires energy in the form of a headloss. The height of the structures must be a little higher, but only by 0.5 - 1.0m. It may be assumed that in the context of a value engineered plant to meet the final quality the elimination of mechanical paddles and adjustment of the intermediate quality improves cost effectiveness.

Experimental work at WRc has shown that hydraulic flocculation in sinuous flow labyrinths (long used in some horizontal flow settlement tanks) reduces the time and improves the efficiency. Floc breakage is undoubtedly less and the shear more uniform than in mechanically mixed tanks. Also the residence time distribution in a labyrinth arrangement will be closer to plug flow and this will therefore increase the efficiency and reduce the time. One patent has described the use of aeration tower packings as a flocculator. While such packings or even gravel can be used for flocculation floc tends to accumulate and some method of backwashing must be added, which removes the advantage. Such systems tend to be more expensive than simple paddles in otherwise empty tanks. Labyrinth flocculators are used in some industrial designs, eg. the Kofta Sandfloat combined flotation/filter. Flocculator design is discussed in Chapter 4.

Most plants hitherto have dedicated flocculators linked to individual flotation tanks, as with horizontal flow settlement tanks. This practice has not been challenged until 1995 when a retro fitted installation was commissioned in N.Wales using a common flocculation system with distribution channels similar to those used in filtration. Discussion at the 1997 conference indicated that other examples exist. There is a danger that if the floc is too large there could be fallout but it is generally agreed that only very small floc is needed in this case. Dahlquist et a. (1966) claim that the floc size entering the flotation section both at 5 and 20 min. flocculation was less than 10μm at which size there should be little damage in passing along conventional channels and penstocks.

The other example, where a common flocculation system is used, is Cocodaf, counter current flotation, where flocculation is purely hydraulic, similar to the contact tanks preceding floc blanket clarifiers. However the contacting conditions are more efficient in this case.

There would seem to be scope for challenging the traditional practice which has become established in the past 20 years.

12.3.3 Air Dissolution

The amount of air to be added to the flow depends on the mass of material to be removed and also the amount needed to make up any lack of saturation. Conversely if the raw water is supersaturated less air will be needed. For clarification of low turbidity water about 5 litres at atmospheric pressure (6g) per cubic metre of flow are normally necessary. Larger amounts appear to be required as the solids concentration in the raw water increases. However as the solubility of air in water at 10°C is 22 N litres per cubic metre a small change in the saturation of the raw water

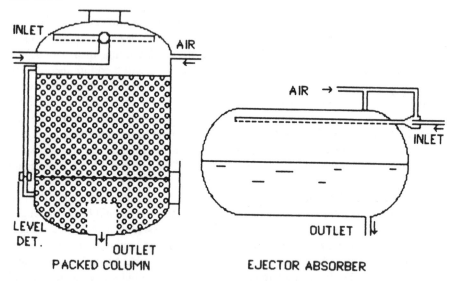

Fig.12.3 Packed column and ejector/separator type absorbers for air dissolution.

will have a marked effect on the air demand. Also as oxygen is twice as soluble as nitrogen diurnal changes in the dissolved oxygen will affect the air demand. In most plants the recycled water is injected at a depth of about 2.0m. The air added is thus close to that required to bring the saturation at that depth up the 100%. Thus one cannot reduce the quantity very much before the bubble blanket disappears. An anaerobic water with the normal nitrogen content would double the air demand. A common design figure is 10g air/m^3, (10ppm).

The dissolution of the air in water is a straight forward chemical engineering process and several options exist. The early literature tended to overplay this aspect. There can be a compromise between cost and efficiency. The packed tower absorber (Fig.12.3.) is the textbook approach. Data on mass transfer coefficients is available from chemical engineering textbooks and efficiencies approaching 100% can be achieved. Simpler but less efficient alternatives include ejector mixing with separator tanks, open spray tanks absorbers and static mixers. Another device uses cross flow over a candle diffuser. All of these use external air compression. The latter options require prototypes for calibration and tend to be used on proprietary plant. The packed tower may not suitable for use with waste water where biological fouling occurs. Direct entrainment of uncompressed air by injectors has been used but the water pressure required to drive the system is twice that of the other methods.

It must be remembered that owing to the different solubilities of oxygen and nitrogen the composition of the gas in the head space in the absorber is close to 8:1 not 4:1 as in the atmosphere. The ratio in the recycled water is of course the same as in the atmosphere to maintain a mass balance. Saturation levels must be based on the 8:1 value. Thus for example at 10°C and 5 bar saturation occurs at 106 litres (free air) per cubic metre. The diffusivity of nitrogen is 76% of that of oxygen and will limit the mass transfer process.

Typical absorbers usually have a specific flow rate of about 80-100m/h and a packing depth of 1.0-1.5m depending on the size of rings chosen, usually 25-38mm. Often the packing is placed on a support grid or tray but this is unnecessary and a carry over from aeration tower design. No gas is vented from the absorber, there is no through flow and the air may be fed in at any point. A simple basket strainer may be used to prevent the packing from passing down the outlet.

Both oil free air compressors and oil lubricated ones are used. The oil content in the air delivered from modern compressors is low enough to avoid compromising the water quality with respect to hydrocarbons at the dilution which prevails. One low energy alternately for compressing the air which has been proved experimentally is the use of a compression vessel which is alternatively filled and drained using a timer operated three way valve or two plain valves and a pair of non return valves (Fig.12.4). The air is compressed by a bleed from the recycled water supply and driven forward to the absorber. Non return valves allow drainage which may be returned to the main flow. A cycle time of 2-4 mins. is easily accommodated as the rate of rise of the air interface in the absorber is about 1.8m/h at 5 bar, or 125mm in 4 mins. The volume of the vessel is that of the free air required in this cycle time. The operating pressure must be at least 3.5 bar otherwise a satisfactory bubble blanket is not obtained. The most common pressure, regarded by some as an optimum is 5 bar but examples of up to 7 bar exist. For a given duty the cross section of the absorber varies inversely with pressure and furthermore the additional flow in the flotation stage caused by the recycle also falls as the pressure increases. The bubble size also appears to fall with increasing pressure as discussed below.

The air carrying capacity varies directly with pressure and directly with flow. Because the pressure with a fixed orifice system varies with the square of flow the air capacity thus varies with the cube of flow and with (pressure)$^{1.5}$. A design based on 5 bar can

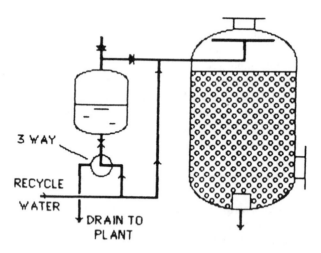

Fig. 12.4 Hydraulic compression of air for dissolved air flotation.

only be turned down by 41% before the pressure falls to 3.5 bar and the performance falls off. At 6 bar the turn down would be 55%.

Some plants run with a fixed recycle and ignore power economy. Some use a higher design pressure and some have multiple aeration manifolds that can be switched in or cut out to vary the hydraulic characteristics if the system. The use of valves in place of orifices enables the flow to be adjusted but with a large number of valves this is an impracticable approach for short term changes.

The recycle flow required to carry the air varies with temperature. Thus a nominal 10% recycle at 10°C would have to be 13% at 25°C but only 8.2% at 2°C. However the air demand for making up any undersaturation will also vary similarly.

A useful monitoring procedure, rarely used, is to measure the air content of the riser section of the flotation tank. Where access is possible this may be done by sweeping aside floating scum and dipping an Imhof cone into the bubble blanket, placing a plate over the mouth and allowing the cone to stand inverted to allow the air to rise to the top. Such a procedure may be adapted for plant use. The recycle may then be set to give a fixed bubble volume concentration regardless of raw water conditions.

The power consumption comprises both water and air components. For example if a 50% margin of safety is allowed over a basic 6 N litres/m³, ie 9 N l/m³, and the design temperature based on 15°C as a maximum, the solubility of the 8:1 N_2:O_2 mixture taken as 118 N litres/m^3 at 6 bar, the recycle ratio will be 9÷118 = 0.076. Thus to treat a main flow of 1m^3/s, 76 litres/s of recycle flow must be raised to 6 bar (60m approx.). Allowing 70% for the pump efficiency the power will be :-

$$Power = \frac{Q.H\rho g}{efficiency}$$

Hence:-

$$Power = \frac{0.076x60x1000x9.81}{0.7} = 63.9 kw$$

The air flow at atmospheric pressure will be 9.0 l/s. For air compression:-

$$Power = 2.49 Q_A P_1 [(\frac{P_2}{P_1})^{0.286} - 1]$$

where P1 etc is measured in Pascals. The theoretical power at 70% efficiency will be 2.12kw, but practical blowers/compressors may not sweep the compressed air fully and may be even less efficienct. Nevertheless the air compression power is relatively small compared with the water power. Some power can be saved if the compressed air is cooled (eg. with a water spray). Compressors fitted with receivers working at a higher pressure than the absorber waste power. Additional savings in power can be achieved if the compressed air is cooled or if a water mist is injected into the compressor to keep the air cool. (A special compressor would have to be selected in this case.)

12.3.4 Absorber Level Control

Three methods of controlling the level in an absorber are available. The most obvious method is to maintain a constant delivery from the recycle pump, at a rate to give the necessary pressure at the nozzles as discussed above. The air is admitted to the absorber via a solenoid or float valve, or an appropriately operated valve working in proportional mode, controlled from the water level which itself may be allowed to vary over a band to reduce the frequency of operation of the inlet valve. Alternatively the compressor speed may be varied. A high water level triggers air admission.

The air flow should not be greatly in excess of the steady demand, again to reduce the frequency of cycling and energy consumption. In one design the air is admitted though a float operated valve set at the intended water interface. Such a valve may be located in an external chamber. This is more convenient on smaller plants.

With a pump with a flat duty curve the admission of air will merely reduce the water inflow and maintain a steady outflow. With a rising duty curve there will be a small increase in outflow still resisted by the square law characteristics of the nozzles but providing that the instantaneous inflow of air is not massive the fluctuations will be within the accuracy of defining the air demand.

Variable speed pumps are required if the air supply is to vary, as discussed above. This method of operation assumes that the absorber efficiency remains constant at a high figure, which will normally be the case, so that the system merely draws in the necessary air to give the level of saturation. The air compressor in this case needs no pressure tank and the delivery pressure will be that of the absorber, which in turn will reduce the compression power to a minimum.

The other common method of control is to set the absorber air pressure and to trim the recycle pump speed to maintain the correct level. This approach requires a calculation of the pressure required to give the correct flow. The recycle pump speed varies only slightly in controlling the level and does not cause significant flow variations in the outgoing flow which remains a function of the internal pressure as set. As usually constructed with a constant high pressure air receiver working well above the absorber pressure the power consumed by the air system is higher than it need be.

It is possible also to control the air <u>flow</u> at a fixed rate as required by the process, from the compressor, and to adjust the water flow to maintain the water level. In this case the saturation efficiency will vary and the water level will vary to give the necessary surface area on the packing to dissolve the air. This system can operate continuously without valve operation except for an emergency low level alarm condition, which may trigger a reduction in the water flow via the variable speed pumps.

The simplest arrangement of all uses a bleed valve or orifice at the intended level which is set by design to weep a very small fraction of the flow which is either run to drain or returned. The discharge of air is discussed in many textbooks (eg. Coulsdon and Richardson (1990). The velocity of air emerging from a plain orifice at 5 bar is 466m/s and at 6bar 481m/s, whereas with water the velocities are 31 and 34m/s respectively. If an orifice is designed to discharge all the air, when exposed the water flow when the level is high, it will be only discharge about 1.2% of the recycle flow (about 0.12% of the main flow). Good control would be expected with half of this. The power penalty is therefore trivial and cost savings significant. Any air reaching the orifice is rapidly vented and if the inflow of air is set marginally above the desired flow. Steady conditions with a wide margin of safety are achieved without an external control system. This system can only be operated by setting the pump speed to give the required flow. The compressor speed must deliver the necessary flow at maximum rate and excess will be delivered at lower rates unless the compressor speed is varied with the recycle pump speed. In this case the power efficiency is high.

It must be remembered that often the absorber may not be installed at the same level as the flotation plant. Air must not be allowed to come out of solution in the main. If it does there will be an eruption of coarse bubbles which will sink the floating scum. Thus for example if the working water level in the absorber is 5m below the flotation tanks the efficiency referred to 5 bar should be not greater than 95%. Likewise if the absorber is above the flotation unit then it is impossible to achieve 100% saturation at the level of the flotation unit and the formation of coarse bubbles cannot occur. However the recycle flow must be increased to compensate for the lower air content.

It is unfortunate that the solubility of gases falls with increasing temperature. Thus a plant designed for say 10% recycle at 10°C would have to be uprated to 11.5% for 20°C or 12.6% at 30°C. If the higher separation rate at higher temperatures is to be exploited the recycle system would have to be uprated even further. Dissolved air flotation is most economical at low temperatures.

For shut down, either planned or on power failure, means must be provided to prevent undissolved air passing through to the aeration manifolds and disturbing the floating scum. Methods used include a power operated outlet valve on the absorber (which must be powered from a standby source), a vent valve held shut electrically to blow off in the event of power failure, or a float operated alarm level air release valve. If the pipe run to the manifolds is longer than the feed to the absorber, and the depth of water in the absorber significant the air may blow back harmlessly through the recycle pumps. Air compressors with receivers aggravate the problem.

If power outages occur infrequently the above measures may be deemed unnecessary. If the scum is disturbed there may be an additional but temporary fallout of scum and a passing additional load on the filters but this will be much less than that of a floc blanket spilling over from a settlement tank.

12.3.5 Air Injection

The recycled aerated water is allowed to discharge though a valve or nozzle which arrangement which optimises bubble formation and also provides reasonably efficient mixing with the flocculated raw water. Thus it is usual to have a grid of nozzles or a row set in a slot with a spacing of typically 250-500mm between them.

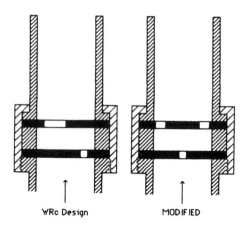

WRc Design   MODIFIED

Fig. 12.5 WRc type nozzle for dissolved air flotation.

Many types of pressure reduction device have been proposed and used. One widely used design is the WRc nozzle (Fig.12.5) or variations of it. Many are similar to the modified version. This comprises an orifice and a target surface close to it so that the emerging jet is rapidly dispersed at very high local shear. In another design two jets emerging from holes in the opposite sides of a short socket tube set in the manifold impinge on each other. This configuration provides a relative velocity twice that of a single jet colliding with a plate. The lower limit for the operating pressure might therefore be less in this case. Some manufacturers and designers prefer needle or globe valves. These have to be set to give an equal flow but offer means of altering the flow/pressure characteristics. Reports from users suggest that they have to be opened occasionally to clear blockages, whereas providing a suitable strainer is incorporated in the line round orifices remain clear. Valves can of course be adjusted to suit the flow rate. Small orifices and slots such as the annulus in a throttled globe valve appear to give better performance than larger orifices. Proprietary spray nozzles have also been used as discussed below.

Most devices are fitted with some type of downstream diffuser, which usually consists of a short length of pipe much larger in diameter than the orifice, to enable the high velocity jet from the orifice to expand to the full bore of the pipe before encountering the floc. Flares and bell mouths have been found beneficial. Excessive length can cause coalescence and the formation of bubbles that may be too large for floc capture and even cause the floating scum to sink . Providing that the essential features are incorporated several designs are capable of a satisfactory performance

Meijers and van Bennekom (1980) and Rykaart and Haarhof (1994) have examined several designs of nozzle and have come to some interesting conclusions as shown in the following table:-

Nozzle	Diffuser	Pressure, bar	Turbidity NTU	Res.Fe mg/l
Bete WL2 120	None	6	1.1	1.1
	A	6	0.53	0.54
	B	6	0.96	0.98
	C	6	0.6	0.66
WRc 3.0mm	WRc	6	1.2	1.19
2.0mm	WRc	6	0.78	0.80
3.0mm	WRc	5	2.4	1.78
Bete TF 8FC	None	6	1.2	1.24
	A	6	0.85	0.87
AKA	None	6	1.04	1.16
	B	6	0.54	0.49
	C	6	0.49	0.47

Bete WL 2 120 is a fan type spray nozzle
Bete TF 8FC is a helical tail nozzle
The AKA is a needle valve.
Diffuser A was a fishtail horn tapering to a slot, B a small 25mm diam. parallel shroud and C a larger 38mm one.
Median bubble sizes varied from 25 to 40μm

Rykaart and Haarhoff (1994) studied several designs and found that bubble formation was complete in 1.7ms. In the WRc type of nozzle the target plate must be no morethan about 5mm from the jet to obtain fine bubbles. Again the bubble size was inversely related to pressure. They also studied the formation of coarse bubbles and found that a tapering outlet following a plain discharge greatly reduces the bubble size and also the number of coarse bubbles.

The nozzle design and operating pressure are therefore most important. However there may be a danger in striving for utmost perfection and in so doing increasing the cost unduly. The main requirement is for a product that can be filtered easily, and a design that is robust and not likely to block.

The depth of injection is usually between 1 and 2m in the riser section. If too shallow there may not be sufficient time or depth for capture, If too deep all the air may remain in solution and not take part in flotation. Excess air may then be required. The nozzles are housed in a riser which brings the flow close to the surface so that the hydrostatic pressure and thus the supersaturation is minimised. In most designs the wall of the riser slopes but there seems to be no good reason for this providing that the flow pattern encourages mixing. Excessive residual dissolved air will aggravate air blinding on any subsequent filter. The velocity over the riser wall is sometimes set to encourage movement of the floating floc towards to sludge weir. In some examples the air is injected into the down comer before the riser. Any coarse bubbles in this case will rise against the flow and may cause problems with scum on a water surface that is not scraped or cleaned.

12.3.6 The Flotation Zone

This corresponds to the settlement zone in a horizontal flow settlement tank. The area if defined by the Hazen Equation, using the bubble rise rate instead of the settlement rate. Most tanks are rectangular but circular radial flow examples exist. As with settlement tanks rectangular tanks nest conveniently when constructed in concrete but in thin steel circular tanks usually prove cheapest. With a given reach from the riser to the far end circular tanks may easily have a diameter twice the length of a rectangular tank.

The separation zone for calculating the surface rate should strictly be the at the base of the bubble blanket, Stevenson (1981), and should exclude the area of the riser in the same way that the area of the flocculation zone in a non recirculation settlement tank is not included in the effective area. However practice which has now become traditional is to include the entire water surface over which scum can collect. This may lead to inconsistencies in comparing different designs.

The flow through the flotation zone is of course the sum of the net flow plus the recycle. Allowance must be made for the latter in calculating surface rates and plant areas. If wash water is recycled to the front end of the works this must also be included.

Based upon the whole water surface rates vary from 5 to 13 m/h, with 10m/h being a fairly common upper limit. Dahlquist et al (1996) claim satisfactory operation at 16m/h but in the 1997 conference he indicated that the limiting feature was not the carry over of solids but air, free from solids, which causes blinding of filters. He extended the upper limit by adding a lamella separator to coalesce and separate the air being carried over. According to Shawcross, rates of up to 25m/h have been explored (Anon, 1997), but the temperature is of course an important factor. The rate is inversely dependant on the viscosity of the water thus is reasonably to expect a design which operates successfully at 10m/h at 3°C to do so also at 16m/h at 20°C, providing that the recycle system is designed accordingly.

Flow patterns in the flotation tank have been studied by several workers, using computational fluid dynamics (CFD). The air blanket stabilises the flow in a manner similar to floc blankets and the pattern changes when air is injected. A plain inverted weir is often satisfactory but as rates rise perforated pipes or a filter below produce a more even flow pattern. The velocity over the wall of the contact zone is important. If too low unaerated water can spill in the reverse direction (Dahlqvist, Anon, 1997). If too high the air blanket at the far end will plunge. This velocity is exploited in many hydraulic sludge removal systems and an optimum must be sought.

The stability of the scum varies. Scums from water containing large amounts of algae may float almost indefinitely whereas predominantly inorganic suspended solids may only float for an hour or so. The part of the scum above the water level tends to drain and dry out so that it becomes top heavy. Patches can invert and part may then sink. The consensus is that most sludges are stable for at least 24 hours. It is always possible the water may contain some solids that are not trapped by the air but always sink. Some tanks are therefore provided with scrapers such as the reciprocating type referred to elsewhere. For the treatment of most waters suitable for flotation the quantity of settleable solids is usually small and they are allowed to accumulate on the floor of the tank and removed occasionally by draining down the tank and flushing out. The floor must therefore slope to a gulley, and a hose connection provided for flushing.

Various views have been expressed about the size of single flotation tanks. If these are scraped the tank size will be governed by the maximin size of scraper that can be installed. Wide experience covers lengths of up to at least 10m (even a 20m centre inlet at Frankley, Birmingham, England) and similar widths, which would produce a unit throughput of about 24 Mld per unit. Purac claim to have built a tank 15m square (Anon. !997). Some argue that beyond a certain length all the bubbles will have risen to the surface but this would run counter to Hazen's concept and such a tank would actually be operating at less than its maximum capacity.

Tank depths are usually at least 2.0m, some are only 1m deep. There is evidence that excessively deep tanks allow undesirable circulation patterns to develop, and baffling may be necessary. A high velocity over the riser cill can cause rolling circulation, and a compromise between this and conveying of the sludge must be sought. A 4:1 length to depth ratio has been suggested,

12.3.7 Clarified Water Collection and Level Control

The density difference between water and the air bubble blanket is much greater than that between the water and floc in a settlement tank. Thus fairly crude arrangements usually suffice for collecting the clarified water (however see the previous section). Very often this is merely a slot at the base of one wall. In some examples perforated pipe collectors are installed close to the floor. If the flotation tank is combined with filtration, the separated water passes directly down into the filter media.

The accuracy of level control is usually dictated by the method of removing sludge. Some arrangements may require control within ±5mm. An alternative to level measurement is to measure the flow over the sludge weir. The level may be allowed to vary considerably if the sludge is accumulated and washed out batch wise when a filter is washed. Scrapers or hydraulic methods require fairly close control of the level against the beach or weir. A full length decanting weir behind a curtain wall at one end of the tank is commonly used. Beyond this any of the systems used to control filters, such as actuated valves or syphons will be applicable. Accurate control is of course only required during the actual scum removal operation and the level may be allowed to vary at other times.

If the flow to individual units is controlled hydraulically, and not by a weir, a changing level in one unit may interfere with the flow to another unit.

12.3.8 Removal of Floating Scum

Several methods have been used depending on the compromise that is required between simplicity and water economy. Some have argued that because the scum can concentrate to perhaps 5% w/w that it should be removed in as dry a state as possible. This has frequently led to the situation where the scum has had to be diluted with flushing water to make it flow away. Others have taken the view that simple non mechanical methods are most cost effective and comparable with the non mechanical philosophy that has grown up with floc blanket clarifier technology. Hydraulic sludge removal has been found in one example to produce overall concentrations on a low suspended solids water of 0.8% w/w, during a 4 week trial,which compares very favourable with floc blanket tanks on such waters. No problems then occur with conveying the sludge to the treatment unit. Thickening together with the wash water from the filters is usually still necessary prior to final disposal.

Mechanical scrapers vary from full chain and flight units, to partial units which make use of the fact that the direction of flow of the water in the tank carries the floating scum forward at last part of the way. The scrapers carry the scum to the end of the tank, up a beach and over to a trough or hopper. In some designs scrapers are carried on a carriage which reciprocates while the blades rise on the return run and fall on the forward run. Velocities are low, usually about 1m/min. The water level remains constant, as the lift up the beach is limited. If the sludge is to be conveyed it must usually be flushed forward with water as already mentioned. Some small plants merely use a narrow hopper against the end of the tank to collect the sludge. Archimedean screws are also used in some plants to convey thick sludge to the end of the sludge weir.

A simpler arrangement is the beach scraper paddle (Fig.12.6), which allows the water flow to carry the scum to the beach and a rubber bladed paddle moves the sum up the ramp and over into the sludge trough. The rubber blade must form a seal against side cheeks on the ramp to prevent entrained water running back into the tank and carrying fragmented floc back into the product water.

Complete hydraulic removal is a further development of the above. Methods include a falling or tilting weir to allow decanting of sludge over the full width of the tank, or a fixed sludge weir with a discharge valve or adjustable down stream weir. With combined filters the water level in the unit is allowed to rise to initiate decanting of the sludge over a fixed full width , sometimes assisted by a travelling blade. There is a growing popularity for such simple systems and some even suggest that the sludge might be as dilute as filter wash water as it does not necessarily affect the size of the sludge thickener, which will often be limited by the underflow flux.

In hydraulic sludge removal the crest height must be about 20mm to clear "icebergs". The technique is therefore not applicable to very short units where the throughput might be less than the instantaneous sludge flow. If the level is raised the rate must be controlled and integrated with the plant control system so that unnecessary surges are not transmitted to the filters. In all of these versions wastage is low and less than that from a typical settlement tank.

In both the hydraulic approach and also to beach paddle design side wall lubrication with a water sparge greatly helps the movement of the scum carpet towards the weir. However in each case it is important that the direction of flow within the unit from the riser to the far wall should be the same as the intended movement of the sludge.

Fig. 12.6 Paddle type scum removal device.

All of the sludge removal methods carry some danger of fallout. Longhurst and Graham (1987) and Schofield (Anon. 1997) report this problem as being particularly common with mechanical scrapers. The hydraulic version may possibly do the least damage as blades do not have to penetrate the surface. Scrapers usually operate intermittently to minimise this problem.

Heavy rain will readily sink the scum and wind will interfere with removal. The scum freezes readily forming a honeycomb plate structure that will either prevent its removal or damage the scraper system. The flotation section (but not the flocculation section necessarily) must therefore be covered and protected accordingly.

Sludge channels should be narrow and have a rounded or benched bottom to provide a reasonable velocity to carry the sludge, and should preferably have an end exit rather than a bottom exit so foam is carried away.

Sludges from flotation plants are generally agreed to be treatable by conventional gravity thickening, although occasionally means for breaking up aerated clumps may be needed. (Pumping can be beneficial in this respect.)

12.4 Post Treatment

The quality of water produced by a dissolved air flotation plant is usually at least as good as that from solids recirculation or blanket clarifier systems and filters used for further treatment are entirely conventional except that the type with shallow supernatant water is probably inadvisable. The water from a flotation plant will be fully saturated at a level corresponding to the base of the bubble blanket (probably corresponding to the level of the riser cill, some 300-500mm below water level). However the aeration in flotation strips out any supersaturation in the raw water (caused by algae for example).

Dahlquist et al (1996) found that water that had been deliberately stripped of excess air tended to show early breakthrough from the filters. However the stripping treatment may have damaged residual floc. Nevertheless weirs used either for level control or flow splitting into filters will tend to remove excess air.

12.5 Design Options

The conventional form of a dissolved air flotation plant is rectangular and horizontal flow as shown in Fig.12.1. In some package plants the flocculation tanks are connected to the flotation section with tubular sections and the recycle stream is injected into these tubes.

It is convenient to combine flotation with filtration (Fig.12.7.). This saves one set of tanks. In this case dedicated flocculation tanks serve combined flotation/filtration tanks. The surface rate of the flotation tank will generally have to be reduced to that required of the filtration section. Some stipulate that this may be no more than 6m/h (all in). The use of sludge removal equipment is then optional and depends of sludge stability. The sludge can be removed at back washing. For sludge removal the filter outlet control system must maintain the level accurately against the rising headloss of the filter. The tanks must be deep enough to accommodate the filters as well as the flotation section but the flocculation tanks are then deeper and can occupy less ground area. The area will be about 4 m^2 per m^3/h. for the combined process steps. Patents on this arrangement have long expired. The disadvantage of this arrangement is that is not possible to add chemicals between the flotation and filtration stages, which may often be necessary for optimum colour and manganese removal.

Most plants have dedicated flocculation sections, each of two stages, which on large plants may comprise two vertical paddles per stage. Four streams require 16 paddles plus drives. As mentioned earlier, one example in Wales now has a common flocculation section serving four flotation units. This is a trend that may spread.
Two early plants (already mentioned) were provided with hydraulic flocculation which simplifies the design and reduces costs. They were able to meet process guarantees without difficulty.

A conversion in South Wales (Fig.12.8.) used existing circular concrete tanks to produce two flocculation stages and a flotation stage stacked vertically with non water retaining diaphragms so that the ground area occupied corresponds to 11m^2/m^3/h. A single flocculator shaft with two paddles was used, and pH correction between flotation and filtration was possible. This design is suitable for circular steel tanks.

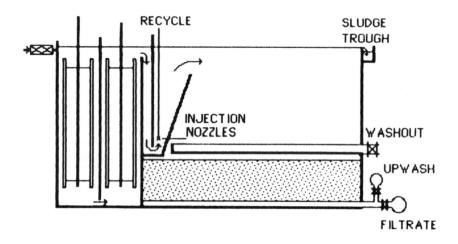

Fig. 12.7 Combined flotation and filtration.

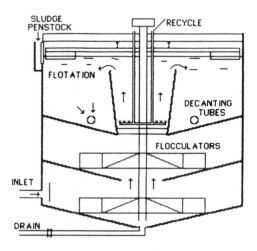

Fig. 12.8 Cantref triple deck flocculation and flotation.

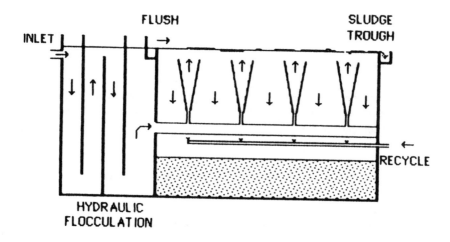

Fig. 12.9 COCODAFF counter current flotation/ filtration.

A more radical departure is countercurrent dissolved air flotation which is featured but not commented on in a patent by Krassnoff and Luthi (1982). The process which has been developed, incorporated into a 200Mld plant and described by Eades and Brignal (1994) (Fig.12.9.) involves hydraulic flocculation in a common section serving several streams for 15 mins followed by distribution of the flow into individual tanks through upstanding diffuser cones arranged in a somewhat similar manner to the inlets in a flat bottomed floc blanket tanks but inverted. The flow therefore descends against a rising bubble blanket produced from low level nozzles. The system optionally has a filter bed below the aeration nozzles. The scum is flooded off hydraulically to maintain mechanical simplicity.

This design would be expected to have certain fundamental advantages in that the countercurrent contacting regime is intrinsically more efficient than cocurrent contacting. The pressure at the nozzles is dissipated into clarified water instead of flocculated water and floc damage should be avoided. Far fewer nozzles are used (2m centres) and the process is modular like flat bottomed clarifiers so that any size may be built. An unexpected benefit was that any floc falling out is recaptured by the bubble blanket, whereas in conventional flotation it would fall to the floor. The treated water quality appeared to be comparable with conventional flotation but very dependant on the contact (flocculation) time. It is difficult to make a true comparison with conventional cocurrent flotation because the objective was the use of non mechanical flocculation and shorter retention times to achieve a lower cost for comparable quality.

12.6 Advantages and Disadvantages

The fact that dissolved air flotation has largely replaced settlement in new plants in Britain over the past 20 years indicates that overall it must be regarded as superior to settlement for the qualities of water that prevail. This is probably for the following main reasons:-

1. The process is able to start, stop and restart immediately without a long build up period when the quality is substandard. (Some are used on a fixed rate, start stop basis)

2. In the event of failure there is no accumulation of past sludge (on which recirculation and floc blanket clarifiers rely for their operation) to pass forward to block the filters.

3. Although settlement tanks can now be operated at rates to produce similar size of plants this involves the use of polyelectrolytes which are subject to strict regulation and many utilities prefer to avoid their use if possible.

4. The capital and operating cost, while slightly more than for settlement, are not excessive compared with the overall cost of the plant and the benefits.

5. The majority of waters to be treated in Britain are stored or are from upland sources and have intrinsically low settlement rates but are suitable for flotation. Flotation is known to be able to handle most of the extremes of turbidity that occur in Britain. In Southern Africa very turbid waters merely have a presettlement stage upstream.

Dissolved air flotation is less appropriate for water with high suspended solids which settle rapidly. As mentioned the power consumption rises with temperature but this does not rule out its use in more tropical areas.

As currently designed dissolved air flotation might be regarded as more complex and less suitable for areas of low mechanical skill but it would seem that there is still scope for adaptation and simplification (omitting mechanical scrapers, and items requiring delicate setting, items of control instrumentation, for example). The hardware is in any case far more robust than lamella systems, which have made little headway in Britain.

Where the incoming water is at a head of at least 35m the power consumption may be eliminated by feeding the absorber direct with uncoagulated water bypassing the break pressure tank. In most cases the impurities in the undosed part of the flow will be absorbed on the floc and the final water quality will not be compromised. (A similar bypassing practice was at one time used at Crownhill works at Plymouth on a direct filtration plant as a means of economising on aluminium sulphate.) The by pass ratio will of course fall as the working pressure rises. There may be limits with highly coloured raw water but the process may be evaluated using the flotation jar test procedure.

The power consumption, equivalent to lifting the main flow through about 6m is trivial compared with the pumping cost for distribution.

12.7 References

Anon. (Purac) (1966). "Method of Water Purification", UK Patent 1,141,675

Anon. (1997) , Dissolved Air Flotation, Internat. Conference Proceedings, Chartered Inst. Water and Environmental Management, London.

Barrett,F. (1975), "Electroflotation", Water Pollution Control, **74**, 59

Bunker, D.Q., Edzwald, J.K. Dahlquist,J. and Gillberg,L. (1994) "Pretreatment considerations for Dissolved Air Flotation", see Eades and Brignal (1994)

Dahlquist,J., Edzwald, J.K.,Tobiason, J.E., Hedberg,T.,Amato.T., and Valade,M. (1996) High Rate Flocculation, "Flotation and Filtration in Potable Water Treatment", Chemical Water and Wastewater Treatment, Eds.Hahn,H.H.,Hoffmann,E. and Odegaard,H., Springer,Berlin & New York.

Eades, A, and Brignal, W.J. (1994), "COCODAF: Countercurrent Dissolved Air Flotation",Flotation Processes in Water and Sludge Treatment, International Association on Water Quality/ International Water Supply Association/American Waterworks Association Joint Conference, Orlando, USA.

Fukushi, K.., Tambo, N. and Matsui, Y. (1994) "A Kinetic Model for Dissolved Air Flotation", See Orlando Conference above.

Janssens, J. (1989) unpublished papers.

Longhurst,S.J. and Graham,N.J.D.(1987), Dissolved Air Flotation for Potable Water Treatment: A Survey of Operational Units in Great Britain", Public Health Eng. **15**, 71

Meijers, A.P. and van Bennekom, C.A.(1980) "The effect of Different Nozzle Types on the Flotation Process", (in Dutch), H_2O,**13**,270

Melbourne, J.D. and Zabel,T.F. (1976) Editors, Flotation for Water and Waste Treatment, Conference Proceedings, WRc. Marlow, UK.

Rubin, E.A. and Gustaf,E (1967) Improvements in and Relating to Installations for the Purification of Liquids", UK Patent 1,184,477

Rykaart, E.M. and Haarhoff,J. (1994) "Behaviour of Air Injection Nozzles in Dissolved Air Flotation", See Eades and Brignal (1994).

Stevenson, D. G. (1981)"The Definition of Surface Rates in Clarifiers", J. Inst Water Eng, Scientists, **35**,179

Valade, M.T., Edzwald,J.K., Tobiason, J.E., Dahlqvist, J., Hedberg, T. and Amato. T. (1966), "Particle Removal by Flotation and Filtration : Pretreatment Effects", J.American Waterworks Assn. **88**, (12), 35.

Zabel,T.F. and Melbourne,J.D. (1980), "Flotation", in Developments in Water Treatment, Ed. W.M. Lewis, p139, Applied Science Publishers, London

Further Reading
Zabel, T.F. (1984)"Flotation in Water Treatment", Chapter in The Scientific Basis of Flotation, NATO Advanced Science Institute Series, Ed. K.J.Ives,Nyhoft, The Hague.

CHAPTER 13.

OTHER TREATMENT PROCESSES

The other prefiltration processes used in water treatment tend to be proprietary and detailed design procedures may not be in the public domain.

13.1 Sirofloc

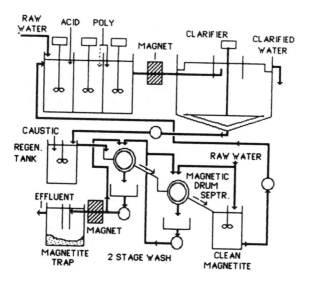

Fig.13.1 Flow diagram for the Sirofloc process.

This process which was developed by the Australian Commonwealth Scientific and Industrial Research Organisation (C.S.I.R.O.) is based on the discovery that magnetite powder (Fe_3O_4) carries a positive charge at low pH values which reverses at high pH. Thus magnetite is able to absorb colour and other organic matter and be regenerated with caustic soda. The second novel feature is that the very fine (micron sized) magnetite which is needed to provide the necessary surface area and thus the absorption capacity normally settles very slowly but flocculates reversibly in a magnetic field to settle rapidly.

The basic patent was published (CSIRO) in 1981 at the same time that a plant was completed at Mirrabooka, W. Australia. Several papers have been written on the subject eg. Anderson et al. (1979). Some British utilities have experimented with the process and three plants have been built, one capable of 60Mld. Only two were built in Australia. Yorkshire Water trials were summarised in a paper by Gregory et al. (1988).

The flow sheet is shown in Fig 13.1. The raw water is mixed with 1-10μm magnetite to give a concentration of about 1% w/w and the pH corrected to 5 with sulphuric acid. The mixture flows through four mixed tanks in series to give a few minutes contact and polyelectrolyte is added about half way through this step. The mixture then passes through a magnet and into a conventional scraped clarifier where it settles at a rate claimed to be in excess of 5m/h.

The settled magnetite is extracted and forwarded to a recovery plant where it is slurried with caustic soda at around pH 12, separated on a magnetic drum, washed and separated again on another magnetic drum before passing to a clean magnetite stock tank from which it is recirculated to the contact tanks at the front end of the plant. The washings pass through a final settlement tank to recover the last traces of the magnetite but a small amount is lost. Magnetite is therefore a consumable chemical, believed to be lost at levels of a few mg/l based on the main flow.

The process does not in fact rely entirely on the charge on the magnetite. To trap suspended solids a cationic polyelectrolyte, usually a liquid type, is added to bind the suspended solids to the magnetite. Thus there is a similarity with the sand ballasted separation processes. These polyelectrolytes hydrolyse rapidly at high pH values and the suspended solids are released from the magnetite. Nevertheless the process is understood to encounter difficulties if coarser grit enters the system.

Maintaining flow in the magnetite circuit is not easy. Plants are provided with pipework which is easily disassembled in the event of blockage. Nevertheless at one works the operators claim (Miles (1993) that the process has proved fairly easy to operate, more so than the previous direct filtration plant. The clarified water has a turbidity comparable with dissolved air flotation plants and possibly better than many settlement plants. There is some doubt about the filterability of residual solids although further dosing with polyelectrolyte would no doubt deal with this. Residual iron and aluminium could prove difficult without pH correction at the filtration stage.

The original Mirrabooka plant had no filters, the clarified water going direct into supply. British plants have all been provided with polishing filtration.

The Sirofloc process has specific applications. The consumption of acid to achieve pH5 would be prohibitive on waters of moderate to high alkalinity. Problems of encrustation will also occur in hard water following reaction between caustic soda and calcium or magnesium in the recovery stage. The effluent is an alkaline concentrate of the colour and the suspended solids extracted from the raw water. This may be difficult to discharge. At Mirrabooka, Australia it was possible to blend the effluent with the alum sludge produced by the first phase conventional plant alongside, and to dry the mixture in lagoons. At Little Hempston, Devon the effluent had to be coagulated with ferric sulphate settled and dewatered by conventional means to produce a filter cake for disposal to landfill.

The principle of magnetic ballasting is interesting and other ways of exploiting it may well emerge when the present patents expire. In its present form the Sirofloc process appears to have specific uses. One attraction at Little Hempston was the avoidance of coagulants such as aluminium or ferric sulphate, although some would regard the cationic polymer with greater suspicion. The process is sensitive to the type of magnetite and will be dependant on a continuing supply of the right type.

13.2 Microsand Ballasted Clarification

Sibony (1981) claims that three Hungarian inventors applied for a patent in 1964 (now expired) describing a process using 20-100μm microsand as a ballasting agent to assist settlement of floc. The rights to this were acquired by the French company OTV. The process was included in a review by Gomella in 1974. The first British example (Simtifier) was constructed in the early 1970s. Proprietary names include Cyclofloc, Fluorapide and Actiflo. The basic concept is the addition of fine sand (3-4kg/m^3) to the coagulated water followed by settlement and recovery of the sand. The principles seem similar to dissolved air flotation but inverted. The volume of sand used is of a similar order although a little less than the air in the latter. The sand size is not quoted but from settlement rates it must be similar again to the bubble sizes in flotation. In contrast to Sirofloc the ballasting agent is widely available and cheap, but it is still a consumable chemical even with recovery.

The original Cyclofloc version involved coagulation, polymer dosing and mixing with sand in stirred tanks followed by settlement in a conventional scraped clarifier. The Fluorapide version uses a hopper shaped tank to hold a fluidised bed with the sand being added to the hopper. The equipment is very similar to the hopper shaped floc blanket clarifier with similar simplicity.

In both cases the floc coated sand is extracted and cleaned in a bank of hydroclones, which unfortunately consume a considerable amount of power and produce a dilute sludge.

The Fluorapide version is also fitted with lamella plates to stabilise the fluidised bed and increase the surface rate. The Annet sur Marne plant in Paris is reported to operate at 9m/h but more recent plants run at considerably higher rates.

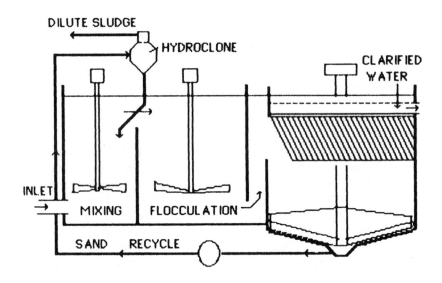

Fig. 13.2 The Actiflo process.

The Actiflo version is a development of the Cyclofloc and not unlike the Densadeg in layout. The raw water is coagulated together with added microsand, flocculated in two stages, and forwarded to a combined lamella scraped tank. The underflow is treated in hydroclones to recover the sand. The "apparent settling rate" ie the yield per unit area of lamella pack is claimed in publicity material to be 40-60m/h, with 8 mins retention in the flocculation section. The average product turbidity is claimed to be less than 2.5 NTU, ie comparable to traditional settlement tanks. The overall depth of the structure is 6.8m.

Sibony quotes experimental data showing that the final turbidity is related to the grain size of the sand, 50-63μm sand giving the lowest turbidity. Alginate polymer was usually required to bind the floc to the sand. The sand consumption is given as 1-2g/m^3 of treated water ie. 50-100 kg/day on a 50 Mld plant, a figure confirmed by experience in Britain. A total of 38 Cylcoflocs and Fluorapides had been constructed by 1981.

No information has been given in the references for the power consumption. However the power consumption on a plant in Wales is reported to be 25.5kw for 38Mld, corresponding to 58kw/m^3/s, or a headloss of about 5.9m, which is very similar to dissolved air flotation. The polyelectrolyte dose is about 0.1mg/l. The sand recycle system must be abrasion resistant and Sibony indicates that pump and hydroclone parts are consumable. It will be interesting to see whether other suppliers develop the technology now that the patents have expired.

13.3 Pellet Clarification

This is a Japanese development that thus far has only been used in commercial applications, mainly because of the quantity of polyelectrolyte used and the reluctance of the Japanese to use of polyelectrolytes. It is an interesting process as it constitutes an extension of floc blanket clarification. Patents were taken out by the Ebara Company (1977) and Ide and Kataoka (1979) have described plants involving 6 m. diameter clarifiers.

The essence of the process is the coagulation of the raw water with aluminium sulphate with appropriate contact or flocculation before the addition of polyelectrolyte in doses above 0.5 mg/l. The process works best on water with some mineral turbidity. The settlement tank is merely a scraped hopper tank, the scraper being required to prevent consolidation and also to encourage the formation small mudballs that can be discharged as pellets. The surface rates without any lamella system are claimed to be between 42 and

120m/h (the latter for a 100mg/l china clay suspension). Apart from the high polymer dose the process is remarkably simple with no external power requirements other than the small amount needed for the scraper and no recycle system. The sludge is granular.

The start up procedure is not disclosed but it may be expected that unless some inert material such as sand can be added as a seed it could take some time before the solids inventory for correct operation is accumulated.

13.4 Membranes

Although they have been used for a long time in desalination membranes are now being considered seriously for the production of potable water from non saline sources. Both the operating and capital costs have hindered their use hitherto but advances in membrane technology, particularly at the interface between reverse osmosis and ultrafiltration have opened the way ahead. The use of membranes for softening is well established in the USA.

The subject is a very large one, beyond the scope of the present book but some general principles can be summarised.

Firstly it is possible to select membranes which give 100% rejection of bacteria and other organisms such as cryptosporidium. This is an attractive feature for operators with doubtful sources. Unfortunately the membranes may have the odd flaw and the performance must be monitored. Moving to low pressure reverse osmosis colour can be removed without coagulation. Some calcium may also be removed but the rejection of monovalent ions is low.

The raw water, possibly after pretreatment, is pumped into the plant at a pressure to suit the membrane and the degree of rejection required. Pressures can vary from 6 bar to 20 bar. It is essential to have a tangential flow over the membrane surface to flush the rejected species forward and to prevent "concentration polarisation". A second pump therefore maintains circulation in a closed loop to give a constant velocity over the membrane surface. A bleed valve enables the discharge of the reject stream in the loop to be controlled while maintaining the pressure in the system.

The percentage recovery with low pressure plants is not high. On softening plants saturation with respect to calcium carbonate will be a limiting factor, even with the addition a sequestrants which increase to solubility. The recovery can be as low as 50%. On plants used for colour removal on small supplies the regulators may prefer the reject stream to have no more than twice the colour of the feed stream.

Membrane plants use tubular cartridge modules to house the membranes. Three types of membranes are in use:-

1. Spiral wrapped assemblies containing the membrane and interleaving separator gauzes which convey the feed across the membrane surface and also the permeate stream away from the surface to a collector.

2. Bundles of hollow fibres, which usually have the flow from outside to inside so that the permeate flows down the core of the fibres.

3. Tubular membranes (6 or 12mm) assembled in bundles like small heat exchangers, in which the flow is from inside to the outside. These have the smallest specific membrane surface area but have a great advantage in that sponge balls can be passed through to keep the membrane surface clean.

One design of hollow fibres employs powdered activated carbon, added to the feed, to provide an ongoing mild scouring action. Cleaning is a vital part of the operation and plants usually include automatic circulation facilities for the cleaning agent. The spent cleaning agent constitutes a chemical effluent for disposal.

Several companies are able to supply membranes systems and rapid development may be expected. However the surface area of membrane is linearly related to the throughput and the cost of a plant may therefore be expected to rise with size more steeply than with conventional technology. Such plants are therefore likely to be installed on smaller supplies. The absence coagulant sludges and the absence of any need to set coagulant doses or to convey such chemicals to the site will also make such plants more popular. A disinfectant after treatment is still necessary.

13.5 Textile Fibre Filters

In recent years two makes of textile fibre filter have achieved some success and further examples are anticipated. These use fibres of polypropylene, nylon or polyester about 15-20µm diameter in a washable configuration. It is reported that they are able to retain 4µm cryptosporidium oocysts and at one works a 60Mld installation has been installed as a long stop after sand filters where such problems have been encountered. The available designs are proprietary and manufacturers should be consulted. One comprises a mop shaped bundle of yarns that is mechanically wrung to remove the collected solids. The other consists of an array of spools mounted on a carrier tube. Cleaning is achieved by a travelling jet that passes between the spools.

13.6 References

Australian Commonwealth Scientific and Industrial Research Organisation (1981) British Patent 1,583,881

Ebara, (1977), British Patents 1,478,337 and 1,478,345

Gomella, C. (1974) "Recent Developments related to Preclarification.", Proceedings of 10th International Water Supply Congress, Brighton, U.K. (IWSA), p1.

Gregory, R., Maloney, R.J. and Stockley, M. (1988), "Water Treatment using Magnetite - A Study of a Sirofloc pilot plant", J. Inst. Water and Environmental Management, **2**,532

Ide, T. and Kataoka, K. (1979) Proceedings of Second World Filtration Congress, p377

Miles, J. (1993) Verbal Communication, Inst. Water and Environmental Management, Scientific Section Visit/Meeting. South West Water

Sibony, J. (1981), "Clarification with Microsand Seeding. A State of the Art." Water Research, **15**,1281

CHAPTER 14.

PRECIPITATION SOFTENING

14.1 Introduction

At one time in Britain calcium and sometime magnesium salts were removed from public water supply on a large scale in order to meet statutory requirements to maintain the composition after new harder sources were introduced. This practice was curtailed after the early 1970's when the association between soft water (low in calcium) and cardiovascular disease was recognised. Softening is now practised on a large scale mainly on industrial supplies and power stations and as a pretreatment for reverse osmosis plants. In one example softening at a reduced level has been continued in order to maintain the output of an existing plant and to utilise existing sludge handling facilities. Many larger plants treat surface water and combine clarification with softening, thereby avoiding a separate treatment stage. The sludge mass is mainly that from the softening process and the calcium carbonate improves the handling of the sludge.

For power station cooling water the main requirement is for the removal of calcium bicarbonate that would be precipitated when carbon dioxide has been stripped off by the combined effect of heating and aeration in the cooling towers. For high conversion reverse osmosis (non sea water) plants the degree of concentration is usually limited by the low solubility of calcium sulphate and silica. Conversion of the calcium to sodium may extend the conversion of water very considerably. The same is true of silica. This is removed by adsorption on magnesium hydroxide. It may be necessary to add magnesium in the form of sulphate to obtain the necessary final quality.

Precipitation softening uses equipment similar to that for conventional coagulation and settlement and may be regarded as special case.

14.2 The Processes Involved

The twin processes of **clarification** and **precipitation softening** are to some extent separate. They operate on somewhat different principles but because there are process units which will can carry out both processes either independently or in parallel it is obviously more economical in capital cost, with little or no penalty in running cost, to combine these steps.

Before going further it is worth reviewing the principles of the processes involved. Firstly **clarification** involves the coagulation of suspended solids, which in

Britain is typically variable from a few milligrams per litre possibly to about 100 mg/l on occasions. Coagulation is achieved by the addition of salts such as aluminium or ferric sulphate (or chlorides). These are hydrolysed to the hydroxides which produce a light voluminous floc precipitate which enmeshes the suspended solids by a process known as sweep coagulation. Water soluble polyelectrolytes are often added in order to strengthen the floc (agglomerates of primary precipitate) so that it forms larger agglomerates and as a result settles faster and produces a lower residual turbidity. The iron or aluminium hydroxides are very insoluble and the reactions are fast. However the slow step in the process is flocculation in which submicron primary particles collide as a result of the motion of the fluid, adhere and so grow into visible agglomerates which settle. This process proceeds at rate which is dependant on the solids concentration and on the shear rate or energy dissipation in the fluid. If the solids from the process unit are collected and mixed with incoming water the solids concentration at flocculation can be increased to 10-20 times the concentration produced by the chemical dose in a non-recirculation situation. By so doing the retention time in the flocculation zone can be reduced very substantially. This concept is employed very widely in almost all the water treatment plants outside the American municipal sector.

It is not possible to increase the energy input per unit volume very much because the same energy that produces flocculation will also break the floc if it is too vigourous.

Precipitation softening differs from clarification in that the precipitate still has an appreciable solubility and the reaction does not go to completion under normal circumstances. With any material capable of crystallising there is a concentration range in which the material remains in solution under all circumstances, ie. below the solubility limit. There is a higher concentration above the solubility limit (the meta stable region) at which it will remain in solution indefinitely if free of suspended solids, but it will deposit on existing suspended solids, particularly of the same material. At even higher concentrations there is a 'labile' region where nuclei are formed spontaneously. Once these form they start to grow at a rate dependant on the difference in concentration between the supersaturated solution and the normal solubility limit. The rate of reaction in a cubic metre of water depends on this concentration difference and also the total surface area offered by the suspension. The latter depends on the solids concentration and the particle size of the suspension, smaller particles having a larger specific area.

In the simple laboratory jar test chemicals are added to a sample of water and a precipitate obtained. However one normally finds that the residual hardness is still well above the expected theoretical limit. If the solids in the process unit are collected and recirculated to concentrate them in the same manner as with clarification a larger surface area can be presented to the supersaturated solution and the degree of supersaturation greatly reduced. Thus in both situations a high concentration of recirculated solids is beneficial.

Unfortunately the settlement of a suspension varies inversely with the concentration, in a relationship of the following approximate type:-

$$u = u_o \, e^{(-10C)}$$

where u is the settling velocity of the suspension, u_o the average single particle settlement velocity, and C the volume concentration.

Thus if all the solids in the process unit are recirculated a point will be reached where the suspension is unable to settle fast enough to release the water and as a result the interface between the suspension and the clear water starts to rise, leading ultimately to carry over of solids. There is a maximum concentration for any particular surface rate. A unit operating at a higher rate cannot tolerate such a high concentration as a slow one (rate being measured in cubic metres/hour per square metre of separation surface, or plain metres/hour.)

Polyelectrolytes can be used to vary the settling capability of a given material by strengthening agglomerates so that they become larger and settle faster.

Softening introduces a specific problem not present in plain clarification, namely that the supersaturation of calcium carbonate causes deposition on all the surfaces exposed to the water. Thus it is essential to add the reacting chemicals into the mixed suspension at a single point and not into the incoming stream where it could block any distribution orifices or pipework. This limits the type of process unit, and in the case of multiple units each unit must have its separate lime dosing line.

14.3 Alternative Processes

Alternative approaches to precipitation softening are base (ion) exchange and membrane softening. However both of these when applied to surface water require clarification and filtration before softening. Both produce a liquid effluent which may not be acceptable to the environmental regulator. Precipitation softening has the advantage of producing a solid by product that in agricultural areas may find a market.

14.4 Process Units

There are three or four types of process among those described earlier unit suitable for combined clarification and precipitation softening:-

Floc blanket -

Upflow hopper tank and variants

Solids Recirculation-

Accelator or Accentrifloc

Reactivator   and variants

Single pass clarifiers are very inefficient and although reference may be found in older books it is unlikely that they would be offered currently. It would in theory be possible to convert a single pass clarifier by installing a sludge recirculation pump but in practice the capacity required would be beyond the hydraulic capability of the existing sludge withdrawal system.

Flat bottomed clarifiers either pulsed or unpulsed are unsuitable because the softening alkali would have to be distributed to all the individual inlet jets, which would be impracticable. A single dosing point upstream of the tank would cause serious scaling up of the distribution system with calcium carbonate. Dissolved air flotation does not have the appropriate contacting system.

Pellet Reactors are another specialised option, which are described below. They produce 1-3mm calcite beads at a very high surface rating. However they do not clarify and are used mainly on borehole water.

Floc blanket clarifiers used for softening have a higher inlet velocity than for clarification and often provision for varying this. One design has an adjustable conical plug valve at the inlet to vary the velocity. Velocities are usually about 1.5m/s. Surface rates for calcium removal have been as high as 5m/h. If magnesium is to be removed the surface rate is nearer to 2.0m/h. However with the decline in interest the technology has stood still. It is likely that these figures could be increased very considerably with the use of polyelectrolytes.

Variants on the hopper tank design include tanks with an ejector type inlet which re-entrains the sludge into the raw water and directs it into a contact zone similar to that of the Reactivator type recirculation tank.

The density of the chalk blanket is much higher than that of hydroxide floc and the headloss though the blanket is therefore significant (it may be 200-300mm) which must be taken into account in the hydraulic design. It can also affect the stability of flow splitting unless weirs are used.

Recirculation clarifiers likewise use higher recirculation rates to maintain suspension. The scraper duty is also higher and this may prove a limiting feature, although the main function of the scrapers is to remove the coarser particles while the finer ones are merely recirculated in suspension by the impeller.

The recirculation impeller pumps many times the net flow. The ratio can be measured by comparing the volume concentration of the suspension at the bottom of the so called draft tube and at the top. Thus 8% for example at the top and 32% at the bottom would indicate a 4 fold reduction or a 3:1 flow ratio. Recirculation would place a much higher demand on the conveying capacity of the scraper if in fact all the solids had to be scraped. Calculation on the standard design however shows that the scraper often can only convey a mass flux of a similar order to that of the reject sludge (ie. equal to the load produced by the added chemicals without recirculation.) Reactivators appear to rely, especially when used for clarification duty, on the fluidity of the sludge pool being maintained, without too much settling out as it is drawn back from the settlement zone to the recirculator intake. There is a danger that if too much settles out then the recirculation of solids would be impaired. On the other hand the coarser particles with a lower specific surface area will settle preferentially and thus be removed preferentially, which is what is required. The finer particles can then continue to grow as they pass through the reaction zone again.

High recirculation rates are beneficial in that the concentration of solids in the reaction zone is maintained at a high level. However as already intimated a point can be reached where the slurry will not settle fast enough. This aspect can readily be quantified by observing the settlement of samples of the liquor from the reaction zone, and obtaining the initial settlement rate over the first 10-20% of the measuring cylinder. This can be repeated with decantation to make up various concentrations to determine the settlement rate versus concentration.

With fast settlement experience has shown that the recirculation rate must be kept high to avoid blockage of the recirculation inlet. If the particles become too coarse they may even fall back in the central draft tube. Too high an impeller speed can in theory cause floc breakage but this is probably less likely with hard calcium carbonate particles.

Sludge Withdrawal

In all of the relevant process units the solids produced by the reactions and those extracted from the raw water accumulate in the sludge inventory within the unit. Solids must be withdrawn to maintain mass balance.

Clarifiers of the floc blanket type (hopper tanks in this application) produce a floc blanket concentration which is related to the `surface rate' of the interface of the suspension and the supernatant water. The added solids cause the surface of the suspension to rise. The inventory of sludge can be controlled by using submerged hoppers or weirs so that the excess can be decanted off.

The Reactivator and similar designs operate without any means of control on the separation level, which normally remains close to the bottom of the centre skirt. Instead the concentration of the solids tends to rise until the separation rate can no longer be attained. The unit must therefore be controlled by continuous sampling, the sludge discharge rate being adjusted either upwards or downwards to maintain a constant suspension concentration and mass balance. As already discussed this target will vary with the flow rate, and can be determined by the settlement measurements.

14.5 Pellet Reactors

These operate in a rather different manner from the above. They are essentially fluidised bed crystallisers and produce hard calcite granules of about 1-3mm diameter. They are also used for other applications such as the recovery of metals from effluents or industrial processes and also phosphate precipitation as calcium phosphate The objective is to obtain growth without nucleation. They are particularly successful in this respect and to maintain the particle numbers artificial nuclei must be added in the form of fine sand, crushed recycled calcite and in one development even garnet.

One of the earlier versions was known as the Spiractor. This consisted of a narrow conical vessel with two tangential inlets for the raw water and chemical injection lances. The surface rate at the top was about 25m/h. Hilson (1970) developed a non spiralling version, with a single inlet, more akin to a narrow hopper tank, which operated on boreholes near Blackpool, in England. More recently the technology has had a boost from Dutch work by Graveland et al (1983), (1986) which followed a Dutch policy of municipal softening. In these units the raw water enters through nozzles somewhat like filter nozzles in a flat steel floor plate in a column which may be up to 2.7m diameter. The softening chemical is added either via lances passing through the floor or through special nozzle assemblies that permit particularly caustic soda the enter from the space between two steel plates. With sodium hydroxide the rise rate may be 100m/h. The product can have a turbidity as low as 1.0 NTU. Sodium carbonate softening does not appear to have been attempted but one might expect it to work as well as with sodium hydroxide. With lime the reaction is slower because the lime suspension must first dissolve. (Lime water is too dilute to be of use.) The surface rate is then about 70m/h and the turbidity may be 50 NTU. However this figure is misleading because compared with hydroxide floc the particles are very much denser and occupy far less volume in filter sand, so that this concentration is still easily filterable at conventional rates. The vessels are about 6.0m high, with a fluid bed depth of about 4.0-4.5m. The fluidised bed of course has a headloss similar to that of a corresponding bed of filter sand

The process cannot be used for clarification. River solids will not stick to the calcite grains. Traces of iron and manganese may be included in the grains but above a limit of about 1.0mg/l the pelleting action starts to break down. Phosphate also interferes although at high levels the process may be used to produce pelletised calcium phosphate. Indeed there are several other materials that can be precipitated in pellet form. It might be expected that conditions which lead to the precipitation of calcium carbonate hexahydrate may interfere, although the seeding may limit the problem.

The inventory of pellets grows as the reaction proceeds.The number remains constant but the particles grow. The bed depth is controlled by regularly discharging the pellets, which flow freely like ball bearings. The bulk density of the pellets is similar to that of filter sand. No sludge treatment plant is required. Fine sand (0.3-0.4mm) is added to maintain the numbers and control the grain size. The sand size is governed merely the settling velocity which will prevent the sand from being washed out. Graveland (1992) has found that garnet is more cost effective because of its higher density, and the larger number of particles per unit weight. In spite of the higher cost per unit weight very much less is used, leading to a cost saving.

The process is started up with 100% fine sand, and can be stopped and restarted without difficulty although the lime or caustic soda lines must be purged on stopping to prevent blockage with calcium carbonate.

The process is an elegant but specific one which in the right circumstances solves both the softening problem and sludge disposal in a very compact manner.

14.6 Process Chemistry

In order to coagulate the suspended solids in the raw water as well as the finer chalk particles ferric or aluminium sulphate are often used (usually the former). Little data on the use of polyelectrolytes appears to be available although experiments on lime saturators showed that the rise rate on these could be increased very considerably. Inorganic coagulant salts increase the consumption of the softening chemical.

$$6\ Ca(HCO_3)_2 + 2\ Fe(SO_4)_3 \rightarrow 6\ CaSO_4 + 2\ Fe(OH)_3 + 6\ CO_2 + 3\ H_2O$$

The first softening reaction is:-

$$Ca(HCO_3)_2 + Ca(OH)_2 \rightarrow 2\ CaCO_3 + 2H_2O$$

100 parts   74parts   200 parts
(as $CaCO)_3$

The above reaction leaves nothing in solution (except supersaturation).
If excess lime is added so that the pH exceeds about 10.7 then magnesium may be removed and replaced by calcium.

$$MgSO_4 + Ca(OH)_2 \rightarrow CaSO_4 + Mg(OH)_2$$

This reaction is reversible but may be slower in the reverse direction once magnesium hydroxide has precipitated. Magnesium hydroxide is a more flocculant material more like ferric hydroxide than calcium carbonate.

The other available reactions are:-

$$Ca(HCO_3)_2 + CaSO_4 + 2\ NaOH \rightarrow 2\ CaCO_3 + Na_2SO_4$$

To make full use of this reaction there must be an equal amount of bicarbonate and non-bicarbonate hardness. Most of the time this is not so, so that some lime would still be necessary to achieve minimum carbonate hardness, unless residual sodium carbonate was tolerated.

$$Ca(HCO_3)_2 + 2\ NaOH \rightarrow CaCO_3 + Na_2CO_3$$

Caustic soda is more expensive than lime and hence is not normally used for this second reaction, although its use for the first reaction is particularly popular especially for pellet reactors. More commonly caustic soda has been produced in situ:-

$$Ca(OH)_2 + Na_2CO_3 \rightarrow 2\ NaOH + Ca\ CO_3$$

The remaining softening reaction is:-

$$CaSO_4 + Na_2CO_3 \rightarrow CaCO_3 + Na_2SO_4$$

This reaction is used to remove permanent or non-carbonate hardness.

An unwanted but unavoidable reaction is:-

$$CO_2 + Ca(OH)_2 \rightarrow CaCO_3 + H_2O$$

The pH values at which these reactions occur are determined by reaction equilibria which are fully described in a book by Lowenthal and Marais, (1976) Computer programmes eg STASOFT by Lowenthal are available to predict the equilibria.

The solubility of calcium carbonate varies with composition of the water and the pH. It so happens that a minimum hardness is obtained if the caustic (p) alkalinity (excess lime) is half the (m) bicarbonate alkalinity. This is generally taken as a yardstick in controlling softening plant. A high p alkalinity and pH implies the presence of free calcium hydroxide, which on its own is soluble to about 1500mg/kg. The above book and all other texts which deal with equilibria appear implicitly to assume that the calcium carbonate is in the calcite form which as discussed below may not be the case.

Newcomers to the field will find that there is a divergence of practice in expressing hardness and alkalinity. Some retain pure chemical equivalents (or milliequivalents) while more commonly calcium carbonate equivalents are used. Thus 1 meq/l is equivalent to 50mg/l as $CaCO_3$. Alkalinities are usually quoted as mg/l as $CaCO_3$. Calcium and magnesium may be quoted as Ca or Mg or as $CaCO_3$ equivalent.

The alkali dose must be controlled fairly accurately because short term variations (more than a fraction of the liquid residence time in the slurry zone, usually about 20mins., will not be blended out. Blending outside the reaction zone will increase the residual supersaturation without the means of reducing it.

Natural river waters vary in composition with the weather and rainfall, and diurnally, if there is much weed growth as carbon dioxide and bicarbonate is taken up or given out by the plant life. This will affect the chemical dose. Thus frequent monitoring and adjustment is required.

14.7 Calcium Carbonate Hexahydrate and Other Interfering Species

This topic was very much a point of discussion in the water industry in the 1960's and 70's when softening of public supplies was at it height. Outbreaks of hexahydrate occurred at many works in Britain as far afield as Maldon, Essex and Bristol. A definitive paper has been published by Slack, (1980), previously Chief Chemist of Essex Water Co. Slack claims that the hexahydrate can form at any temperature below 15°C, with increasing likelihood down to 8°C when the probability is 100%. In the presence of hexametaphosphate the hexahydrate forms even at 21°C. Soluble organic matter is also suspected as an additional contributing factor.

The formation of the hexahydrate is triggered by a number of factors. To quote from Holden (1970) "More recently attention has been focused on the difference in solubility of crystalline forms in which calcium carbonate is precipitated. The common forms are calcite and aragonite and to these must be added calcium carbonate hexahydrate, the occurrence of which in softening practice has now been established by Slack (1964). The hexahydrate is considerably more soluble than either aragonite or calcite. The latter is the stable form to which the others revert according to conditions: the formation of hexahydrate is favoured by temperature below 10°C, but temperature is not the only factor as the crystals may include small amounts of other materials, the nature and amount of which may influence the rate of change to calcite. It is known from the alkali industry that the formation of hexahydrates is encouraged by the presence of polyphosphates (Brooks,Clark and Thurston, 1950) and Slack (1964) found that a trace of phosphate (0.08% as P_2O_5) was present in the hydrate produced in softening." Also from Holden:- "Organic matter tends to retard the softening process (Janzig 1952)". "Condensed phosphates from detergents have been shown to inhibit precipitation of calcium in lime soda softening, in concentrations as low as 1mg/l (but not magnesium), irrespective of temperature (Malina and Tiyaporn, 1964)."

From thermal analysis curves published by Slack it appears that the hexahydrate reverts to calcite at a temperature above 32-36° C.

Jar tests at room temperature may be misleading. Slack was able to obtain hexahydrate in jar tests run at the correct temperatures. The presence of phosphate levels above about 1 mg PO_4/l interferes with pellet reactor performance. A paper published in the 1920's mentions that is it precipitated from milk of lime when sugar is added, which confirms that organic materials may trigger its occurrence. This of course raises the possibility of polyelectrolytes interfering with crystallisation. While this is only speculation the monoclinic crystal structure of the hexahydrate would offer a larger surface area for deposition than the cubic calcite and may therefore grow faster under adverse conditions.

Apparently in spite of the literature Graver a specialist contractor in the USA (the designers of the Reactivator who were consulted by a user in Britain in 1995 who encountered such problems) disclaimed all knowledge of such a problem. Indeed there appears to be no reference to it in the current American Waterworks textbooks. As Holden has observed, the production of hexahydrate may explain some of the inferior results often obtained when softening takes place in cold weather. Water abstracted down stream of sewage outfalls would be particularly likely to demonstrate the problem.

Because of the water of hydration in the lattice the hexahydrate will produce 108% more sludge mass than calcite, and the bulk density of the monoclinic mass would be expected to be less than that of the granular calcite. The overall effect on the volume of sludge to be conveyed could therefore be even greater.

Crystals tend to grow under conditions of supersaturation (the metastable region) by following the lattice patterns on existing solid material. Thus where two forms are stable the existing one will tend to determine the form of later production.

The hexahydrate, according to Slack, has a solubility of 95mg/l as $CaCO_3$ as against 17mg/l for calcite. Thus when the calcium carbonate precipitates as the hexahydrate there is a very considerable fall in the efficiency of the process and the expected residual alkalinity cannot be achieved.

To prevent the formation of the hexahydrate three approaches were considered by Slack. Firstly precipitation of phosphate preferably to reduce the PO_4^{3-}

level to 0.2mg/l which required 7mg/l Fe (25mg/l Ferric sulphate) which must be dosed before the process unit. Secondly, Slack was using both sodium carbonate and lime. He found that if these were mixed together before entering the main flow (thereby precipitating calcite first) then this calcite provided nuclei which influenced the form of subsequent precipitation. Lastly and in the case at Essex Water Co. calcite suspensions (commercial whiting) were deliberately dosed to seed the clarifiers (Accentriflocs and Hopper Tanks.) This together with the ferric dosing was successful.

14.8 Post Treatment

As already indicated precipitation softening units are not 100% efficient and often produce a water with residual supersaturation (positive Langelier Index). If this water is filtered the media will become encrusted and may cement together.

If partial softening is required then it is normal to soften only part of the flow and to blend the product with a by pass stream. Calculations based on carbonate equilibria will show whether the blend is stable, as it often is. If full softening is required the product may have to be stabilised either by the addition of carbon dioxide or acid to lower the pH and alkalinity to the equilibrium point. It is possible to add hexametaphosphate or an organic phosphonate sequestrant to prevent precipitation and cementation but these chemicals interfere with particle capture in filtration and high levels of residual coagulant will probably be encountered. Where the water is to be used without filtration in a cooling tower for example this will not matter.

14.9 Lime Saturators

In earlier days and in locations where pure grit free lime is not available lime saturators were often used to produce a clear lime solution for pH correction. The equipment used was similar to the hopper clarifier. The flat bottomed clarifier is suitable for this application and the pulsed variety has in fact been used. The surface rate is low, usually about 1-1.2m/h. The solubility of lime is about 1500mg/l, varying with temperature. Thus although pellet reactors operating on lime would benefit considerably from using lime water instead of lime cream the saturators required would be vastly larger than the reactors. Experiments with polyelectrolyte have shown that the surface rate may be increased considerably but the solubility will be unchanged.

Lime saturators may either be charged batchwise or topped up continuously. In the former case the initial concentration of the suspension will be high and because the settlement rate varies with concentration in a similar manner to floc blankets the surface rate must be reduced. Continuously fed saturators may run with a dilute suspension at a higher rate. Any carbon dioxide or bicarbonate in the water will cause calcium carbonate to precipitate and saturators must be cleaned out regularly.

Improvements in the quality of lime have made saturators obsolete except in exceptional circumstances. Overseas, where locally made lime is used they still persist but are not popular.

14.10 References

van Ammers, M., van Dijk, J.C., Graveland, A. and Nuhn, P. A. N. M. (1986) "State of the Art of Pellet Softening in the Netherlands", Water Supply, **4**, 223

Brooks, R., Clark, L. M. and Thurston, E. F. (1950) Phil. Trans. (A) **243**, 145.

Graveland, A., van Dijk, J.C., de Moel, P.J. and Oomen, J. H. C. M., (1983) Developments in Water Softening by Means of Pellet Reactors", J. American Waterworks Assn. **75**, 619

Graveland, A. (1992) Personal Communication

Hilson, M. A. and Law, F. (1970),"Softening of Bunter Sandstone Waters and River Waters of Varying Quality in Pellet Reactors", Water Treatment and Examination, **19**, 32

Holden, W. S. (1970) "Water Treatment and Examination", p386, Churchill, London

Janzig, A.C. (1952) Title not available. Water Works Eng. **105**, 1045

Lowenthal, R. E. and Marais, G. v.R. (1976) "Carbonate Chemistry of Aqueous Systems.", Ann Arbor Science,Mich.

Malina, J F. and Tiyaporn, S. (1964) Title not available. J. American Waterworks Assn. **56**, 727

Slack, J. E. (1964), Title not available. Proc. Soc. Water Treatment and Examination, **13**, 156

Slack, J. E. (1980) "Calcium Carbonate Hexahydrate: Its Properties and Formation in Lime Soda Softening", Water Research. **14**, 799-804,

CHAPTER 15.

SLUDGE TREATMENT AND DISPOSAL

15.1 Sludge Yields

Most British works built since 1983 and using coagulation with ferric or aluminium salts have followed the general guidance given by Warden (1983) who undertook trials on small scale thickeners and established some design criteria. In earlier days it was not uncommon for sludges to be discharged back to rivers. A general flowsheet for water recovery is given in Fig. 15.1.

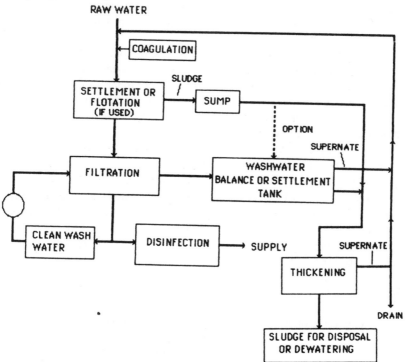

Fig. 15.1 General flowsheet for water recovery.

Mass balance calculations are an essential first step in appraising the problem. In arid areas subject to storms and where rivers carry glacial melt suspended solids in the raw water may rise to several thousand mg/l. At 1000mg/l a 100Mld plant would remove 100 tonnes of solids per day. In such situations it is not unreasonable to return the solids to the river. The coagulant added would be trivial by comparison with the solids load in the river.

Unfortunately for the plant designer the disposal of treatment sludges is usually constrained by regulations framed in a wider context and there may be limits to what might be regarded as sensible for the particular situation. In Britain sludges are treated as industrial waste and usually have to go licensed tips. On rare occasions water works sludges find uses, such as an additive for brickmaking and ferric sludges are also being use in horticulture and silviculture. Charges include transport, disposal and landfill tax. These are based on the weight dumped regardless of water content. Thus there is an incentive to dewater as far as reasonable.

The largest mass of solids in the sludge arising from a conventional plant is discharged from the clarification stage. If presedimentation is used then this is likely to be even larger than the main clarification stage. There are limits to the concentration that will flow easily (about 10% w/w for river solids) and in some very silty areas a very considerable proportion of the abstracted water may in fact return direct to the river to carry the solids away. If wash water is returned to the front end of the plant then even the small amount of solids going forward to the filter will eventually be discharged from the clarifiers. On most plants, treating moderate water qualities, the wash water will comprise the largest volume but the concentration of solids will be only 200-400mg/l.

The design must be based on the maximum sludge yield likely to arise, which for river abstractions is difficult to define. It is necessary to examine historical data. Extremes may be unrepresentative and a 90 percentile provides a reasonable basis for the suspended solids for design purposes. The plant will have some internal capacity within the thickeners to cover peaks lasting a few hours.

The solids load is calculated from the following conversion factors, which can be adapted to other coagulants.

Coagulant –Highest applied dose $\quad$ X 0.234 as $Al_2(SO_4)_3 \cdot 18H_2O$
$\qquad\qquad\qquad\qquad\qquad\quad$ X 0.535 as $Fe_2(SO_4)_3$
$\qquad\qquad\qquad\qquad\qquad\quad$ X 1 as Hydroxide.

Suspended Solids	X 1 as Solids
	X 2.5 as NTU
Colour	X 0.2 as °Hazen (CoPt Standard)
Polyelectrolyte	X 1 as 100% solids

The sum gives the total solids arising in mg/l, (g/m3). This, multiplied by the daily throughput, gives the daily sludge yield.

Volumes

Conventional high level sludge concentrators in floc blanket and recirculation clarifiers give concentrations which vary with the water quality. 0.2% w/w is a general average on a low turbidity water. With low suspended solids and algae it may be even less. On the other hand turbid river waters can easily give 1%. Tanks which rely on a scraped floor will usually give twice to four times these figures.

Dissolved air flotation often gives 5% with scrapers but the frothy scum does not flow easily and must be diluted back possibly to half of this figure to convey it to the next stage of treatment. One plant treating a low turbidity coloured water where the sludge was floated off over a weir was guaranteed to give 0.5% w/w and in a 4 week trial averaged 0.8%. This flowed very easily to the thickeners.

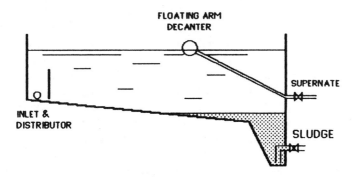

Fig.15.2 Wash water settlement tank.

The sludge mass in wash water is small compared with the primary stage sludge and hitherto it has made sense to return it either after settlement (Fig. 15.2) or direct to the inlet of the plant via a balancing tank to provide a steady return flow. The identification of cryptosporidium oocysts as a problem has caused this practice to be reviewed. It appears that these organisms tend to separate from the sludge and return, thus forming closed loop so that the efficiency of the filters is swamped. Various procedures have been considered depending on local circumstances. In some cases the settled wash water has been filtered (with further treatment if necessary) before return so that the closed loop is broken. The filter only has to handle the wash water flow which may be only 1.5% of the plant throughput. In some designs the filter is steam heated to inactivate the oocysts before back washing. Alternatively the backwash water from this filter can be pasteurised or treated with high doses of ozone.

Sludge lagoons were once common. In drier climates the sludge will dry out and can be removed to a landfill site. However with pressure on land utilisation and also from safety consideration thickening and dewatering have become normal.

15.2 Gravity Thickening

As already indicated Warden's work has provided the definitive approach. The sludge, after blending in a sump to provide a consistent concentration, is dosed with a polyelectrolyte (usually a long chain polyacrylamide), mixed at an appropriate shear and forwarded to a thickener which is merely a scraped settlement tank. The mixing shear is critical for good thickening. Warden gave a guideline of 5m/s through an orifice for alum sludge and 11m/s for an iron sludge. The polyelectrolyte dose was found to be in the region of 2.5kg/Te dry solids. The sludge transfer and dosing system is operated at a constant rate on a start/stop basis to avoid the need for proportional dosing.

The theory of thickening is discussed in Chapter 3. The thickener is designed on the basis of the larger of either the underflow or overflow limiting areas. The overflow is normally not in excess of 1.5-2m/h depending on the type of suspended solids in the raw water and the underflow 2-4kg dry solids/m². Iron sludges may be treated at the higher loading. The scraper is designed to convey at four times the design volumetric underflow rate.

Warden's intention was to operate the thickener at a constant solids level and to transfer the sludge as it rose to a buffer hopper, using a sensor to detect the level. The buffer hopper then supplies the dewatering press and must have sufficient capacity to deal

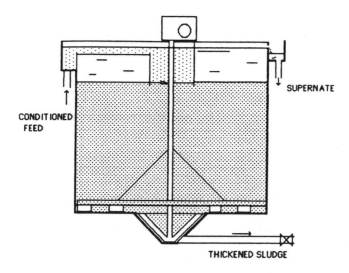

CONDITIONED
FEED

SUPERNATE

THICKENED SLUDGE

Fig.15.3 Gravity thickener for waterworks sludge.

with the variable flow into such presses. Centrifuges receive a steady flow.

In contrast to previous practice and to sewage sludge practice Warden found that pickets (vertical rods) are not needed with water works sludge (Fig.15.3). His thickeners are shallow, (1.8m depth). In some cases the buffer capacity is built within the thickener by making it deeper but ensuring that it always has the required minimum depth. This is usually more cost effective. Standby thickeners are also used as sludge storage tanks with suitable interconnecting pipework on larger plants where larger hoppers would otherwise to expensive to build. Utilities often require sufficient sludge storage to enable the dewatering section to work only 5 days a week for normal operation.

Supernatant water from thickeners is normally returned with the filter wash water supernatant. It may sometimes cause problems because conditions in the thickener can become anaerobic, causing taste problems and even producing soluble manganese. Residual polyelectrolyte has been known to cause problems on filters. The amount used in sludge treatment often exceeds that used in clarification. Chlorination of the supernatant usually solves such problems and on plants where preozonisation is practised this should also provide a remedy.

15.3 Dissolved Air Flotation Thickening

An alternative to gravity thickening that appears to have been explored in continental Europe is dissolved air flotation. This has also been used for sewage sludge thickening , where it has been the subject of papers by Bratby and Marais eg. 1977. The simplicity and effectiveness of gravity thickening using polyelectrolytes has however prevented dissolved air from make any process on water treatment sludge in Britain.

15.4 Sludge Freezing

When sludge freezes the colloidal structure is broken down. On thawing it reverts to a fast settling granular material which can run on to lagoons, allowed to dry and transferred for landfill at about 40% dry solids without further treatment. In the colder parts of the USA natural freezing has been used extensively to modify hydroxide sludge in this way.

In the 1960's before polyacrylamides were available freezing of sludge was introduced at four plants in Britain (A.E.R.E Harwell, Daer, Fishmoor and Stocks). The power consumption for refrigeration was significant (about 60 kwh/m^3) although the thawing section absorbed heat from the freezing section. Panel heat exchange surfaces used in the process suffer damage from the stresses of the expanding ice and maintenance was high. At the Daer plant the problem was overcome by using cooling coils in tanks lined with glycerine filled bags to cushion the expansion. For success the rate of freezing had to be fairly low which precludes the use of commercial flake ice machines. The pelleted ice machines now available would seem to have some potential in this direction. Such mechanical freezing is still used in the USA.

Kawamura and Trussel (1991) refer to freezing of sludges at water treatment plants in Japan, which is apparently still being practised. The thawed sludge is even pressed to achieve 40% dry solids for land disposal. The power consumption is quoted as 1,100kwh per tonne dry solids (55kwh/m^3 for 5% thickened feed). The feed is sludge gravity thickened without polyelectrolyte and would be expected to be only about 2% dry solids which would suggest only 22 kwh/m^3 .

A project at A.E.R.E. Harwell in the early 1970's using direct contact with butane where drops of sludge froze at the top and thawed at the bottom of a column, exploiting the different hydrostatic pressure, proved unsuccessful because of problems with emulsification of the butane with the product. The work was an extension of a desalination programme using a similar principle.

Freezing is applied to the thickened sludge to minimise the volume to be treated. It would not be practical to attempt to freeze centrifuged sludge.

15.5 Recovery of Coagulant

Many attempts have been made to recycle aluminium hydroxide back to sulphate for reuse. A crude approach is merely to acidify the thickened sludge with sulphuric acid and to allow the product to settle or to filter it. Some success has been achieved but the recycling both of organic matter and metal impurities can be a problem. The acidified mixture settles and filters much more easily than the original sludge but the residue must be neutralised for disposal.

A considerable amount of effort has been devoted to solvent extraction (so called liquid ion exchange) to recover the aluminium, using carriers such as tributyl phosphate in kerosene solvent. This has been successful technically but the extractant is not of a type to generate enthusiasm with operators and activated carbon polishing is essential in order to avoid contamination.

A process of the author's which proved promising in pilot trials and which solved the main problem with direct acidification involved concentration (or thickening) of the sludge followed by acidification of part, say ⅔rds, and rejection of the remainder. The acidified portion was mixed with fresh coagulant and dosed back to the incoming flow. The reject fraction is of course already neutralised by the water being treated. Providing that excess sludge is in contact with the acid the latter is fully utilised and the residual undissolved material can often be beneficial and assist flocculation. The coagulant dose might therefore be considered to be based on the equivalent of acid and the fresh coagulant. However it was found that with aluminium hydroxide the acid would form a soluble monobasic salt. Indeed fresh sulphate would also dissolve a one third equivalent so that there are benefits in adding the fresh coagulant to the acidified mixture. Jar testing to determine doses must therefore be based on the recycled mixture. (There were indications that bicarbonate might be involved in the reaction.) This process eliminates the need for lime for neutralisation, but the residual sludge is similar to normal material but reduced in volume.

In another set of trials involving counter current washing of the acidified sludge it emerged that a significant fraction of the suspended solids present in the source was calcium carbonate. This of course dissolved preferentially and greatly diminished any chance of economic success. Such a problem may be widespread, even in reservoir waters which produce saturated waters from algal blooms.

Indications from reviews suggest that most attempts at recovery in Japan and the USA have been abandoned because of doubtful economic benefit and concerns about recycling of impurities.

15.6 Sludge Drying

In arid areas with ample land around the works the sludge can be dried to a powder. Normally artificial drying would be uneconomical because in contrast to china clay production for example the end product has little intrinsic value, even less value than sewage sludge. There may be situations however where artificial costs of landfill disposal may lead to a reconsideration of the situation. Should this be the case the dry material may find uses in new treatment processes. Drying would only be economic on centrifuged or pressed (and possibly frozen and thawed) sludge, to reduce the water content from 80-85% to nearly zero.

15.7 Dewatering

Plate type filter presses and centrifuges are used extensively for dewatering the thickened sludge. The presses are batch machines which consist of stacks of vertical recessed plates covered with suitable filter cloths. They depend on the use of polyelectrolyte to achieve viable filtration rates and to produce coherent cakes that will peel off the cloth when the cakes are discharged. Plate sizes on full size units vary from 1m square up to about 1.2 X 1.8m. Cake thicknesses are between 25 and 40mm. Some designs have rubber membranes to squeeze the cake before it is discharged. Many presses are provided with an automated discharge system. The cake falls into a trough from where it is conveyed to a skip for disposal.

The polyelectrolyte dose applied for thickening is usually satisfactory for pressing. Operating pressures are usually about 7 bar and the thickened sludge is delivered either via pressure vessel with compressed air above the sludge, using a progressive cavity transfer pump, or with a hydraulically operated rare pump that is able to maintain constant pressure against the declining flow. The filter cloth must be chosen with care and different sludges require different cloths. These have to be hosed down at intervals.

The time cycle for filtration varies with the cake thickness but is usually between 8 and 24 hours. Discharge of the cake is not reliable and an operator is usually present. The cake has a solids content of 20-25% or higher on silty water. Small pilot presses are used for the evaluation of cloths and process conditions.

The main competitor for filter presses is the scroll centrifuge which is able to wind the sludge up a conical beach within the machine. and out directly into a skip. Extra polyelectrolyte may be required. The centrifuge has the advantage of being truly continuous and of not needing the attention necessary with filter presses when they are discharged. The machines are much smaller and lighter than the presses for the same duty. On the other hand the power consumption is much higher and the dryness of the dewatered sludge less (15-20%). Whereas a filter cake may be like chocolate and can be walked on, the centrifuged sludge is a thick greasy consistency which will not stand a man's weight. Centrifuge manufacturers are able to hire out small machines for evaluating process conditions.

Belt type filters are also used, but they give a much wetter sludge, eg. 10-11% solids, compared with 15% on a centrifuge with the same sludge. Heavy doses of polyelectrolyte are needed. Better results are obtained with softening sludges.

A full description of both filter presses and centrifuges together with theory and methods of evaluation is given by Purchas and Wakeman (1986).

Where a less sophisticated dewatering procedure is required it is possible to use drying beds with polyelectrolyte treated and thickened sludge. A bed of fine sand is laid on graded gravel support layers with a land drain system below. The sludge is run on to the bed and allowed to filter and dry out. It is then skimmed off with some of the sand. WRA (WRc) extended the principle to porous asphalt as used for tennis courts, in which case front end loaders can be used. The principle was taken further in the so called Rapid Sludge Dewatering Process in which porous ceramic tiles were laid over an underdrain which was connected to a vacuum system to form a large version of the laboratory Buchner funnel. The sludge could be removed by loader after several sequential fillings with conditioned sludge. However after emptying the tiles had to be cleaned with high pressure jetting and the labour involved has limited the development of the technique.

15.8 References

Bratby, J. and Marais, G.V.R. (1977) "Flotation" in Solid/Liquid Separation Scale up. First Edition only, Uplands Press, London.

Kawamura, S. and Trussel, R.R., (1991) "Main Features of Large Water Treatment Plants in Japan",J. American Waterworks Assn. **81**, 56.

Purchas, D.B. and Wakeman, R.J. Eds.(1986) Solid/Liquid Separation Equipment Scale Up. Uplands Press, London and Filtration Specialists.

Warden, J.H. (1983) "Sludge Treatment Plant for Waterworks", Water Research Centre, Report TR 189.

Further Reading:-

Cornwell, D.A. and Koppers, H.M.M. (1990), Slib, Schlamm, Sludge. American Waterworks Association Research Foundation, Denver CO, USA. ISBN 0-89867-532-4.

Part C

Filtration

CHAPTER 16.

THE STRUCTURE AND HYDRAULICS OF GRANULAR BEDS

16.1 Introduction

Before discussing the processes that occur within a granular bed during filtration it is necessary to understand the behaviour of fluids flowing through in clean media. As in most technologies there tends to be a set of specific technical terms well used and understood by those in the field but confusing to outsiders who may use such terms to describe totally different parameters. For this reason a glossary has been included in this book. In Chapter 30 in which the specification of filter materials (media) is discussed some of the concepts are also explained.

A granular bed as used for filtration comprises indivisible grains which rest on each other. They do not fit accurately as bricks in a wall but leave voids in between through which the water flows and in which the dirt removed during filtration collects. This leads to the term 'in-depth' filtration which is used by some advertisers as if it were something new. Solids may also collect on the surface of the bed without penetration, producing 'surface filtration'. In this case the water has to pass through the 'cake' or surface mat. This has a high resistance to flow and steps are normally taken to avoid this happening.

Granular media filtration of water is primarily an interception process and close inspection shows that it relies on leading edges to capture the suspended particles in a similar way (but by a different mechanism) to that by which snow in a breeze is captured on trees and telegraph poles. A granular bed provides a mass of leading edges and space for the deposit to accumulate. The dirt particles retained on the grains are no longer in the line of flow but are rather like the sediment on the bed of a river and have less effect on the hydraulic gradient than in the case of surface filtration.

16.2 Sources and the Nature of Media

This chapter is concerned with the characteristics of the bed in its settled or working condition. Sands are most readily obtained from natural deposits laid down by seas, rivers, glaciers etc, and thus they tend to be naturally water worn and partially rounded. The natural processes also tend to classify such materials into size bands and the sands of interest for filtration are not particularly plentiful. Any walk along a seashore will reveal coarse shingles or fine silver sand that has been ground down too far by wave action with little in the 0.5 to 6mm range.

Filter media is also obtained by crushing massive material such as anthracite, basalt, slag etc, but it is difficult to control the process to produce the relatively narrow size band required for filtration, and the yields are not high. A supplier will usually serve several industries so that reject material from one can be sold elsewhere. If material is produced locally, specifically for a particular works, there is an incentive to use a wider size cut to improve the yield. A well established supplier with many outlets can however produce close cuts without a major price penalty. Material such as chopped extruded plastic can of course be produced virtually without wastage. The cost of such materials is normally much higher unless suitable scrap can be found.

Granular materials such as silica sand, basalt, flint, anthracite, pumice, limestone etc. are all relatively cheap compared with fibrous materials, and this has made granular media filtration a very competitive process for large scale work. As a result it has hitherto withstood the onslaught of 'higher technology' processes. For this purpose, the chemical nature of the grain is relatively unimportant, providing that it does not dissolve. Indeed in practice even its physical properties are not of great concern as it does not have to withstand great forces. Relatively weak materials such as granular activated carbon are used successfully.

Traditionally, filter sands were rounded and many textbooks (and indeed some suppliers) still perpetuate the myth that rounded grains are beneficial. Conversely other suppliers complain that some customers still insist on rounded media. In fact the reverse is the case. A possible explanation for this discrepancy lies with the fact that the backwash rate, as discussed in Chapter 20, is very dependent on the voidage of the bed, and a filter designed for rounded media may not provide a sufficient backwash rate if angular media is substituted.

In spite of the claims of some plant suppliers, filter media can never be mono sized unless they are produced, for example, by an extrusion and chopping process. Sands tend to be produced in a normal size band of 1.5:1 to 2:1 or even 3:1 (between the 5 percentile and 95 percentile sizes). Some friable materials tend to break down and produce an even wider size range. In some cases, a wider range may be chosen deliberately to reduce the cost of the media.

16.3 Bed Voidage

This is defined as the space between the grains expressed as a fraction of the total bed volume and is some times referred to as bed porosity. (The term porosity is retained in this book for the internal porosity of the grains.) It varies slightly depending on the previous history of the bed, for example whether it has just been poured in, or backwashed several times, or indeed whether it has been air scoured (ie.

cleaned by blowing air through it.) Voidage is important because it affects the pressure loss produced by fluids flowing through the media and it also provides space for the suspended solids removed from the fluid.

A bed of uniform ball bearings will settle to give a voidage of 41%. Indeed the particles used need not be perfectly round or of steel. McGeary (1961) reports that lead shot, steel shot, glass beads, sand, beans and poppy seeds all give voids in the region of 36%-41%. This figure is an average because if individual pores are considered, it will be found that the arrangement of individual particles varies from an open cubic pattern (Fig. 16.1) with a high voidage of 47.6% to a hexagonal rhombohedral pattern (Fig.16.2) with a voidage of only 25.95%. The former is unstable and tends to collapse if there is any vibration. The latter is completely stable and cannot be compressed further. This condition will occur rarely within the bed because in practice, with a range of particle sizes, the individual grains do not fit into each other perfectly. Overall there will be a random pore configuration between the random grain sizes.

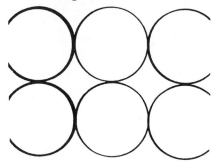 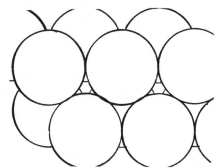

Fig. 16.1 Cubic packing of spheres   Fig. 16.2 Rhombohedral packing of spheres

Real filter media comprises irregular grains, often with rounded edges but the surfaces are not polished (Fig. 16.3). They do not readily slip into a close packed form but tend to prop the pores open. The voidage under any set of conditions increases with the angularity (ie, the converse of the sphericity) thus anthracite, which is a crushed material, has a higher voidage than rounded sand eg, 45% as against 38% after gentle vibration or tapping. Very sharp materials such as crushed flint give an even higher voidage.Mixtures of particles of different size always have a smaller voidage than mono-size materials of the same shape because the smaller grains can pack into the spaces bewteen the larger grains. McGeary (1961) has plotted a curve (Fig.16.4) for the effect of size ratio of a binary mixture of steel shot on the voidage. The reference voidage of 37.5% falls by 2.5% for each unit size ratio, thus a 2:1 ratio

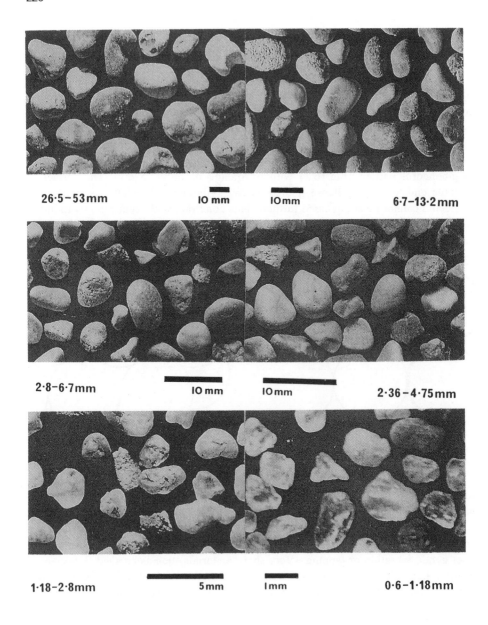

Fig. 16.3 Typical samples of media enlarged to a similar size for shape comparison.

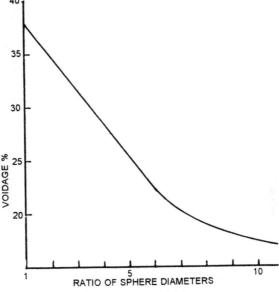

Fig. 16.4 Voidage of binary mixtures of spheres of different diameter ratio (Mc Geary)

only reduces the voidage from 37.5% to 35%. A practical sand will have continuous range of sizes so the overall effect will probably be about half of this. Following on from the above, one might postulate that wider size ranges would be more acceptable for filtration with angular materials than with rounded material, because in the former the higher initial voidage will offset the reduction in pressure loss resulting from the wider range. The width of the size cut however can have an adverse effect on the bed behaviour, for example, when washing with water alone the sand bed will stratify so that the surface layer where most of the solids are retained during filtration will be finer than intended. Also the size range affects the filtration behaviour itself as discussed later in this book.

The pores (Fig. 16.5) in a bed of filter media can therefore range from propped open cubic ones between the largest size of particles present to rhombohedral close packed ones between particles of the smallest size. No attempts appear to have been made to assess the pore size distribution in real beds. The distribution of pores will however vary according to the history of the bed. In studies such as McGeary's work, the beds were vibrated to achieve a final stable packing density. A filter bed going into service after a backwash is in a fairly open condition as the particles come to rest after expansion of the bed. This condition is fairly stable and the relatively small applied pressure eg. 2 m. head, uniformly applied, seldom causes any further consolidation. However, air scour ie. the bubbling of air through the saturated bed causes compaction and reduces the volume of the bed typically by 4-5% of the height.

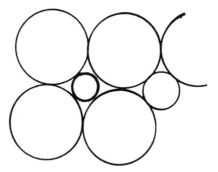

Fig. 16.5 Propping of packed spheres by smaller ones.

Thus if air is applied prior to backwashing, expansion during fluidisation will start from a lower datum point than otherwise and the bed depth after washing will differ from that which prevailed immediately before the wash flow started.

For the laboratory evaluation of filter media a simple test to compare new samples with reference samples is required. It is usual to compare the volume as poured and after tapping the measuring cylinder on the bench until no further change in the volume occurs. Commercial rounded sands usually have a voidage of 41%-42% as poured and 38% after tapping. Anthracite may be 45% as poured and 42% after tapping. A sand in service after fluidisation typically has about 43% voids. Some very rounded materials can have a voidage as low as 33%. These will be easy to wash, but will have a low solids holding capacity and a higher clean bed headloss.

16.4 Hydraulics

(Note that the symbols used in this book are listed in a table at the front.)

At filtering velocities the flow is mainly streamline (laminar) and the head or pressure gradient is proportional to the flow or velocity. This behaviour is known as Darcy's Law. Darcy's observations were however only the beginning and the theory of flow through porous media has been developed progressively by Kozeny, Carman and Rose. A good digest of the types of flow and the derivation of the equations will be found in Coulson and Richardson's textbook on Chemical Engineering (1990), which is widely available. It gives a derivation for the Kozeny equation. To achieve a practical filter design the headloss (pressure loss expressed as the height of the corresponding column of water) must not be excessive, particularly if the filter is to housed in an open tank (a gravity filter). This has a major impact on the choice of the filter media.

A granular bed is characterised by the surface area and voidage. An analogy is usually drawn between the behaviour of the bed and a bunch of parallel circular capillaries of hydraulic diameter d_m. The term hydraulic diameter is in common use in the calculation of the behaviour of non-circular pipes and channels and is 4 times the cross section of the flow divided by the wetted perimeter.

The surface area of the wall of each capillary wall is:-

$$a_c = \pi d_m l$$

where l is the length of the capillary. The volume of the capillary is:-

$$v_c = \frac{\pi d_m^2 l}{4}$$

If the voidage of the assembly is e the number of capillaries in a unit cross section of the bed will be:-

$$N_c = \frac{e}{V_c} = 4 \frac{e}{\pi d_m^2 l}$$

Hence the specific surface area (per unit volume) will be:-

$$S_c = N_c \cdot a_c = 4 \frac{e}{d_m}$$

$$d_m = \frac{4e}{S_c}$$

The velocity in the capillary pores $\mathbf{u_c} = \mathbf{u/e}$ where u is the external approach velocity.

The specific surface area of the bed (S_c) per unit of bed volume is related to the specific surface area of the grains (S), per unit volume of grain material, eg. **S= 6/D** for spheres, by the expression:-

$$\mathbf{S_c = S(1-e)}$$

Thus:-

$$d_m = \frac{4e}{S(1-e)}$$

Note that S is in m^2/m^3.
The Poiseuille equation for capillary flow states that:-

$$\Delta p = \frac{32 u_c \mu l}{d_m^2}$$

where Δp is the pressure loss across the length l of the capillary, μ is the absolute viscosity and u_c is the mean velocity in the capillary. Hence substituting for d_m:-

$$\frac{\Delta p}{l} = \frac{2 u_c \mu S_c^2}{e^2}$$

However, **u_c= u/e, and S_c= S(1-e)**,
therefore:-

$$\frac{\Delta p}{l} = 2\mu u \frac{(1-e)^2}{e^3} . S^2$$

This is the usual form of the Kozeny Equation except that a constant K is substituted for the factor 2 which only relates to straight capillaries. A porous bed comprises a random mixture of large and small pores, the larger ones being the most conductive. Practical mixtures with a voidage of around 40% have been shown empirically to have a tortuosity or effective pore length of about 2.5 times the bed depth, so that the constant generally used when the Kozeny equation is applied to granular beds is 5.

For spheres **S= 6/D** (where D= spherical grain diameter) hence:-

$$\Delta \frac{p}{l} = [180\frac{\mu}{D^2} . \frac{(1-e)^2}{e^3}] . u$$

The term between the square brackets constitutes the hydraulic resistance, the pressure loss being this multiplied by the velocity.

One may well question the relevance of the diameter of spherical grains when practical grains are usually irregular in shape. However since this diameter, which is used extensively in several chapters, is a more convenient surrogate for specific surface area it may be defined as the diameter of the sphere which would have the same 'exposed' surface area as the irregular grains in question. This is much the same concept as the hydraulic size discussed in Chapter 30. The term 'exposed' implies the exclusion of cracks, crevices and internal porosity.

In water treatment the term headloss (ΔH) is commonly used. This is related to the pressure loss by the equation:-

$$\Delta p = \Delta H \rho g$$

where ρ = the density of water and g = gravitational constant.

Interestingly, it appears that all the discussions so far have only considered a capillary model. One can of course consider a parallel plate or lamellar model in which case the equation for headloss is:-

$$\frac{\Delta p}{l} = \frac{24 u_c \mu}{w^2}$$

where w = the width of the lamella channels. The number of lamellas in a cube of unit volume will be e/w. The surface area of the bed S_c will be 2e/w, ie. $w = 2e/S_c$.

The conversion between areas of the particles and that of the bed, and between the lamellar surface area and the bed surface area remains as before. Hence in this case:-

$$\Delta \frac{p}{l} = 6 \mu u \frac{(1-e)^2}{e^3} \cdot S^2$$

This constant is much closer to the generally recognised experimental value of 5. Indeed the network of pores in a granular bed bears a closer relationship to the network of lamellar films in a foam than to a bundle of parallel capillaries.

The above discussion assumes that all the capillaries or lamellas are of uniform bore or thickness. Obviously in practice this is not so. If one considers a mixture of two equal quantities of materials of equal voidage but two different sizes, it is necessary to consider the conductance in each part in a manner akin to electrical resistances in parallel. Hence for such a mixture with particles differing in size by 2 to 1, overall resistance for two half beds in parallel will be related to $1/2.5D^2$, whereas based upon the average diameter ie, $1.5D$ the term would be related to $2.25D^2$. Thus the conductance of mixtures will tend to be somewhat greater than that predicted by the Kozeny Equation. However, the error in the above case in only about 10%, which is probably much less than the errors created by the uncertainty of voidage data, which as discussed below have a dramatic effect on the resistance of the bed.

At high filtration rates with coarser media and possibly within the so-called worm holes in a clogged filter bed (which are discussed later), the flow tends to become transitional ie. partly turbulent, and the Carman equation as discussed by Coulson and Richardson must be used.

$$\frac{R'}{\rho u^2} = 5Re^{-1} + 0.4Re^{-0.1}$$

where:-

$$\frac{R'}{\rho u^2} = \frac{e^3}{S(1-e)} \cdot \Delta\frac{p}{l} \cdot \frac{1}{\rho u^2}$$

or

$$\frac{\Delta p}{l} = (\frac{R'}{\rho u^2}) \cdot \frac{S(1-e)}{e^3} \cdot \rho u^2$$

and

$$Re = \frac{u\rho}{S(1-e)\mu} = \frac{Du\rho}{6(1-e)\mu}$$

Re is the modified or particle Reynolds Number, R' is the average shear stress at the grain surfaces. The other terms are as before.

As an example with particles of 1.5 mm hydraulic size and velocities of 20 m/h, the $0.4Re^{-0.1}$ term is 18% of the $5\ Re^{-1}$ term.

The above equations rely upon a measurement of the specific area of the media. This can be derived from a sieve analysis by calculating the hydraulic size of the media ie. the area average size. The procedure is described in Chapter 30. It is debatable whether one should use the area average size or the area squared average size. However, Sakthivadivel (1972) has confirmed that the Carman/Kozeny equation gives reliable results usually with K=5 over a wide range of particle sizes and shapes. As will be discussed later, the above equations cannot be used to predict headlosses in a clogged filter bed because the terms e and 1-e describe different factors.

The 1-e term does not change whereas e does.

Strictly speaking the hydraulic size should be multiplied by a sphericity term to give the true surface area for non spherical grains. This parameter is rather difficult to measure. For rounded grains it approaches 1. Angular particles produce a higher voidage and as already indicated this has a much greater effect on the outcome.

It should be noted the $(1-e)^2/e^3$ term makes the headloss very sensitive to voidage.(Fig.16.6) A 1% change in voidage, ie. 2½% relatively, will produce an 11% change in pressure gradient, whereas a 2½% relative change in grain size will only produce a 5% change in headloss. Unfortunately, it is easier to change size than voidage, a producer can do very little about the voidage of the material that he is selling, but he can adjust the size. Thus it is sometimes necessary to adjust the size to compensate for uncontrolled changes in voidage. (Such changes are difficult to make within the constraints of a contract).

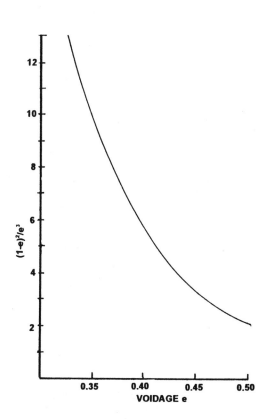

Fig. 16.6 The effect of voidage on the relative hydraulic resistance of a granular bed.

The voidage can change after a bed has been put into service. The media as placed will be well mixed but washing with water alone (or with separate air scour) causes some stratification, which enables smaller grains to migrate from the pores between larger particles and form a separate layer on their own, higher up in the bed, leading to a drift in hydraulic size from a larger value at the bottom to a smaller one at the top. This will reduce the headloss although it will also cause the bed to occupy a larger volume. This in turn can cause problems with sand loss. Thus it is always advisable to underfill a bed somewhat initially, and adjust after the bed has been backwashed a few times.

Overall, the Carman/Kozeny equations are probably more precise than the available data which characterises the media. The specific surface area is dependent somewhat upon the shape. This is itself difficult to quantify. Measurements can be made on single particles but it is too tedious to make such measurements on large

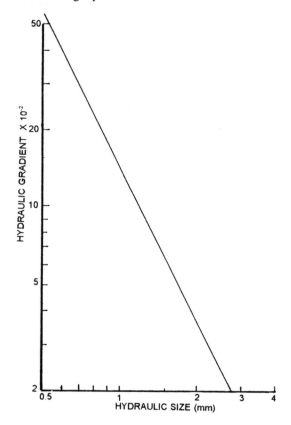

Fig. 16.7 Hydraulic gradient versus hydraulic size at 1mm/s flow rate for 40% voids and water at 10°C.

numbers of particles. The voidage of the material varies with its recent history and as already mentioned, the condition of the bed must be specified. Thus it is not surprising that some workers prefer to measure headloss directly rather than rely upon calculations.

In the next chapter the Kozeny Equation is applied to the blocked or partially blocked filter bed. Here uncertainties arise from the heterogeneity of pore blockage. One can have the same approach velocity into a uniform blockage or into a mixture of fully blocked pores and open clean worm holes with the same apparent open voidage. The headlosses are very different, as will be discussed later.

16.5 Airflow

The flow of air in a dry porous bed is also defined by the Carman/Kozeny equations. However, once wetted, the behaviour is dictated by surface tension effects and buoyancy. Air tends to penetrate in capillary trains (Fig. 16.8) with a leading end reaching out to the next smallest pore. The pore diameter (ie. the size of rod that could be inserted) between three equal spheres is 0.15D. The maximum bubble pressure is defined by:-

$$p = H\rho g = \frac{4\gamma}{D_p}$$

where γ = surface tension and D_p = pore diameter.

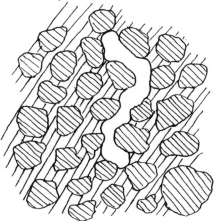

Fig. 16.8 Air capillaries within a saturated granular bed.

Thus a rising capillary will break into vertical elements with a length corresponding to the penetration head. The surface tension of water is 0.071 N/m, thus the head for 0.8 millimetre hydraulic size sand will be:-

$$\frac{4 \times 0.071}{0.0008 \times 9810} \quad = \quad 36 \text{ mm}.$$

One may expect capillary trains exceeding this length to rise up through the bed or else maintain a continuous airflow, but elements shorter than this will remain trapped. The water will part to make way for the air, and the rate of progress will depend upon the airflow through the train of pores. If air is introduced from an orifice, an interconnecting matrix bubble with a height equal to the penetration head is likely to form before the capillary trains set off.

With fine media or near the surface the air may be unable to penetrate against the surface tension and lens shaped voids with an arched roof form. The roof eventually breaks allowing the air to pass upwards and possibly form another lens shaped cavity higher up.

Rising bubbles in sand diffuse horizontally very little and nozzle patterns have been observed on the surface of a sand bed as much as 2.8 metres above the floor. Free bubbles rising from a single orifice through a reasonable depth of water will spread laterally to some extent, and a cross section through the plume will reveal a normal distribution to the air flux.

However, a mass of air bubbles rising from a flooded completed filter floor before placement of the media will tend to congregate in swarms causing a rolling action, which is very apparent in aeration lanes in sewage treatment plants. For this reason observations of the air distribution from a newly installed filter floor before the media is installed can be very misleading.

16.6 Simultaneous Air/Water Flow

If air and water are applied simultaneously to a bed of granular media, the air is propelled to some extent by the hydraulic gradient created by the flow of water. This has been studied to some extent by Amirtharajah and is discussed in Chapter 20. However, he was concerned with the threshold for movement of grains and not the hydraulic gradient through the bed nor with the air/water voidage ratios. This appears to be a somewhat under researched topic as far as fine granular media is concerned, although much work has been done on manufactured high voidage packings for gas/liquid mass transfer.

From the foregoing it can be assumed that for low co-current flows, the capillary trains will be shortened by the ratio of the hydraulic gradient to the air buoyancy and the air will penetrate more readily. However it is difficult to quantify the air hold up and therefore the local velocity of the water. Headloss measurements in specific cases have shown that the air reduces the pressure loss in an upflow situation in a manner analogous to an air lift, but precise data linking air and water flow rates generally with hydraulic gradients for given sizes of media do not seem to be available for low voidage systems. The subject is discussed further in Chapter 20, where one approach seems to give data which correlates with experience.

Some designs of biological aerated filters provide another interesting example where air is required to flow upwards counter to water flowing downwards. In conventional fine sand beds eg. hydraulic size around 0.8 mm. any air coming out of solution within the filter during down flow operation effectively shuts off the permeability of the bed towards water. The air will only escape from the bed when the water flow has ceased.

Biological aerated filters use media with a hydraulic size of about 2.8 mm. The capillary trains will therefore will only be about 10 mm long and therefore close to the grain diameter. There is a much greater probability of a large pore being available nearby to allow the air to escape upwards with an even shorter capillary train. Thus there is probably a fairly sharp lower limit for the grain size in such counter current aerated filters. A parallel situation exists where a filter is to be refilled after draining down and dry bedding. A fine media will remain air locked and must be filled from below. A coarse bed can be filled from above without backwashing because the air can escape.

16.7 Dispersion of Flow

Fluids do not pass either through pipes or porous media in a sharp piston fashion. If a sharp pulse of dye is injected into the stream this pulse will diffuse both in the direction of flow and laterally, and the dye concentration in the outgoing stream if measured and plotted against time will show a normal distribution in the statistical sense to which a standard deviation can be assigned. This behaviour has been characterised and is useful in considering the displacement of dirt from a filter bed during backwashing, for example, and also the extent of longitudinal mixing of the flow during filtration.

Some striking photographs have been published by Hiby (1962). These illustrate the way in which a transverse stripe of dye injected into a viscous fluid becomes elongated as it flows through a labyrinth of cylinders and emerges as a very ragged front. (On reversing the flow, providing only a small distance has been

238

travelled, the original stripe is restored). Hiby also published photographs showing how a stream of marker dye entering a labyrinth of cylinders on the centre line between two cylinders where the flow is highest, subsequently becomes a boundary flow of low velocity around the next cylinder downstream. Here the similarity between capillary flow and granular beds assumed in the derivation of the Kozeny equation falls down. In a single capillary there will be little transverse mixing of the boundary element into the main stream and hence a wide dispersion of any input trace will be observed at the outlet. However, in a granular bed, the surface flow creeping off one grain inevitably becomes the mainstream of the next layer. In a granular bed as in any other flowing system a sharp pulse or change in concentration at the inlet disperses to form a normal distribution or S-shaped curve, respectively, at the outlet (Fig. 16.9). The shape of the curves can be characterised in statistical terms, eg. variance. Levenspiel (1972) has studied such systems in detail and has reported a correlation for packed beds:-

$$K_D e/uD = 2.0$$

where K_D = the dispersion coefficient and D is still the grain size. u and e are still the approach velocity and voidage.

NO OF STAGES, J 16

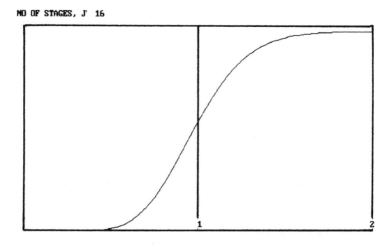

Fig. 16.9 Dispersion of a step change (eg. in concentration) on passing through a granular bed based on the example on page 239. X = units of apparent transit time in the bed.

The variance of the distribution curve is given by:-

$$Variance = \sigma^2 = 2\frac{K_D}{uh} + 8 \left(\frac{K_D}{uh}\right)^2$$

σ=standard deviation, h= bed depth hence K_D/uh is related to K_D/uD by the number of grain diameters in the depth. If a step change in the composition of the ingoing stream is made, 95% of the change ie, the 2.5 percentile and the 97.5 percentile are within two standard deviations of the mean.

Thus for a 1 metre deep bed of 0.8 millimetre media and a voidage of 40%, with a flow rate 2 mm/s (7.5 m/ hr):-

$$K_D = \frac{2.0 \times 0.002 \times 0.0008}{0.4} = 8 \times 10^{-6}$$

Hence σ = 0.06. Thus 95% of the change occurs between 0.78 and 1.12 times the mean time of 200 s, ie. 156-224s. There is therefore not a long transition time for the filtrate quality when the input quality changes, and likewise during backwashing, one would expect a fairly sharp displacement of dirt from the bed if the deposit has been fully loosened before backwash commences.

16.8 Multi Media Filters

This term is used in the water treatment industry to descibe filters which are charged with more than one type of filter media in the same bed without any intervening barrier, for example anthracite on top of sand or anthracite-sand-garnet. There are filters with two beds (like bunks) possibly both of sand but of different sizes. These would not be dual media filters but double deck filters. As will be discussed later, there are advantages in providing an increasingly finer media in the direction of flow. Such a bed offers the benefits of a deep coarse bed with a large holding capacity combined with the efficiency and quality of fine media. Such a coarse-to-fine gradation can be achieved by using two of more different types of media, with different densities so that coarser grains of lower density will rest in a stable manner over finer but denser grains. Such beds are described as dual or multimedia bed. Under laboratory conditions at least five layer beds have been achieved. The selection of media grades is governed by the behaviour during the backwash and hence a detailed discussion is to deferred to Chapter 20. As far as pore characteristics and hydraulics are concerned, the other types of media behave like sand of equivalent voidage and size. The only area of concern is the interface where fine particles of the dense media can penetrate up into the coarse particles of the lighter media. This would be expected to reduce the local voidage and increase the headloss. There is however evidence which is discussed in Chapter 20 that this does not matter and that the dispersion of the fines from the dense material up into the coarser media avoids the blinding effect of an unskimmed layer of fines on the top of the lower bed. There does not appear to be any data to confirm this, but it would appear possible. Nevertheless, most operators like to see a sharp interface. Sharp interfaces are usually possible with sand/anthracite were there is usually a 4-fold ratio in the submerged densities between the two media but with sand/garnet the ratio is barely 1.7 and considerable intermixing at the interface is usual.

Anthracite is the most common alternative to sand as a filter material. This is available in a number of densities ranging from 1400 kg/m^3 to 2000 kg/m^3, the common varieties being 1400 and 1600 kg/m^3.

Activated carbon has also been used as a lightweight material for filtration as opposed to absorption (ie, it is not regenerated). The fact that these materials are used indicates that hardness is not a critical parameter. The load on the base of the filter is often two tonnes per square metre in a typical filter plus the static submerged weight of the bed. Far more damage is caused to the media by handling and placing, or by abrasion during backwashing. Porous grains are likely to block internally fairly quickly when put into service, and the internal porosity is not likely to contribute to the flow. Thus only the external voidage should only be used in the Kozeny equation.

16.9 References

Coulson, J.M., Richardson, J.F., Backhurst, J.R. & Harper, J.H. (1991), Chemical Engineering Vol.II Ch.4, Pergamon.

Hiby, J., (1962) "Longitudinal and Transverse Mixing during Single Phase Flow", Proc. Symposium on Interactions between Fluids and Particles,. Inst. Chem. Eng. London.

Levenspiel, O., (1972), Chemical Reaction Engineering, Wiley.

McGeary, R.K., (1961) "Mechanical Packing of Spherical Particles", J. American Ceramics Soc. **44,** 513

Sakthivadivel, R., Thanikachalam, V., Seetharaman, S. (1972), "Headloss Theories in Filtration", J. American Waterworks Assn. **64,** 233

CHAPTER 17

PROCESS MECHANISMS

17.1 Introduction

Granular media filtration is a complex process in which the particles to be removed must be conditioned chemically so that they may adhere to each other and to the filter media. The particles are many time smaller than the grains of media and capture involves one of several mechanisms the least important of which is straining, sieving or surface filtration which predominates in most forms of filtration found in chemical engineering. The deposit is retained in the pores of the media while the water flows past the deposit rather than through it. In most cases the chemical treatment is an essential and integral part of the process. Indeed Cleasby et al (1992) go as far as saying that chemical treatment prior to filtration is more critical to success than the physical facilities at the plant. Granular beds have a limited internal volume and filters are often preceded by a primary stage which removes excessive loads of solids, leaving the filter to perform a polishing role. These aspects are discussed in Chapter 19. In this chapter the way in which the particles to be removed arrive at the surface of the grains, the pattern of deposition and the resultant effect on the flow through the filter are considered.

Several detailed reviews of the mechanisms involved in granular media filtration and the mathematics that go with them have been published eg.Ives (1980). Ives also (1975) edited the proceedings of a NATO workshop on the Scientific Basis of Filtration and Tien (1989) has fairly recently produced a very detailed book on the subject. Slow or biological sand filtration is a different process which operates by mechanisms other than those discussed here. It is discussed in Chapter 27.

It is not the intention here to analyze the published equations in detail but to examine the processes from first principles. The discussion below presents some new aspects of the process but it does not explain the behaviour entirely. One is left with the conclusion that there are features still to be discovered.

Many attempts have been made to model granular media filtration. Iwasaki (1937) was perhaps the first to propose an exponential decay of concentration with depth. However more than 50 years later papers on the subject are still appearing, for example Choo and Tien (1993), who compared five models, and also Ojha and Graham (1992). Tobiason and Vigneswaran (1994) showed that their model did not predict the hydraulic gradient very well but their experimental data accords with some of the present author's predictions. These recent papers have already reviewed the literature available to date in some detail.

Models may be divided on the one hand into mathematical correlations of experimental data which provide no little or no insight into the mechanism of the process and on the other hand into attempts to analyze the fundamental processes involved in terms of established principles, using for example a series of cell units with sequential time increments. Unfortunately the latter have not been particularly successful and various artificial features have had to be introduced. The situation is undoubtedly complex but Stevenson's work (1997) appears to make significant progress in this direction by considering the filter media as a mixture of different voidage and size elements, and the suspension also as a mixture of sizes.

17.2 Random Mixed Media

Most workers have assumed that filter media are essentially homogeneous materials with characteristics corresponding to the average of the relevant parameters. Voidage has tended to be ignored. When one examines the bed on the scale of a few grains it becomes obvious to an observer that the media are far from homogeneous. A typical filter sand has particles whose size varies by a factor of at least two. Fluidisation during washing causes little stratification of the size fractions of a given media, particularly in full size filters which do not have the stabilising influence of the walls in a laboratory column. Al Dibouni and Garside (1979) found that significant segregation only occurs where the fluidisation threshold of the fractions differs by a factor of 2 or more. Most filter media is graded to within a range of less than this figure. When washed with combined air and water a wide size range will mix, often including support gravel.

The grains do not rest against each other in an ordered manner like pipes stacked in a bundle but in a random manner so that the pores in between the grains can vary from a close packed triangular (rhombohedral) pattern between three of the smallest grains to an open square (cubic) arrangement between four of the largest grains (The voidages being 25% and 47% respectively). Neither are the grains spherical. It is possible to have a variety of pore shapes as well. All of this leads to a wide variety of individual pore hydraulic resistances which form a three dimensional matrix within the bed.

It is a feature of random systems that cells of a given characteristic can form clusters. A low resistance pore is not necessarily followed by a high resistance one. Low resistance channel strings form within the bed and meander through the structure.

This behaviour can be illustrated in 2 dimensional form by the following table of random numbers in which the low resistance paths have been highlighted:-

```
1 1 5 4 1 3 8 8 6 4 0 4 3 6 4 5 4 3 8 4 7 2 9 8
3 8 7 1 4 3 6 6 8 5 5 5 0 5 9 0 9 0 3 2 4 3 1 7
6 8 6 6 4 5 8 2 3 3 5 2 7 0 6 7 2 7 9 3 9 8 1 3
5 4 6 5 6 0 1 7 5 1 1 1 4 1 3 5 9 0 8 8 9 4 6 0
5 8 2 0 4 5 2 3 2 4 1 3 4 2 8 2 3 2 5 1 3 3 9 3
2 8 1 1 4 0 4 0 6 6 5 7 0 1 8 6 1 0 1 8 4 3 4 7
```

The average total through these 6 rows should be 27 but there is one string which has a total of 6, one of 7 and two of 9. The same behaviour will occur on a 3-dimensional basis in a bed of granular media where the chance of finding a low resistance neighbour will be even greater. Inspection of stacked tables, allowing movement to any nearest neighbour, suggests that such low resistance strings continue without end.

The same argument may be applied to the suspended solids, which in practice are not of a uniform size but spread over one or two orders of magnitude. As already noted, Iwasaki suggested that the residual concentration decayed exponentially with depth into the media and others have concurred with this. This is probably true for idealised monosize particles passing through monosize media laid down in an ordered manner. However it is also widely agreed that the decay constant or filtration coefficient is a function of the particle size and in practice a set of exponential decays must be summed. Such a sum is not exponential but tends to be logarithmic with slopes that change with the size range. This can easily be demonstrated by summing a set of different exponential decay curves.

In Stevenson's (1997) work therefore the local heterogeneity was taken into account and the process considered to proceed in a parallel matrix of different sizes and voidages. Likewise the particles to be filtered are considered as a set of various size individuals. Computational limitations have restricted the study to five of each species, but without limit to the size range.

17.3 Bed Hydraulics

As discussed in the previous chapter Kozeny (1927) was the first to establish the relationship between pressure loss, voidage and specific area when a fluid flows through granular media.

$$\frac{\Delta p}{h} = 5 \cdot \frac{(1-e)^2}{e^3} \cdot S^2 u\mu$$

The Kozeny equation, above, is satisfying in that it may be derived directly from the Poiseuille equation governing the pressure loss produced by flow in a capillary (taking the porous material to be equivalent to a bundle of capillaries), or more accurately by flow though a stack of parallel plates. If one assumes the capillaries or plates to be parallel to the direction of flow the constants are 2 and 3 respectively. Interestingly, if the parallel plates are set at an angle of 39° which will extend the effective path length then the constant will be 5 as found by Kozeny. Such an angle is not an improbable average for the path between random particles in a bed of media.

The specific area, S, in the Kozeny equation is that of the solid material, which for spheres is 6/D. Some workers include a sphericity term but to simplify the treatment this will be ignored here. D may therefore be regarded as the equivalent diameter.

If one considers small isobaric layers of a real bed it must follow from the above that there will be a considerable local variation in velocity through the pores, some even allowing transitional or turbulent flow. If the above equations are applied a ± 6% range used in the model described later a 13 fold range of velocity would be expected even without taking the grain size variation into account. No data is available on the extent to which low resistance pores penetrate into a bed or whether the wormholes are eventually blocked. As already indicated such runs appear to continue indefinitely.

The (1-e) term (which occurs as a square) in the Kozeny equation and several of the equations quoted below, is in fact a conversion factor for the specific area per unit volume of grain material to the specific area per unit volume of the bed. When one considers the nature of the deposit it is questionable whether the effective surface area does change much as the deposits are laid down. Indeed the snow caps observed in practice appear to take the form of aerodynamic fairings and are located on points of minimum shear at the poles of the grains (in a model in which the flow comes from the polar axis). The voidage term which appears as a cube in the equation defines the interstitial flow velocity and varies with the deposit. Thus it is suggested that a distinction must be made, in the context of filtration, between the clean bed voidage and the residual voidage so that the Kozeny equation when used to describe filtration should become:-

$$\frac{\Delta p}{h} = 5 \cdot \frac{(1-e_o)^2}{e^3} \cdot S^2 \cdot u\mu$$

Carman (1936) extended the theory to cover transitional flow, and his equations, adapted to distinguish between the two types of voidage are given below:-

$$Re = \frac{u\rho}{S(1-e_o)\mu}$$

$$\frac{R'}{\rho u^2} = 5Re^{-1} + 0.4Re^{-0.1}$$

$$\frac{\Delta p}{h} = \left(\frac{R'}{\rho u^2}\right) \cdot \frac{S(1-e_o)}{e^3} \cdot \rho u^2$$

As will be apparent later the velocities in clogged media rise well into the transitional region. At low Reynolds numbers the above becomes identical to the Kozeny equation.

The above equations are discussed in detail in chemical engineering textbooks such as Coulson et al. (1991).

17.4 Surface Shear Stress

Equation 6 may be rearranged to derive the shear stress from the head or pressure loss:-

$$R' = \frac{\Delta p}{h} \cdot \frac{e^3}{S(1-e_o)} = \frac{\Delta p}{h} \cdot \frac{e^3 D}{6(1-e_o)}$$

It follows that with a range of grains and pores in parallel, experiencing a common pressure loss across a section of the bed the largest and most open pores are subject to the highest shear stress. If material is scoured from such pores under high shear stress, the voidage will increase further increasing the shear stress until the pore has blown clean. This would appear to explain the formation of wormholes which are commonly observed in filters. These seem to be the highways by which water is able to penetrate the clogged upper layers of a filter, in which little filtration occurs, down to the lower working layers.

For the above equations to hold it is implicit that the deposit (e_o-e) must be uniformly dispersed. If half of the clean bed voidage is filled uniformly then the headloss will rise eight fold (from e^{-3}), but if half of the pores are clean and the other half fully clogged then the headloss will only be doubled. This point has profound implications in filtration where one can observe clean wormholes passing through the clogged media. From the above observations about the clustering of low resistance elements in a random mixture one may expect a pattern of highways and byways, the highways are capable of being swept clear and the byways clogging up. It is also apparent that although one can predict the headloss from a known pattern of deposition one cannot predict the pattern of deposition nor the residual voidage from the headloss because there are too many unknown parameters. The deposit may be expected to withstand shear stresses only below a given limit otherwise it will not adhere to the grains. There is also likely to be another higher limit above which existing deposits will be blown away.

17.5 Interception

If one considers particles of floc being carried along a streamline which approaches a grain from the polar direction any particle which approaches on a streamline passing closer than one floc radius to the grain will graze the surface of the grain and potentially stick to it (Fig.17.1).

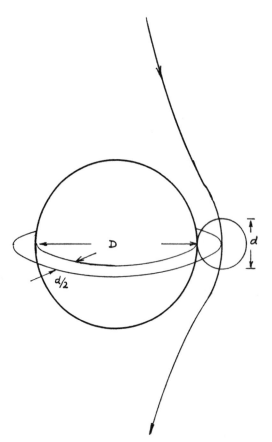

Fig. 17.1 Interception model.

There will thus be an annulus surrounding the grain corresponding to a 'capture cross section', to borrow a term used by nuclear physicists. Particles passing through this area will be captured if the capture efficiency is 100%. This efficiency depends on the chemical conditions but for the present purpose all collisions are assumed to lead to capture. If d<<D, the area of the annulus around each grain which includes all streamlines on which the centres of suspended particles move towards collision will therefore be:-

$$A_c = \frac{\pi}{2} . D . d$$

The volume of each grain is $\pi D^3/6$, thus in a bed of unit area, voidage eo and height h the number of grains present, N, will be:-

$$N = \frac{6}{\pi D^3} . (1-e_o) . h$$

The total capture cross sectional area per unit plan area of bed will be:- a per unit of plan bed area will be:-

$$A_h = N A_c = 3 \frac{d}{D^2} . (1-e_o) . h$$

However velocity of the fluid passing over the grains is not uniform but increases initially linearly with distance from zero at the surface (again for d<<D). The velocity is also proportional to the shear stress and inversely to viscosity. At a short distance y from the surface the grains (compared with the grain diameter) the velocity will be:-

$$u_y = R' . \frac{y}{\mu}$$

where R' is the shear stress and μ the viscosity. As this is a linear function the average velocity through the annulus will be half of this. y corresponds to the limiting streamline which passes the equator of the grain at a distance of d/2 from the surface. Thus the flow passing through the capture cross sections around each grain in a unit area of bed of depth h (by combining the above equations)is given by the equation overleaf.

$$Q_c = \frac{3}{4} \cdot \frac{d^2}{D^2} \cdot (1-e_o) \cdot \frac{R'}{\mu} \cdot h$$

where Q_c = the flow from which particles will be removed (per unit area of bed and depth h). R', the surface shear stress, is defined earlier.

$$R' = \frac{\Delta H}{h} \cdot \frac{\rho g e D}{6(1-e_o)}$$

(In the water industry "head" (water gauge) (ΔH) is preferred to pressure)

$$\Delta H = \frac{\Delta p}{\rho g}$$

At the high flow rates in residual open pores the flow will be turbulent.

The filtration coefficient is derived from the exponent of the ratio of the flow from which particles are removed to the total flow per unit area, ie. Q_c/u, thus:-

$$c/c_o = \exp\left(-\frac{Q_c}{u}\right)$$

Two correction factors need to be introduced to bring the simple model more into linewith the actual situation. KF is a calculated correction factor for the increased roughness or extended silhouette of the surface after the first floc particles have been deposited as illustrated in Fig.17.2.

It can be demonstrated geometrically from this figure that the capture cross section is increased by a ratio of $\theta/\sin\theta$ where $\cos^{-1}\theta = d1-d2/d1+d2$, for the section occupied by a particle, d1 being the diameter of the resident particle and d2 that of the arriving particle. For example KF is 4.4, 1.57 and 1.06 for values of d1/d2 equal to 0.1, 1 and 10 respectively. The benefit tails off exponentially as the vacant spacesaround the grain are filled at random. Larger particles thus have a particularly beneficial effect on the capture of small particles, as first described by Mackie (1989).

K_M is also a calculated geometrical correction factor which reduces the capture efficiency at high shear values (ie high velocities). It is based on the assumption the particles only deposit in the polar region of each grain, above a latitude corresponding to cos-1 (limiting shear/actual shear), (Fig.17.3). In practice this

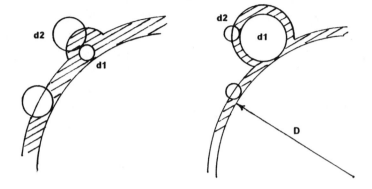

Fig. 17.2 Modification of the capture cross section by captured particles.

Fig. 17.3 The reduced capture cross
section at high flow rates.

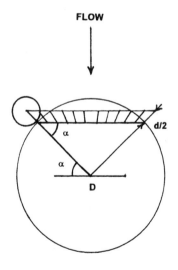

latter refinement has little effect
because pores tend to revert to either
a fully blocked or fully open
condition and the intermediate
condition is transient. Thus if these
corrections are included, the fraction
remaining from interception is defined
by the statement:-

$$C/C_o = \exp\left(-\frac{K_F . Km . d^2 . e . \rho . g . \Delta H}{8 . D . \mu . u}\right)$$

At low Reynolds Numbers ΔH may be calculated from the Carman Equation so that account can be taken of the turbulent flow in the more open pores. If flow is laminar ΔH can be replaced by the Kozeny Equation and Equation 15 reduces to :-

$$C/C_o = \exp\left(-\frac{45}{2} \cdot K_f \cdot K_m \cdot \frac{d^2}{D^3} \cdot \frac{(1-e)^2}{e^2} \cdot h\right)$$

The d^2/D^3 term is similar to that in the equation derived by Tambo (1979) for floc capture within floc blankets. It is better however to use Equation 16 together with the Carman equation because in a practical situation some cells in the model have flow velocities in the transitional region.

17.6 Settlement

While interception requires a flow to carry the particles to the capture zone round each grain settlement like diffusion operates independently of flow. Settlement becomes more significant with dense particles such as clay and silt and is exploited particularly in roughing gravel filters which are used in rural areas in developing countries.

The area of tank (Fig.17.4) required to provide complete settlement of particles settling at a velocity of us from a flow of Q is given by the Hazen Equation:-

$$A = \frac{Q}{u_s}$$

A granular filter can be considered to resemble a stack of miniature shallow settlement tanks or a lamellar settler. (Fig. 17.5)

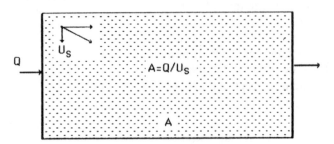

Fig. 17.4 The Hazen model for settlement.

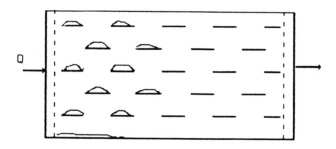

Fig. 17.5 Settlement area presented by slats.

The projected or shadow area Ap of the grains in a unit volume of bed will be:-

$$A_p = \frac{\pi D^2}{4} \cdot \frac{6}{\pi D^3} \cdot (1-e_o) = 1.5 \frac{(1-e_o)}{D}$$

However the settlement area is limited by the angle of repose so that the effective diameter of each grain is less by a factor corresponding to the sine² of the angle (see above). Thus the effective settling area in a depth h is:-

$$A_p = 1.5\sin^2\alpha \cdot \frac{h}{D} \cdot (1-e_o)$$

Thus for an angle of repose (α) of 45°, a 1.0 mm sand would have an area of 300 m²/m³.

Each grain offers only a small fraction of the area for complete settlementand the residue mixes with the supernatant which passes forward. Thus there will be an exponential decay of concentration with depth. Taking a unit area of bed the flow Q becomes equivalent to the approach velocity u. However to obtain the relevant interstitial velocity this must be divided by the net voidage e to give u/e. The removal can therefore be defined as follows:-

$$\frac{C}{C_o} = \exp\left(-A_p \cdot \frac{eu_s}{u}\right)$$

The settlement rate of a single particle in laminar flow is given by the Stokes Equation overleaf:-

$$u_s = (\rho_s - \rho_L) \cdot \frac{gd^2}{18\mu}$$

where ρ_s and ρ_L are the densities of the solid and liquid, and d the particle diameter. Thus, by combining the above equations, the overall removal of particles by settlement will be as follows:-

$$\frac{C}{C_o} = \exp\left(-\sin^2\alpha \cdot e(1-e_o) \cdot (\rho_s - \rho_L) \cdot \frac{gd^2h}{12Du\mu}\right)$$

Thus for example a 1 metre deep bed of rough surfaced 0.5mm sand will remove 97% of clay particles of 5µm diameter and density of 2500 kg/m3 from an approach velocity of 10 m/h. Hydroxide floc particles however are much less dense and sedimentation is less likely to play a significant role. Observations with transparent sided columns and even with a single wire placed across the flow using a feed of derived from conventional coagulation with aluminium or iron salts have shown that floc collects on the upstream face of the grains or wire even in an upflow situation. Interception would appear to be the dominant action in typical practical situations, especially for polishing filtration following settlement or flotation.

17.7 Diffusion

Particles below 1-2µm diameter are captured primarily by diffusion in the same way as species in solution in ion exchange on resins, where the theory has already been developed.

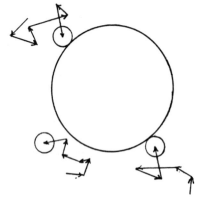

Fig. 17.6 Concept of diffusion to the grains of media.

The diffusivity K_D of a small particle is defined by the equation found in many textbooks on physical chemistry:-

$$K_D = \frac{K_B T}{3\pi \mu d}$$

K_B is the Boltzmann Constant, ie. 1.38×10^{-23} J/deg K., T the absolute temperature, d the particle diameter and μ the absolute viscosity.
Snowdon and Turner's (1969) equations developed for diffusion in ion exchange kinetics have been adapted to the present situation.

$$Sc = \frac{\mu}{\rho K_D}$$

$$Re = \frac{uD\rho}{(1-e)\mu}$$

Snowdon and Turner have shown empirically that:-

$$Sh = \frac{0.81}{e} . Re^{1/2} . Sc^{1/3}$$

The mass transfer coefficient K_L and the Sherwood Number are related by the equation:-

$$K_L = Sh . \frac{K_D}{D}$$

The concentration change produced by exposing the area Sc in a unit volume to an approaching flow of u is:-

$$\frac{\delta C}{C} = K_L \frac{S_c}{u}$$

Sc is as before the specific area of the bed. For a bed depth of h and unit plan area.

$$Sc = 6(1-e_o)h/D$$

254

Thus the concentration of monosize particles will vary with depth exponentially as follows:-

$$C = C_o \exp\left(-\frac{6(1-e_o)\,hK_L}{uD}\right)$$

Combined interception and diffusion

The above equations for interception and diffusion may be combined by adding the exponents. The result plotted for various particles sizes in 0.5mm media at 10°C at 7.5m/h filtration rate is shown in Fig. 17.7. It will be noted that a minimum exists at about 1µm.

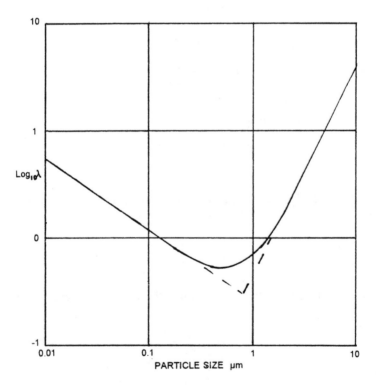

Fig. 17.7 The variation of filtration coefficient with particle size for interception plus diffusion for 0.5mm media (hydraulic size) at 10°C and 7.5m/h.

17.8 Surface Filtration

As indicated earlier surface filtration in which the media merely acts as a sieve is something to be avoided. If the media consisted of uniform size close packed spheres the pores would pass spheres less than 0.155 of the diameter of the spheres.

Thus with reasonably rounded media there is a good chance that all particles larger that 0.155 of the smallest fraction of the media will be trapped before penetrating very far. Thus with 0.5-1.0mm sand surface filtration is possible with particles of 77 μm and larger. This would trap several species of filamentous algae. As soon a one pore becomes blocked by a bridging particle the size of subsequent particles which will be retained will fall. A population of algae will start to create the equivalent of a filter paper which will itself become blocked by finer particles and the resistance to flow will multiply, effectively shutting down the filter. Surface filters using precoat powders, cloths, papers or membranes, used extensively in the chemical industry and also for small water supplies operate at very much higher pressure differentials than granular media filters. Their solids holding capacity is more often limited because of the hydraulic resistance of the deposit. Granular media filtration by contrast does not require the flow to pass through the deposit, but around it as it resides on the grains or in blocked pores. In a treatment plant solids which block the surface of filters are therefore removed first by microscreening, settlement or flotation. Dual or triple media filters provide a partial solution by increasing the size threshold which may cause surface filtration.

Often the surface of a filter bed is covered with the solids being filtered out and one can have the impression of surface filtration. The 'cake' can be several millimetres thick. This layer is however a porous blanket and probably comprises larger floc particles. Such floc is however able to remain porous as it would have been in a fluidised floc blanket in a clarifier. If it is free of blocking components such as filamentous algae it may also rest on the dead sections of the media and permit the wormholes that have formed in the bed to suck material from above, thereby forming wormholes in the floc layer. Close inspection of such a floc layer will often reveal the entrances to such holes.

17.9 Inertial Capture (Fig.17.8)

This is the process whereby particles carried in a fluid stream are unable to follow the streamlines over the surface of the grains and continue along their previous trajectory to collide with the grains.This is a significant mechanism in air filtration, where the viscosity is low. With liquids the much higher viscosity causes particles to be swept round the grains. Ives (1975) considers that this mechanism may be ignored.

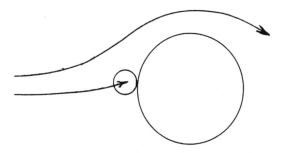

Fig. 17.8 The concept of inertial capture.

17.10 Internal Trapping (Fig.17.9)

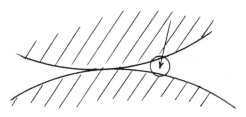

Fig. 17.9 Internal trapping in the crevices between grains. (Flow is normal to the paper.)

The last mechanism to be considered is that where the flow passes through the wedge shaped crevices close to the points of contact between the grains. Over the band where the crevices are narrower than the particle diameter capture can occur by mechanical straining and it is therefore efficient for those particles that move on such a path. It is difficult to model the trajectory but it would appear from a consideration of laminar flow between parallel plates that the proportion of the flow passing through such narrow gaps compared with the flow passing through the capture halo around the surface of a grain will be very small. Nevertheless this is a mechanism that has yet to be quantified. Visual observation would suggest that the role is not a major one.

Internal trapping of a different type will occur with dual or triple media filters at the interfaces where percolation through the upper layer is stopped by the change in grain size.

The overall change in the concentration will be the product of all of the above removal mechanisms. However for modelling purposes only interception, settlement and diffusion are considered to be significant.

17.11 Conditions in a Clogged Filter

From the foregoing it will be understood that filtration proceeds until a limiting condition occurs when the deposit resolves itself into blocked pores and wormholes. The latter provide the highway through which flow continues to the layers below. The headloss in these wormholes provides most of the headloss through the clogged bed, and as the depth of the clogged zone increases so the headloss increases fairly linearly, as is found in practice. Conditions are static in the wormhole zone with no further filtration occurring. However there is a danger that any increase in flow may cause the limiting shear stress in the larger blocked pores to be exceeded causing what may be described as a 'blow out' or avalanche which will in turn produce a turbidity spike. This situation can occur for example when the flow increases as a result of one filter in a set being taken out of service for washing. At higher flow rates a larger proportion of the bed appears to be devoted to wormholes in order to maintain the limit on the shear stress. It is also possible to have avalanches in clogged beds when a section of the deposit breaks off, causing blockage of a wormhole and disturbance of a larger zone. These appear to be worse at low flow rates where the wormholes may be far apart.

Anyone who has run experimental columns will know that the headloss on restarting after an overnight shut down is somewhat less than that before stopping. This is probably the result of floc compaction and shrinkage, a natural feature that was discussed and investigated by Francois(1987). It is possible that the deposit has some natural resiliance and is compressed by the flow. When the flow stops there may be some recovery and swelling, particularly of hydroxides and organic deposits. Such swelling would render the deposit weaker and more likely to shed on restarting. Any local flow reversal on stopping will tend to dislodge deposits. Also with a mixture of Reynolds Numbers the flow distribution will change as the flow changes. All of these effects can cause particles to be dislodged on starting and stopping. This problem is well recognised in the Badenoch Reports (Anon. 1990 and 1995), which strongly recommend washing of used filters after a stoppage.

In this model the initial filtration coefficient as such is not sensitive to the flow rate. The latter affects the amount of floc that can be retained in the bed and the breakthrough pattern. This is confirmed certainly by experimental data. With higher rates the maturation period is considerably reduced and once the limiting shear stress has been reached in some pores the filtration efficiency will start to deteriorate. Overall the average quality declines.

17.12 Flocculation

Flocculation is the process whereby small particles suspended in the water collide either as a result of molecular bombardment or from the outcome of fluid shear. Such contacts lead to agglomeration. The process is discussed in more detail in Chapter 5. Flocculation has been considered as a possible component of filtration for many years and has again been examined more recently by Graham (1988). Undoubtedly it is a plausible process with direct filtration where the coagulant is added immediately upstream of the filter. Floc will not have reached its equilibrium size and with the ample shear available the freshly destabilised particles will grow. However the older mature floc from a previous treatment stage is more likely to be broken down, particularly in the wormhole section of a bed.

Furthermore the concentration of solids lower down and particularly in a polishing filter is lower than in a typical flocculating situation, for example within a settlement tank. It also falls as the water passes through the bed, making flocculation even less likely. (Flocculation in this case is controlled by the Smoluchowski Equation (discussed in Chapter 5) and the rate is a function of concentration and shear rate.) More significant are perhaps the observations of Ginn et al (1992) that solids which are shed tend to be coarser than those entering the filter. These may be pieces of displaced deposit rather than the product of flocculation of the suspension within the pores. Compared with new floc displaced deposit is less likely to adhere further down the bed. It would be a daunting task to include the age characteristics of the floc in any one cell in a mathametical model. Overall flocculation would seem to play a minor role in filtration.

17.13 A Mathematical Model (MFLX-Multilayer Flexible Model)

A filtration simulation programme has been built around the above theoretical basis by treating the bed as a matrix of five different grain sizes in a defined size range by five different voidages (arbitrarily ±6% about the nominal voidage for the bulk media), see Fig.17.10, instead of a single homogeneous medium. The suspended solids to be filtered were also considered as five discrete fractions. The concentrations of each size fraction and also the deposit volumes of each within each of the 25 cells in each depth increment were all recorded for each time increment.

Computing limitations restricted the number of depth increments to 50.

The effluents from the 25 cells in each depth layer were blended to produce an average composition to feed to the next layer. This was deemed to produce a crude simulation of the random mixture of large and small pores already discussed.

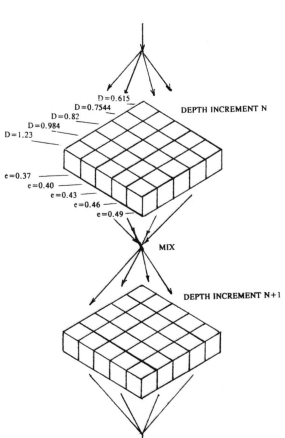

At each cell calculation of the average shear stress (R') on the surfaces was derived and if this exceeded a stated limit then deposition was not allowed. If this exceeded the limit by a defined factor then the deposit present was stripped and added to the ongoing flow into the next layer. The output concentration at the bottom of the bed was the sum of those of the individual fractions. Ideally one would wish to use an even larger matrix, and smaller time increments. Rolling averages would smooth out the resultant curves and make the model correspond to the many vertical cells which comprise a real filter.

Fig. 17.10 The multilayer 5 X 5 matrix model of MFLX simulating a disperse filter bed.

Facilities were included to allow the calculation to switch to three different types of media at specified depths to allow dual or triple media filters to be explored. The input variables were therefore:-

Media depth   }
Hydraulic size of media } for each layer
Grain size range   }
Voidage

Flow/time profile.
Volume concentration of solids in the feed.
Mean size of the suspension
Size range of the suspension
Limiting shear stress (determined by calibration)
Ratio of deposit to shedding shear stress limits

Limiting time. Limiting headloss. Depth increment
Time increment. Fluid density. Fluid viscosity

All of these are real parameters not arbitrary coefficients.

The size fractions for the media and the suspended solids were assumed to be present in equal amounts but a size distribution facility could easily be added.

There is a general consensus that small particles produce a stronger deposit. Thus one could include a term to vary the limiting shear stress with the particle size or average particle size of the deposit in any cell. However this refinement has not been incorporated at this stage.

No account has been taken of flow dispersion in the residence time within the bed. Thus any dislodged material appears as a sharp spike whereas in practice there will be some smoothing. However the time increments used are long compared with the internal retention time of most filter beds and the effect will be minor.

The floc shear strength is the main parameter which must be determined by trial and error to make the predictions fit the data. Extrapolation to other situations is then possible.The distribution of the voidages also must be an estimated but this will be between known and defined limits.

17.14 Experimental

While most of the work involved programming and calculation some experimental data was available. Much more data to support the general predictions of the present work will be found in published literature. Simple filter columns filled with various sizes and depths of sand were run using a chalk river with low turbidity (normally 2-5 NTU, 220 mg/l alkalinity and 250 mg/l calcium hardness (both as $CaCO3$). The river water was dosed with ferric sulphate to give 5 or 7mg/l Fe. Unfortunately the experimental work predated the refinement of the model and was not specifically designed to test the conclusions.

Measurements have shown that for a Leighton Buzzard sand with a packed voidage of 38%, the voidage after fluidisation, ready for service was 43%. The size used in calculations was the hydraulic size as defined by Stevenson (1994) and in the B.E.W.A. Standard (1993), eg. centreline 0.6-1.18mm = 0.82mm.

17.15 Results

Figures 17.11, 17.13, and 17.15 to 17.17 show examples of the results obtained with the present programme. The erratic curves are caused by the finite size of the increments and the fact that only a single vertical section is considered. In a real filter such fluctuations would usually be averaged out. (It would of course be possible to take smaller time increments and calculate rolling averages but this would have extended the run times (about 90min.) even further).

Fig.17.11 is based on a 10 fold particle size range, which is not unrepresentative of actual suspensions.(Fig. 17.12).This is repeated at twice the flow rate in Fig.17.13, while Fig.17.14 gives some experimental curves, which could be supported by countless other published examples, eg. Bucklin et al.(1995). Fig.17.15 shows a calculation for smaller monosize particles which demonstrates some of the features of the model more clearly. Figs. 17.16 and 17.17 illustrate the flow/time profiles of filters in a plant with equal flow and declining rate regimes.

17.15.1 Headloss vs Time

Filters running at constant rate produce a linear increase in headloss with time. This characteristic is reproduced by the present model. Most of the headloss occurs within the fully clogged part of the bed and as this proceeds forward so the headloss increases.

Fig.17.15 (monosize suspension) however shows an initial diversion in this linearity which is clearly visible in curves published by Tobiason and Vigneswaran (1994) based on actual experiments with monosize suspensions. The curve is caused by the headloss following the Kozeny Equation as the deposit is laid down up to the point where cells start to shed deposit. In the disperse size suspension the overlapping behaviour of the various size fractions masks this effect, as shedding of large floc starts much earlier in the run.

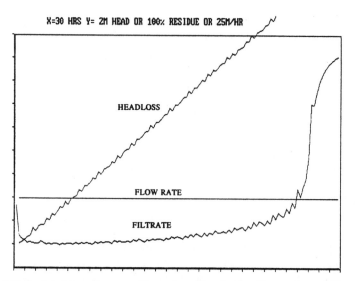

X=30 HRS Y= 2M HEAD OR 100% RESIDUE OR 25M/HR

HEADLOSS

FLOW RATE

FILTRATE

Fig. 17.11 Calculated headloss profile and residue in the filtrate from 900mm of 0.6-1.18mm media, 2:1 size range, 43% voids, 7.5m/h filtration rate, 5μm mean particle size 10 fold size range, 0.2% vol. concn. 2.5N/m2 shear limit.

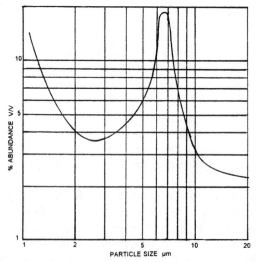

Fig. 17.12 Particle size distribution from a floc blanket clarifier (River Thames) feeding filters.

X=30 HRS Y= 2M HEAD OR 100% RESIDUE OR 25M/HR

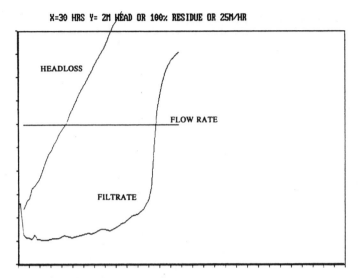

Fig. 17.13 As Fig. 17.11 but at 15m/h.

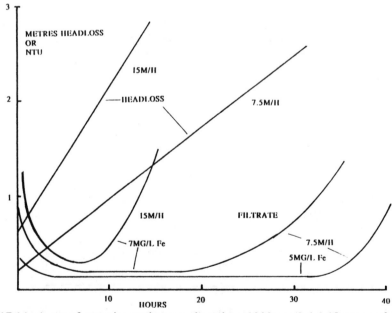

Fig. 17.14 A set of experimental curves based on 1000mm 0.6-1.18mm sand 5mg/l Fe as ferric sulphate.

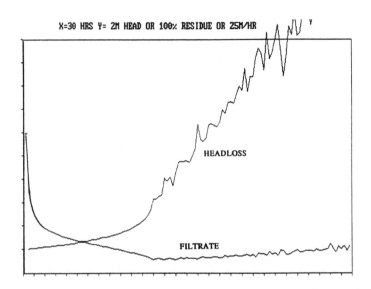

Fig. 17.15 As Fig. 17.11 but based on 2μm monosize particles, 1m/h, 0.1% vol. concn. 10N/m² shear limit.

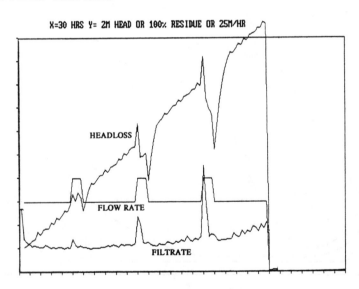

Fig. 17.16 As Fig. 17.11 but with 33% step increases to simulate the flow profile with four filters in a plant.

X=30 HRS Y= 2M HEAD OR 100% RESIDUE OR 25M/HR

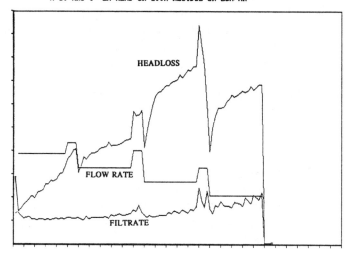

Fig. 17.17 As Fig. 17.16 but operating on a ±30% declining flow profile.

17.15.2 Voidage and Maturation

In Table 17.1, which shows the cell velocities and deposit volumes in the top depth increment, it will be noted that initially all the cells attract a deposit. The flows vary widely but those with the highest flow, which also have the lowest efficiency tend to clog more rapidly so that the flow falls whilst the lower flows rise. This produces a new type of maturation which is very clear in Fig.17.15 and lasts for a considerable proportion of the run. This is separate from that caused by a roughening of the grain surface discussed above and which only affects the first hour of the run. The process rapidly ceases when the shear in the most open cell reaches the limit and detachment commences.

17.15.3 Flow Changes

In the example in Table 1 the initial flow was 7.5m/h and only one of the 25 cells was blown clean to act as a wormhole. On switching to 15m/h three other cells discharged initially although one subsequently started to filter again. Ideally the model should have more than 25 cells to illustrate the effect more clearly, especially at low flows where in this case only a single cell remained clear.

The effect of flow changes experienced by filters when others in a plant are taken out of service for washing is shown in Fig 17.16. No other models, so far published, appear to be able to reproduce spontaneously this sort of behaviour, which is well known to plant operators. The tendency to breakthrough increases with the load on the filter. On gravity filters the high headloss often cannot be sustained and the filter slows down towards the end.

Fig.17.17. illustrates declining rate filtration, long advocated by Cleasby and reviewed by him in 1993. The lower headloss at the end of 24 hours and the reduced turbidity spikes are also reproduced spontaneously.

17.15.4 Effect of Flow Rate (Steady Conditions)

At low flows the efficiency appears to be unaffected by the flow, a point which the present experimental data confirms. However once the limiting shear is reached the efficiency becomes flow dependant. The situation is complex as illustrated by the curves and breakthrough occurs much earlier at high flow rates. Figs.17.11 and 17.13 compare well with Fig.17.14.

17.15.5 Effect of Bed Depth

As indicated earlier, the exponential decay in solids concentration with depth assumed in many theoretical papers does not hold with practical media and solids. Fig. 17.18 shows a set of experimental data and calculated data. The slope of the calculated curve varies with the particle size range, wider ranges giving flatter curves. This will be self evident in that there will be a larger amount of fine material which is difficult to filter with the wide range, for a given mean size. Again the model reflects practical data.

The size of the depth increment used in calculation is a compromise between speed of calculation and accuracy. In practice the effect on accuracy falls with increment size and test runs are necessary.

17.15.6 Media Size

The above equations as applied to a single particle size and monosize media suggest that the efficiency (filtration coefficient) should vary with the inverse cube of the grain size with interception, inversely with size for settlement and also inversely with diffusion. However even with interception the cube effect is greatly diluted when calculations for a range of particle and grain sizes are superimposed. Mohanka (1971) concluded that the exponent was -1.5. Other authors are less precise eg. "the size effect was overstated". The present experimental work and the calculations indicated that the exponent is nearer -1 but somewhat variable depending on the conditions

Table 17.1.
Example of Flow and Voidages in the 5 X 5 Matrix
Hydraulic Size 0.82mm (0.6-1.18mm) Size Range 2. Voidage 43%
Limiting Shear Stress 2.5N/m². Nominal Particle Size 5 μm. Particle Size Range 10.
DEP= Deposit (fraction of bed volume) U= cell velocity m/h

First Iteration 0.25 hours

DEP	U	DEP	U	DEP	U	DEP	U	DEP	U
0.009	1.69	0.008	2.54	0.008	3.01	0.007	4.34	0.006	6.77
0.010	2.33	0.010	3.52	0.009	4.16	0.008	6.00	0.006	9.35
0.012	3.18	0.011	4.81	0.010	5.69	0.009	8.20	0.007	12.72
0.014	4.31	0.012	6.51	0.011	7.70	0.010	11.08	0.008	17.11
0.015	5.79	0.013	8.75	0.013	10.34	0.011	14.84	0.009	22.79

After 1 hour

0.116	3.41	0.133	4.16	0.139	4.55	0.149	5.73	0.152	8.43
0.141	3.99	0.159	4.84	0.165	5.30	0.173	6.77	0.174	10.35
0.168	4.57	0.186	5.54	0.192	6.09	0.199	7.95	0.196	12.65
0.197	5.14	0.215	6.25	0.220	6.92	0.225	9.28	0.218	14.43
0.228	5.70	0.245	6.99	0.249	7.81	0.251	10.81	0.240	18.84

After 1.5 hours

0.223	0.96	0.243	0.92	0.250	0.91	0.265	0.89	0.279	0.90
0.256	0.98	0.277	0.93	0.284	0.92	0.298	0.90	0.278	2.38
0.291	0.98	0.311	0.93	0.318	0.92	0.332	0.89	0.311	2.46
0.326	0.98	0.345	0.92	0.352	0.90	0.329	2.35	0.321	4.25
0.362	0.96	0.380	0.91	0.347	2.37	0.364	2.35	**0.000**	**154.6**

After 2.5 hours, flow raised to 15m/h.

0.279	0.20	0.294	0.17	0.299	0.17	0.259	0.92	0.276	0.86
0.312	0.19	0.301	0.42	0.308	0.40	0.291	0.95	0.298	1.21
0.346	0.19	0.334	0.42	0.308	1.04	0.324	0.97	**0.062**	**53.47**
0.350	0.48	0.333	1.09	0.341	1.05	0.347	1.30	**0.00**	**109.3**
0.346	1.20	0.368	1.09	0.365	1.39	**0.090**	**55.8**	**0.00**	**140.8**

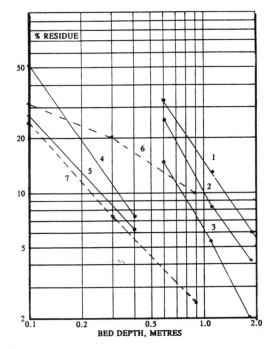

Fig. 17.18 Residual particle concentrations ot turbidity removal as a function of bed depth. Experimental data - 1. 1.18-2.8mm sand 20m/h 5mg/l Fe. 2. Ditto, 15m/h, 3. Ditto 8m/h. 4.).6-1.18mm 15m/h. 5. ditto. 12m/h. Calculated data - 6. 0.6-1.18mm 7.5m/h. 10 fold particle size range. 7. Ditto. 3 fold size range.

and the size range of the suspension.

17.15.7 Media Size Range

Some suppliers of media and contractors claim that close cut or monosize media gives superior results that justify the additional cost. No published evidence has been located. The present programme however indicates that a 1.5 fold size range instead of 2 fold gave a 40% reduction in the residual solids concentration. However the filter run to a given headloss or to breakthrough was about 10% shorter.

17.15.8 Floc Strength

Although not illustrated, the chemical environment and the presence of polyelectrolytes will affect the limiting shear stress. This parameter has physical significance and can be calculated for clean media. In using this model it is the only parameter that is not defined measurable by independent means. Thus its value can be estimated by trial and error to make the calculated performance match experimental observations.

As the floc strength and limiting shear stress rises so does the headloss gradient. The overall efficiency is improved, but not necessarily the initial efficiency unless the chemical changes the effective particle size of the solids being filtered. It has already been mentioned that the shear strength probably varies with the particle size of the suspension.

17.16 Summary

The disperse particle size range, disperse voidage, disperse suspension size model described above appears to be able to predict many features of real filters in a real time manner in a way that has not hitherto been achieved in previously published models. Perhaps most surprising is the prediction of wormholes through the more open pores which has been a matter of concern to some (Anon 1990). It is satisfying to be able to predict the behaviour in this way from an analysis of measurable characteristics of the system and a priori argument.

The model predicts a shedding and rearrangement of the particles of deposit within the advancing front, a feature proposed by Mintz (1966) and which was the subject of considerable controversy at the time.

From the above work it would appear that the application of the concept of 'filtration coefficient' to real filters is perhaps dangerous. It is only valid for idealised monosize systems.

The relevant equation is more of the type:-

$$C_h = C_1 . h^{-x}$$

Where C_1 = concentration at unit depth, and x is an exponent which will vary somewhat but appears from experiments to be about 1.5 in at least one representative situation. This obviously does not hold for very small depths. A similar equation can be produced by summing a range of exponential decays. Indeed the decay of the mixed fission products from a nuclear explosion, each of which decay exponentially, also follows such a relationship, with x equal to 1.2.

There is ample scope for further development of the principles and the programme would benefit from greater computing power to allow the use of more than five size fractions. There are many aspects of filtration (declining rate operation for example) that can be explored and optimised by this model. The above examples are only samples to demonstrate the capability. The model could of course be used in parallel with a real filter to pace its condition and forecast its behaviour should conditions change.

The main area deserving experimental study is the characterisation of the 3 dimensional bed random matrix. The present model is very much a simplification. Measurement of local flow velocities on a grain size scale would be useful.

17.17 References

Al Dibouni, M. R. and Garside, J., (1979) "Particle Mixing and Classification in Liquid Fluidised Beds", Trans. Inst. Chem. Eng., **57**, 94.

Anon. (1990), "Cryptosporidium in Water Supplies", (First Badenoch Report), H.M.S.O., London.

Anon. (1993), "Standard for the Specification, Approval & Testing of Filtering Materials", British Water, 1, Queen Anne's Gate, London,SW1H 9BT

Anon. (1995), "Cryptosporidium in Water Supplies", (Second Badenoch Report), H.M.S.O., London.

Bucklin, K., Amirtharajah, A. and Cranston, K.O. (1995) The Characteristics of Initial Effluent Quality and its Implications for the Filter to Waste Procedure. American Waterworks Association Research Foundation, A.W.W.A., Denver.

Carman,P.C.(1937) "Fluid flow through Granular Beds", Trans. Inst. Chem. Eng, **15**, 150-166.

Choo, C.U. and Tien, C."Transient State Modelling of Deep Bed Filtration", Water Science and Tech. (1993), **27**, (10), 101-116.

Cleasby, J.L., Sindt,G.L., Watson, D. and Bauman, E.R. (1992) Design and Operation Guidelines for Optimisation of High Rate Filtration Processes. Plant Demonstration Studies. American Water Works Association Research Foundation. A.W.W.A. Denver. (ISBN089867-604-5).

Cleasby, J.L. (1993) "Status of Declining Rate Filtration Design". Water Science Technology, 27,151-164.

Coulson, J.M., Richardson,J.K., Backhurst,J.R. and Harper,J.H.(1990) Chemical Engineering,Vol.II,Chapter 4 Pergamon, London

Francois,R.J.(1987), "Ageing of Aluminium Hydroxide Flocs",Water Research, 12,523

Ginn,T.M., Amirtharajah,A. and Karr,P.R. (1992) The Effect of Particle Detachment in Granular Media Filtration", J.American Water Works Assn. 82,66

Graham,N.J.D.(1988), "Flocculation in Pores", Water Research, 22,(10)

Ives,K.J. (1975) A Scientific Basis of Filtration, Noordhoff, Leyden

Ives, K.J. (1980) "Deep Bed Filtration: Theory and Practice".Filtration and Separation (March/April), 157

Iwasaki, T. (1937), "Some Notes on Sand Filtration", J. American Waterworks Assn. 29, (10), 1591-1602.

Kozeny, J. (1927), "Uber kapillare Leitung des Wassers im Boden", Sitzb. Akad. Wiss., Wien, Math-Naturw. Kl. 136 (Abt.IIa), 271-306.

Mackie, R. I. (1989), "Rapid Gravity Filtration-Towards a Deeper Understanding", Proc. Filtration Society. Jan.Feb. pp32-36

Mintz, D. M. (1966) "Modern Theory of Filtration", Special Subject No. 10, Proc. 7th Congress, Internat. Water Supply Assn. 1, 3-29

Mohanka, S.S. (1971)," Multilayer Filter Design", Water and Water Eng. 4, 143-147

Ojha, C.S.P. and Graham, N.J.D.(1992), "Appropriate Use of Deep Bed Filtration Models", J. Environmental Eng., 118,(6),964-980

O'Melia,C.R. and Ali,W.(1978) "The Role of Retained Particles in Deep Bed Filtration", Prog. Water Tech.,10,167

Rajapakse, J.P. and Ives, K.J. (1990) Prefiltration of Very Highly Turbid Water Using Pebble Matrix Filtration". J. Inst.Water Environmental Management,4,140

Snowdon, C.B. and Turner, J.C.R. (1967), Mass Transfer in Liquid Fluidised Beds of Ion Exchange Resin Beads, Proc. Internat. Symposium Fluidisation, Ed. Drinkenberg,A.A. , Netherlands University Press, Amsterdam.

Stevenson, D. G. (1994) "The Specification of Filtering Materials for Rapid Gravity Filtration", J. Inst. Water & Environmental Management,8,(5),527-533

Stevenson, D. G. (1997) "Flow and Filtration through Granular Media - Effect of Grain and Particle Size Dispersion", Water Research, **31**,310.

Symons,J.M. (1993) Letter, J.American Water Works Assn.,**85**,(3),4

Tambo,N. and Hosumi,H (1979) Physical Aspects of Flocculation II Contact Flocculation, Water Research,**13**,441-448

Tambo,N. and Watanabe,Y. (1979), "Physical Aspects of the Flocculation Process.I Fundamental Treatise",Water Research **13**, 429.

Tien, C. (1989) Granular Filtration of Aerosols and Hydrosols, Butterworth, London

Tobiason, J.E. and Vigneswaran, B.(1994)"Evaluation of a Modified Model for Deep Bed Filtration", Water Research. , **28**, (2) 335-342

Yao,K.M., Habelian,M.T. and O'Melia,C.R. (1971) "Water and Waste Water Filtration Concepts and Applications", Environmental Sci. and Tech.,**5**,(11),1105

273

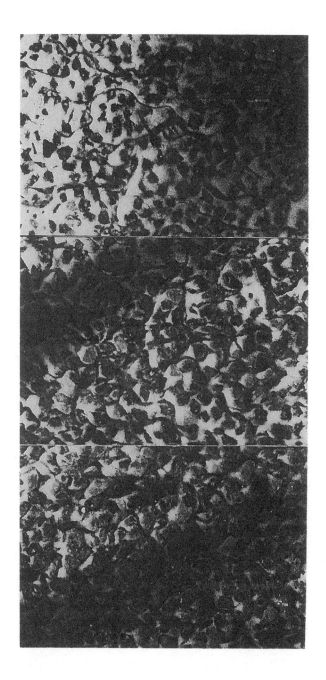

Fig. 17.19 Photographs of floc accumulating within a sand bed during filtration, showing the formation of snow caps and wormholes.

CHAPTER 18.

PROCESS DESIGN

18.1 Introduction

In spite of the mass of literature on the subject, it will be apparent from the previous chapter that granular filtration has not been defined by rigorous mathematical equations in the same way as hydraulic calculations for example. The size and size range of the suspended solids varies with the location and also depend on the nature of the pretreatment used. The chemical composition and concentration of dissolved species affect the probability of capture and also coagulants and coagulant aids will modify the solids. Thus in contrast to hydraulic calculations where one relies on the defined density and viscosity of the fluid the input data for filtration is intrinsically variable. The designer therefore is left with qualitative guidelines, some data on standard practice and sometimes the option of pilot plant trials. There are often constraints of tradition to be overcome and as Cleasby (1990) comments the regulators may not approve of any 'adventurous' designs.

Firstly, it is necessary to define the duty in terms of feed quality and filtrate quality. An acceptable run time and headloss must also be included with the objectives. Other parameters such as media size, size range and the choice of mono or multi media, bed depth and filtration rate follow from the duty requirement. Having fixed the process parameters, other design matters such as backwash rates and the selection of an appropriate floor have to be considered. If quality requirements are very stringent, then the method of operation may need special attention. It is implicit that the quality of the water fed to the filter must be suitable for such a treatment. If the turbidity and suspended solids are excessive, pre-treatment will be necessary as discussed in Chapter 19.

It is generally accepted that the feed to a conventional fine media polishing filter should not exceed about 3 NTU although this figure is based upon 'standard' designs. 3 NTU is a figure that can be guaranteed with most pre-treatment processes. For direct filtration ie. without pre-treatment, the limits are more open ended. An AWWA committee reported by Cleasby (1990) published 'Perfect Candidate' limits of 40° Hazen colour, 5 NTU turbidity, 20 cells/ml algae, 0.3 mg/l iron, 0.05 mg/l manganese. However, there are several examples where these limits are grossly exceeded. In one particular case at Wahnbach in Germany, (Bernhardt and Schell, 1982), turbidities of up to 80 NTU, and algae up to 100,000 cells/ml were treated satisfactorily on a filter designed for such a duty. Much will depend upon the coagulant dose because hydroxide floc is voluminous and occupies far more space than clay particles. The Wahnbach example above does not constitute a limit but provides encouragement for further exploration.

Filtration is generally the final point of removal of suspended solids and hence the filtrate quality may be the subject of legal regulations. For potable applications 95 percentiles and absolute maxima may apply eg. 1 NTU and 5 NTU, more usually 0.4 NTU. The AWWA goal is 0.1 NTU.

0.2 NTU which corresponds to about 0.5 mg/l of suspended solids still corresponds to about 25 kg of suspended solids per day entering a 50 Mld supply. Increasingly, attention is being turned to particle counting as a means of assessing filter performance because low levels of turbidity are difficult to measure. Even at 0.1 NTU there may still be very high counts of detectable particles. The incidence of *Giardia lamblia* and *Cryptosporidium* has provided a further incentive to scrutinize residual particles in filtrates. 0.5 mg/l corresponds to 78,000 5µm particles per cubic centimetre. The figure rises to 1,000,000 at 1µm.

Industrial requirements vary widely, some being more stringent than potable water standards, whereas in other cases, only roughing filtration is needed.

Filter run times have traditionally been designed for multiples of 24 hours because of manual operation. However, many polishing filters have been capable of 96 hour runs. This would indicate a fair degree of over design with an automated plant. There is normally an incentive to achieve cost optimised designs and if, as is common, wash water is recycled the consumption of wash water is immaterial providing that the recovery plant is designed as part of the system. It is possible to wash a pressure filter and return it to service in 5 minutes, so that the outage time is only 8.3% with a 1 hour filtration cycle. Such a short cycle would greatly increase the solids loading that could be applied to the filter. Alternatively higher rates could be handled.

Maximum headlosses in gravity ie. open tank filtration are determined partly by the structure. A total headloss through the bed of 3 metres or more is common with pressure filters, but 2 metres is more usual with open gravity filters. Some continuously washed designs have a maximum headloss of only 0.3 metres. There are certain losses that have to be accepted in pipework and valves if these are to be kept to a reasonable size and cost. There is little point in keeping the filtration headloss much smaller than other losses. It is also claimed that running a filter to a high headloss makes backwashing more difficult. Whether this is true or not is debatable as there are other variables to consider. It is possible that backwash problems are caused by other factors that also lead to high headlosses such as overrunning of pressure filters and excessive use of polyelectrolyte or sand that is too fine.

18.2 Efficiency of Filtration

Filters used for potable water treatment with a clarification stege are expected to remove about 80-90% of the incoming turbidity. A settlement stage would be expected to produce a turbidity of 2.0-2.5 NTU or better, while a filtrate of 0.4 NTU or better (eg. a target of 0.1NTU) will usually be demanded. On the other hand filters in a direct filtration situation may have to produce the same 0.4 NTU from a feed of 10 to 20 NTU so that 98% removal or better is necessary. It may be better to accept a somewhat higher cost and separate the roughing or primary role from the polishing one. The chemical conditions can be adjusted between the two stages and the first stage with a shallower bed washed more frequently, while the second stage matures. A recent emerging trend is a change from turbidity as a quality standard to the onset of breakthrough of micron (μm) size particles, which can occur well before normal turbidity limits are reached. A granular media filter is therefore not a standard device but a system that should be adapted to meet specific needs. Unfortunately many users do not appreciate this and expect an off-the-shelf solution which must then either be over conservative or carry risks that could be avoided with a little experimentation. There are some good examples where experimentation has paid handsomely.

In some cases only a low efficiency is possible because of the presence of considerable amounts of unfilterable material. This is the case often with primary filtration upstream of slow sand filters and tertiary filtration of sewage where no pretreatment is applied.

Unfortunately with the variable nature of the water to be filtered and of the solids to be removed quantitative mathematical relationships cannot be established and pilot plants trials are generally needed if the feed or the design is to depart significantly from previous experience.

18.3 Bed Depth

The previous chapter refers to the exponential or logarithmic decay of particle concentration with depth into a clean filter bed. In radioactive decay the term half life has been used to portray a more readily appreciated picture of the process. In filtration one can equally describe the decay of concentration in terms of the halving depth. This depth is $0.69/\lambda$ where λ is the filtration coefficient. In two halving depths the concentration falls to one quarter and so on. Stevenson (1997) has shown that in practical situations with a range of sizes the summation of individual exponential curves tends to be more logarithmic. The decay with depth is then less than would be expected from an exponential profile. In both models however a better quality can be achieved with a greater depth.

In practice the residual turbidity or suspended solids concentration seldom decays away to zero. There is usually an unfilterable residue, usually of very small particles which penetrate almost to an infinite depth. Hence there is a point at which additional depth is of little benefit. It may be necessary to recoagulate these residues and increase their effective size. On the other hand very much finer media can achieve the same end in much shallower beds. Curiously many designers and experimenters seem to consider depth to be a sacrosanct constant.

The depth necessary to obtain a satisfactory quality in a clean bed is usually called a penetration front (Fig. 18.1). Behind this is as previously discussed, is the clogged or wormhole section of the bed. A **deeper bed** will allow a **greater** holding volume behind the front and hence a **longer run** length or **higher rates**. However, the headloss gradient in the clogged section of the bed is fairly high and furthermore the deposit may not be very stable. Fixed rules such as offered by Kawamurs (1975), such as a depth of 1000 times the effective size can be misleading and do not take account of differening depths for the penetration front and clogged zone, which will vary with the water and method of operation. Avalanches can occur, particularly if the flow changes. The actual depth required depends upon the grain size, as discussed below, and in theory the same quality should be obtainable with an appropriate depth of any size of media. In practice the depth of a coarse media may be too great to be viable.

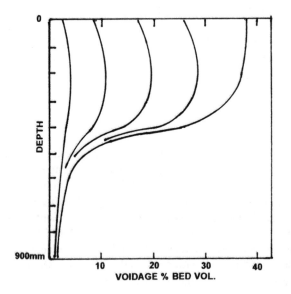

Fig. 18.1 Penetration front in a granular bed filter. (Plotted from data generated from the same run as Fig. 17.11, at 12 hours into the run. The five curves represent the cumulative volume occupied by the 5 size fractions, the smallest being on the left. Note that the smallest fraction constitutes the major component of the deposit at the leading edge.)

Also it appears that coarser media, particularly without polyelectrolyte may be more prone to avalanche and produce an intermittent breakthrough, particularly when the flow changes.

Cleasby (1990) has provided a summary of American practice. Tien (1989) quotes the 'Ten States Standards' that governed design in the USA at one time. Such regulations inhibit advances in the technology. Some typical depths are given in Table 18.1. However in specific cases where pilot studies have been run very different combinations have been selected, such as Los Angeles Aqueduct where 1.8m of 1.5mm effective size anthracite was used on its own.

Filters intended to handle **high** loads of solids must be **deep** in order to provide the necessary volume to hold the solids, unless very short runs are acceptable. There is no theoretical limit to the depth and one with, say, 5m depth would not present a design problem. Sand especially is cheap. The limits depend on the economic cost of structures, but a deep bed usually implies a smaller floor cost, which with the associated washing equipment is usually the main component.

18.4   Media Size

There is a divergence of view on the effect of media size, ranging from a cubic relationship (Stein, see the review by Ives and Scholji, 1985) to something very small, "the effect is overstated", (Omelia and Ali, 1978). Some data presented by Tien (1989) suggests that the relationship lies between a square and cube relationship between size and filtration coefficient. The formation of chains and dendrites discussed by Tien will probably affect the behaviour. Clark, Lawler and Cushing (1992) working with polymer dosing in direct filtration found what they described as a 'marked effect' of grain size with virtually no filtration occurring on grains of 3.68mm diameter. Mohanka (1971) has published some clean bed filtration coefficients based on ferric hydroxide in London tap water. The data can be described by the equation:-

$$\lambda_o = 10.3D^{-1.5}$$

This was at a filtration rate of 5m/h. The constant was 7.0 at 10-20m/h. D is in mm. Tien also provided information on filtration coefficients.

Stevenson's model (1997) based on particles covering a range of sizes, discussed in the previous chapter showed that decay was more logarithmic than exponential and varied with the size range. For 7-10:1 the exponent was about -1, more in line with Mohanka's observations.

The designer must ultimately fall back of experimental measurements on the actual suspension to be filtered. The headloss both in the clean and clogged state will be very much lower with coarse media if they provide a satisfactory filtrate. However fine media provides better security. Stevenson (1996) found that very efficient removals of 2μm particles of china clay could be achieved from hard river water without any chemical treatment using 100mm of sand with a hydraulic size of 150μm. The headloss was of course high by normal standards. Thus there is a continuum between granular media filtration and precoat filtration in which the media (usually diatomaceous earth with a high voidage) is applied to a tubular support in depths of only a few millimetres, and is discarded when clogged.

Typical sands used in Great Britain are listed in Table 18.1.

Table 18.1
Typical mono-media filter bed depths

Media	Depth
0.5-1.0mm	600mm
0.6-1.18mm	600-700mm
0.85-1.0mm	900mm
1.0-2.0mm	1200mm
1.18-2.36mm	1500-2000mm

Anthracite and to a lesser extent 'garnet' are also used. (Garnet is a name given to some five minerals of various densities and one needs to be specific. The same is true of anthracite which can vary from a specific gravity (sg) of 1.3 to 2.2. British material has an sg of 1.4.)

Water utilities, consultants and specialist contractors have proprietary data from past trials and experience on which they call for plant designs.

18.5 Filtration Rates

The filtration rate has a direct effect on the size of a filter. As a first approximation, the headloss in a clogged filter is a direct function of the adhesion shear stress.

$$\frac{\Delta H}{h} = \frac{6R'(1-e_o)}{g\rho eD}$$

where $\Delta H/h$ is the hydraulic gradient, R' the adhesion shear stress, ρ the density of the fluid, D the grain diameter and e and e_o voidages as already discussed.

In this equation, already referred to in Chapter 17, e in the wormhole region equals e_o. Thus the gradient is directly related to R' which will be a characteristic of the particles and their chemical condition. Hence it is possible to nominate some loads per square metre per run. Such figures depend upon the ratio of turbidity to coagulant dose because clay particles, as already mentioned, are far denser than hydroxide floc. In fact for many waters the loading can be quoted in terms of hydroxide only.

Likewise, dense particles such as mill scale can produce very high loads by comparison. A typical loading of say 0.5 mg/l aluminium (1.4 mg/l aluminium hydroxide) might produce a 36 hour filter run at 6 m/h on a mono media filter. The load would then be 300 g/m^2 per run. Suspended turbidity could raise this to nearly 2 kg/m^2 without affecting the filter run. Cleasby (1990) quotes a range of 550 g to 5.5kg/m^2, 1600 g/m^2 being a more common average. Algae, being mainly water, will tend to reduce these figures which are based upon dry weight.

Loading data must be used with caution. If a more concentrated suspension is applied, the runs will be shorter but also a higher percentage efficiency will be necessary to maintain the same filtrate quality, thus the bed should be deeper. However the particles sizes present in more turbid watersare likely to be larger so that the efficiency may be higher. If a particular suspension filters less efficiently at a higher rate as may be the case if diffusion plays a part in the mechanism, a deeper bed will be necessary. However, polyelectrolytes are often used to strengthen the floc, in which case, a deeper bed can be avoided. On the other hand, the higher headloss produced by the stronger floc would indicate the need for a coarser media or a move to dual media.

Thus the picture that emerges from the USA, where some consensus has been reached, is that the traditional filtration rates of 5 m/h established by Fuller in the last century can be safely raised to 10 m/h and with the aid of polyelectrolyte 15 metres per hour can be safely reached. Specific examples which have been developed and approved using pilot scale tests, have successfully achieved rates of 24-33 m/h. However, where no experimental work is possible and quality is paramount, many insist on rates not exceeding 6 metres per hour and accept the cost penalty. The move towards higher rates of filtration with coarser media has been encouraged in the USA by the publication of an AWWA Research Foundation Report (Cleasby, Sindt, Watson and Baumann, 1992), where the use of polyelectrolyte is reported as being virtually essential. In this work high rates are defined as greater than 10m/hr.

In Britain a reluctance to exploit polyelectrolyte has limited progress in this direction and some utilities limit their filters to 6m/h (based on 700mm of 0.5-1.0mm sand. The American study was preceded by a survey of high rate plants, Cleasby, Dharmarajah, Sindt and Baumann, 1989), using the same definition of high rate and restricting the list to plants producing less than 0.2NTU Turbidity. 21 plants are listed. Interestingly one of the conclusions was that the operators must be motivated to achieve a good quality at high rate and the chemical treatment which requires good supervision must be kept under control. The importance of rapid mixing is again highlighted. However this study for the most part considered plants which used filtration after settlement and remarks about flocculation were primarily aimed at the optimisation of the latter stage. Dual or triple media were used in all but one (Los Angeles Aqueduct, which uses deep anthracite). Particular attention was paid to operating strategies such as running the first filtrate to waste, slow start and slow rates of change which are discussed in Chapter 23. The majority of the plants reviewed operated at around 10m/hr, with a few up to 15m/hr.

18.6 Dual Media Beds

The use of anthracite with sand in granular media filters goes back at least half a century. The lower submerged density of the anthracite enables coarser grains to remain in a stable manner on top of the finer grains of sand. This is discussed further in Chapter 20. It follows from the Kozeny equation and other equations presented in Chapter 17, that a greater quantity of floc can be accommodated within anthracite and other coarse media for a given headloss. An additional benefit of anthracite is the greater clean bed voidage of the crushed material which provides more accommodation for the dirt. The finer sand is able to produce the same quality and resistance to disturbance as before and at first sight, a greater capacity should be achievable for same headloss.

Unfortunately, dual media sand filters are often constructed to the same total depth as conventional sand filters and this can interfere with the filtrate quality, especially if taken to the same headloss which tradition requires. Ideally the sand depth should be sufficient to guarantee the quality at the end of the run while the anthracite provides a good storage capacity at relatively low headloss. The safest course is to add the anthracite to the existing sand. Existing structures may be unable to accommodate the extra depth but modifications are often possible and the supplier of the filter might be consulted with benefit.

The lower headloss provided by dual media enables filtration rates to be raised without reducing filtration runs. Many conversions and new plants have been built on such lines but often margins of safety have been eroded. This is an area where computer models, even if only approximate, can provide a useful tool to explore combinations of the two media, and reduce experimental work.

On the contrary side, anthracite is somewhat more expensive than sand. There will be a problem if combined air and water washing is proposed because anthracite is easily lost. The solution most frequently used is the rising wash technique discussed in Chapter 20 which wastes more water.

Several papers on the optimisation of dual media filters have been published. They are unfortunately specific to given circumstances and the definition of optimisation also varies. Some define it as the achievement of the maximum volume per run rather than minimum cost.

Common examples of dual media filters use 150 mm of 1.6 mm hydraulic size anthracite over 550 mm 0.8 mm hydraulic size sand. 300 mm of anthracite provides more capacity but 400 mm of sand underneath is probably marginal if a good filtrate quality is to be maintained.

18.7 Triple Media Beds

In theory, many more layers can be added. For example, Mohanka (1971) and Fox and Metcalf (1974) explored five layer filters. Several papers (reviewed by Cleasby 1990) suggest that the filtrate quality is not significantly better. The clean bed headloss is greater and as a result the runs are shorter. Cleasby notes that in most cases, such filters have been used to achieve higher rates and controlled comparisons are not very plentiful.

It has been claimed by one utility that the triple media filters with fine garnet provides better protection against penetration by aquatic organisms but offers little in the way of lower turbidity or particle penetration. A properly designed dual media filter is probably able to remove most filterable materials leaving nothing for the fine garnet.

It is easy for researchers to construct multimedia beds using carefully sieved materials of close size range. It is not economical to use very closely sieved materials on the full scale. One supplier of triple anthracite/sand/garnet filters has claimed that mixing is beneficial. However measurements of hydraulic size through the bed have revealed that this parameter does not necessarily fall with increasing depth because the larger sand particles can fall down into the garnet layer and fine garnet grains rise into the sand. To maintain segregation the ratio of the submerged densities should be at least as large as the square of the size range. Other dense materials which have potential for this application are ilmenite, magnetite and possibly pyrolusite. Powdered metals may also find application in special circumstances where cost is secondary.

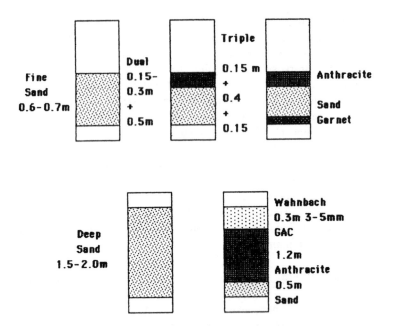

Fig. 18.2 Examples of filter media configurations.

When selecting a triple bed filter it is wise to construct a specimen bed, and to sample and analyze backwash and check the hydraulic size with depth. It is possible not to achieve the desired gradation from coarse to fine, and to end up with a flat profile with coarse sand sinking well into the garnet layer.

Another triple media system that has not achieved the recognition that it deserves is the Wahnbach design mentioned earlier (Bernhardt and Schell, 1982), which combines quality with a high flow rate. This uses conventional sand, high density anthracite and a low density granular carbon as a media, not as an absorbent. The configuration is shown in Fig. 18.2. The coarse carbon layer provides an even lower headloss layer on top of the anthracite and discourages surface filtration during algal blooms. An overall depth of 2.0 m provides adequate holding capacity and quality. This particular plant processes up to 432 Mld (432,000 m^3/d).

The Wahnbach configuration is or course compatible with the garnet based triple media filter and a quadruple bed would be viable. It provides scope for high capacity and efficiency.

It should also be noted that the gravel under bed in upflow filters provides a similar multi media configuration. Unfortunately the coarse gravel is not easily cleaned and tends to remain clogged.

18.8 Optimisation

Filter designs can be optimised from several viewpoints depending upon the intentions of the designer. There is no one optimum design. Firstly, it is necessary to define the objective in terms of acceptable quality, available headloss, cycle time etc. Too many past attempts have made traditional assumptions. The optimisation must take the entire plant into consideration. Minimum overall cost will usually be the motivation behind any exercise of this nature, although with existing plants the ability to uprate without acquiring land or even building new structures is often attractive or even essential.

A filter run will terminate on headloss (ie, the flow can no longer be maintained), breakthrough or turbidity (or any other relevant parameter such as particle count) or time if it is a manual plant and cannot be allowed to continue to the next day. Most operators would wish to see the run terminate on head or time leaving a margin of safety for breakthrough but if there is a gross mismatch then the design cannot be the most efficient and cost effective. The plant must of course be able to handle the worst quality likely to arrive, and particularly in designing direct filtration plants historical data must be consulted. Unfortunately, such data is seldom collected with treatment in mind, so that little information on doses of coagulant for example, may be available. Many sampling regimes are planned for the convenience of staff in demonstrating compliance with water quality objectives rather than to produce a full picture of the variability of the source. Bad weather may deter sampling just at the time that the source reaches a significant peak. The use of a clarification stage upstream of a filter greatly simplifies filter design. Such stages are more tolerant of changes in raw water quality and produce a consistent product for filtration.

Filters in a set may require washing more frequently than is possible, eg. 6 filters may take 6 hours to drain down, wash and refill. The processing of used wash water may prevent the next filter from being washed.

The chart on the next page indicates corrective actions to move the design towards an optimum situation.

Fault: Filtration Runs Terminated by Headloss too Quickly.

Action:
a)   Reduce filtration rate
b)   Use a coarser deeper bed
c)   Use dual media
d)   Add a pre-treatment stage
e)   If polyelectrolyte is being used to reduce the dose or chlorinate
f)   Improve the wash water treatment system

Fault: Early Breakthrough

Floc may penetrate the filters while the headloss is still low.

Action:
a)   Check chemical treatment
b)   Increase bed depth
c)   Use finer media
d)   With dual media, reduce the anthracite and increase the sand depth
e)   Check the absence of `hunting' (frequent over adjustments to the flow) in the control system
f)   Use a polyelectrolyte
g)   Introduce or adjust the slow start system, or consider declining rate filtration
h)   Accept the situation and wash more frequently.

The most radical move towards a cost optimised design involves a break with tradition and complete scrutinisation of plant design. Cost information on the various sections of a plant will of course be required as well as data on the effect of filtration rate on run length and filtrate quality. The following is an example:-

Consider a 50 Mld plant (2083 m^3/h). Experimental data from a real source indicates that filter runs vary with filtrate rate as follows (with a fixed media size):-

Rate	8m/h	Run	24 h	Depth 0.6m	
	12		14		0.9
	16		9		1.2
	20		6.5		1.5

In each case the run terminated on headloss.

The data indicate that the quality within these limits was not greatly affected by the depth except that at 20 m/h, the deepest bed (1.5 m.) gave the best assurance. Thus for consistency it was assumed that depth should be as indicated ie, giving a constant sand inventory. It is assumed that 30 minutes is required to wash a filter and the wash volume for design purposes is taken as 4 bed volumes. Six filters were selected in this exercise to avoid excessive flow changes when a filter is withdrawn for washing. Washwater is recycled and returned to the head of the works after settlement. The total filter area is indicated in the following table.

Rate		Recycle		Net Yield/cycle m³	Filter area
8m/h		1.2%		189.8	263m²
12		2.1		139.2	135
20		4.6		124.0	100

The volume of water used in a given wash is the same in all cases because the sand inventory is the same. In this particular example, the size of the washwater holding/balancing tank remains the same (4 bed vols.), but the turnround time is greatly reduced at the high rates. Even at 6 hours however, the settlement rate demanded is not excessive.

Undoubtedly, the 20 m/h option would be the cheapest in this situation. Although the media inventory remains constant, the floor and upwash equipment, which are the most expensive components are considerably reduced. Operating costs also remain fairly constant if the limiting headloss is unchanged. The backwash power consumption is small by comparison because the pumps and blowers run for such a short time compared with the continuous duty of low lift pumps. The peak power demand is also reduced because of the smaller size of the filters whereas the upwash rate per square metre remains the same.

If washwater is not recovered, then the approach will be entirely different and the cost of water lost will have to be taken into account. This can vary widely and over the life of a plant it may well be that a low rate, low wastage, plant will be the more competitive choice. A typical amortisation period for concrete works is 50 years. A 50 Mld plant will produce about 1×10^9 cubic metres in this period. Every 1% reduction in wastage would generate £100,000 for each 1p that the water costs.

Few plants have been designed by optimisation procedures although the high rates being employed in some recent plants in the USA suggest that this philosophy has been pursued over there. The above exercise can be extended to different sizes of media, dual media etc. In practice the traditional 24 hour minimum wash cycle is a common limitation in plant optimisation, the excuse being that a margin of safety must still be applied. However, such margins can still be incorporated in plant designed for 12 or 6 hour cycles. Wash water recovery plants are cheaper than filters and the recycle flow should be no deterrent.

Some authors have considered optimisation in terms of maximising the run or the yield from the filter in terms of cubic metres per square metre or the yield per cubic metre of washwater used. An engineer or an accountant will mainly be interested in minimising the cost for given quality standards and reliability.

Optimisation and the filter configuration must take account of the chemical treatment conditions. The best design without polyelectrolyte will not be the same as that with polyelectrolyte and certainly in the United Kingdom at the present time, there is an ambivalent attitude toward these materials.

18.9 Pilot Plant Experimentation

Where ever time, finance and facilities exist an experimental assessment of the configurations for granular media filtration is the most reliable approach towards achieving an optimum design. Data from one environment can of course be transferred to similar situations and this is how the majority of experienced designers proceed.

A pilot filter must be provided with a representative feed including any pre-treatment that may be anticipated. A pilot filter may be quite small down to a diameter 50 times the largest grain size likely to be used. Thus 150 mm is a typical size of column for most work and for fine sand 100 mm is adequate. Transparent columns are preferable, particularly for ensuring that washing is satisfactory. The depth of the column should exceed the maximum bed depth likely to be used, plus a reasonable margin for expansion during washing and also, with open gravity columns, a sufficient height must be provided for the driving headloss (eg, 2 metres).

The simplest way of ensuring that the media remains covered in an open gravity column is to provide a syphon break in the outlet (Fig. 18.3). This should be level or slightly above the surface of the media in its working condition. As the headloss through the filter increases, so the level in the column will rise. The incoming water should be fed into the column at a low level, preferably down a descending pipe and not allowed to cascade from a height otherwise the character of the floc may change (the latter is common practice but is nevertheless undesirable). Alternatively the column may be closed and operated under a slight pressure.

The flow must be controlled and measured. It is best if flow meters are inserted in the filtrate line so that they are less likely to block. Where chemicals are added to the feed, an orifice should be incorporated immediately downstream to provide mixing at a representative headloss, say 300mm.

A small column need only have a mesh floor or a mesh with some support gravel. Alternatively a proprietary filter nozzle may be used. A mesh or support gravel provides a more distinct bottom boundary to the working media. A backwash supply is required and possibly an air supply if air scouring is to be explored. A single centre air injection point will suffice on small columns. Backwash conditions on small columns intended for process evaluation differ from those in full size filters. The narrower column tends to inhibit convection patterns and higher rates tend

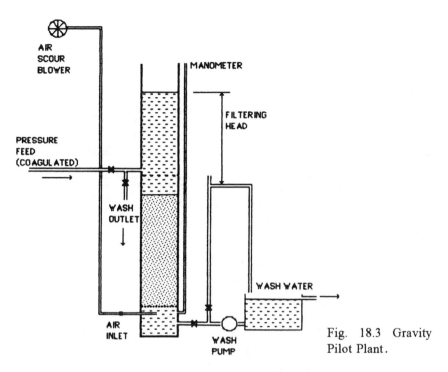

Fig. 18.3 Gravity Pilot Plant.

to be required. Also floc has to be flushed upwards rather than sideways as in most full size British and European filters. Most experimental filters are equipped with multiple manometer tubes but these are of questionable value. One can usually see where the deposit is. The depth needed to ensure adequate filtration filtrate quality is considerably greater than that over which the deposit is visible. A basic filter only needs a single pair of manometers to measure the overall headloss.

Often it is difficult to dose chemicals to a small flow and some pre-treatment processes such as sludge blanket settlement tanks cannot easily be scaled down. In such cases, a large flow must be taken and treated upstream and the majority discarded in a weir splitter box with only a small proportion being taken to the experimental filter.

Maintaining a constant flow while the filter clogs can be a problem. Many allow the feed to cascade into a rising level filter but as mentioned already, this can cause floc breakage and give misleading results. With a pressure filter, the easiest arrangement is to use a fairly high pressure source. For instance, 5-10 times the maximum headloss of the filter and to throttle the flow with an appropriate metal seated valve before adding any chemicals. A solution for open columns is a float

290

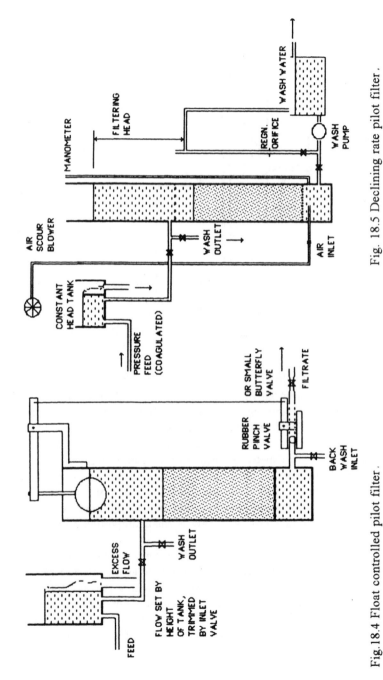

Fig. 18.5 Declining rate pilot filter.

Fig.18.4 Float controlled pilot filter.

operated mechanism which shuts off the outlet as the level falls, an arrangement commonly used in full sized plants. A domestic ball valve float can be located in a 150 millimetre diameter column and linked to a rubber pinch valve (Fig. 18.4).

Manual correction of flow inevitably causes step changes that interfere with the performance and towards the end of a filter run these can produce turbidity spikes. **It is strongly recommended** that the step changes that occur when other filters are taken out of service for washing should be simulated in the pilot plant runs. The turbidity spikes that occur in practice will otherwise be missed and the results may therefore be misleading. Unintentional changes should however be avoided.

If a pilot filter is required to simulate declining rate operation, a constant level header tank with an overflow will be needed (Fig. 18.5). The filter can then be run under constant head and the flow will decline as the filter clogs. The actual height of the head tank above the syphon break outlet and the setting of the throttling valve orifice can be calculated from the following equations.

Let the flow range from 1+R to 1-R where R is the ratio of decline (eg. if R=0.2 the flow will range from 120% to 80% about the design flow).

In the clean condition $\Delta H_{tot} = \Delta H_m(1+R) + \Delta H_t(1+R)^2$

in the dirty condition $\Delta H_{tot} = (\Delta H_m + \Delta H_c)(1-R) + \Delta H_t(1-R)^2$

R, ΔH_m and ΔH_c, the media and clogging headlosses,are known and hence the throttling loss ΔH_t, which must be added, can be calculated. The above headlosses are those for the design or average flow rate.

Instrumentation used in a pilot filter will vary according to the resources of the sponsor. Manual recording is possible but tedious. The first priority must be a recording turbidimeter. Flow may be known and should not vary rapidly. Headloss normally increases steadily and manual reading is possible whereas turbidity can change rapidly. Coagulant residuals will often be of concern and regular sampling may be required. It is possible to automate and add instrumentation up to or exceeding the standard of a full production works.

If any chemical treatment is likely to be involved eg, with direct filtration or even after pre-treatment if a polyelectrolyte is to be added, it is best to start with shallower bed depths and higher rates than ultimately expected in order to produce measurable residues in the filtrate. The chemical conditions can then be more easily optimised. A very low or zero residue could indicate an excessive bed depth. If a deep bed is used the residue may not be sensitive to the chemical conditions.

There is little point in continuing filtration runs to the end point during this phase. A series of short runs long enough to establish the initial quality and headloss gradient, will enable optimum chemical conditions to be established more quickly. Polyelectrolytes tend to condition the media and the benefit of a high dose may be retained for a period after it has been stopped. Doses should be explored in increasing steps rather than decreasing. Strong hypochlorite will destroy the effect with many polyelectrolytes.

Having established the chemical treatment with a bed shallow enough to leave a finite residue the normal course will be to explore a coarse matrix of bed depths and rates, possibly media size and dual media according to the objective.

Ives (1986) has introduced the concept of a filterability number:-

$$F_L = \frac{\Delta H_L C}{C_o u t}$$

Where ΔH_L= increase in headloss in time t. C and C_o are filtrate and feed concentrations or turbidities and u is the filtration velocity.

The apparatus described by Ives is small and only uses 1 litre of sample. It therefore serves the same purpose as a jar test. It is not intended to assist the selection of the media configuration but only the best chemical treatment.

An other rough method of assessing the correct chemical treatment is the standard Jar Test described by Stevenson (1986). However with filtration one is not looking for settleable floc. The flocculation time is usually shortened and the product filtered through a No. 1 Whatman filter paper rather than being allowed to settle.

Experimental studies with pilot plants are expensive but on large schemes they can provide substantial savings if the design can be matched to the duty and large margins of conservatism reduced, possibly allowing stages to be eliminated. On the other hand for small schemes standard designs with in-built safety factors will be more appropriate. This is where previous experience of similar situations is invaluable.

18.10 Turbidity and Quality Assessment

The performance of filters tends to be assessed primarily on turbidity removal and also to some extent on the residual coagulant. These are pragmatic measurements that are easily made with relatively low cost instruments. The turbidity of a suspension is however very much dependant on the particle size. Also traces of macromolecules such as polysaccharides that may be present in the raw water and which are difficult to precipitate will scatter light. Turbidity may be measured by absorption of light (eg the Jackson Candle, JTU) or by light scattering (Nephelometry, NTU). Although both methods may be calibrated by the same standard suspension of formazin the results produced different methods on a given sample may differ.

Nephelometry is essential for filtrates where the absorption is low. It is known that the particle size distribution of material penetrating granular media filters changes either by flocculation or filtration followed by detachment. Larger particles scatter light less and hence it is possible to *reduce the turbidity without actually removing any solids*.

Turbidity measurement has limits and increasingly particle counting is being employed for high quality filtrates. This assesses both the numbers and the size of the particles penetrating the bed. Even this technique is limited to particles above 1-2 μm. This is sufficient to identify the presence of particles of a size similar to organisms such as *Cryptosporidium* but useless for colloidal residues.

Particle counting can be misleading because the mass of the particles varies with the cube of the size. For example 8 times as many 2μm particles have the same mass as a given number of 4μm particles. Breakthrough curves based on numbers give a very different picture compared with the particle volumes involved (Fig.18.6). The numbers involved with submicron particles probably make counting impracticable in this range.

It has to be remembered that some unfilterable material (ie. filter penetrating material, is usually present in most practical situations. An infinitely deep bed is not likely to yield a zero residue. The system requires further study. These may be uncoagulated particles present in the raw water, colloidal fragments broken off from the deposit within a filter, post precipitation and residual soluble material. Aluminium and iron coagulants often leave unprecipitated, possibly complexed, traces. The chemical conditions must be optimised and as discussed in Chapter 19 it may not be possible to optimise for all the species present in one single stage of treatment. The unfilterable fraction must be deducted from the model if any attempt is made to simulate the process. A common quote is that "it is only possible to filter that which is filterable". The chemical conditions are as important as the rest of the design. Not infrequently designers and contractors are required to guarantee the performance of plants in situations where they have no control over the raw water and where no pretreatment is available or expected.

18.11 References

Bernhardt, H. and Schell, H. (1982) "Energy Input Controlled Direct Filtration to Control Eutrophication", J. American Water Works Assn. **74,** 261

Clark, S.C., Lawler, D.F. and Cushing, R.S.(1992) "Contact Filtration, Particle Size and Ripening," J. American Water Works Assn. **84,** (12), 61

Cleasby, J. L. (1990) Water Quality and Treatment, Ch.8., Filtration, American Water Works Assn. Ed. Pontius, F.W., McGraw Hill, New York.

Cleasby, J. L., Darmarajah, A.H., Sindt, G.L. and Baumann,E.R., (1989), Design and Operation Guidelines for Optimisation of the High Rate Filtration Process: Plant Survey Results, American Waterworks Research Foundation, Denver, CO., USA.

Cleasby, J. L., Sindt, G.L., Watson, D.A. and Baumann, E.R., (1992), Design and Operation Guidelines for Optimisation of the High Rate Filtration Process: Plant Demonstration Studies, American Waterworks Research Foundation, Denver, CO., USA.

Fox, G.T.J. and Metcalf, S.M. (1974), "Design of Multilayer Filters for Use in the Water Treatment Industries", Proc. Filtration Soc. (July/Aug), 383.

Ives, K.J. (1986) "Deep Bed Filters", Solid /Liquid Separation Equipment Scale Up, Ch.8., Ed. Purchas,D.B. and Wakeman,R.J., Uplands Press, London.

Ives, K.J. and Scholji, I. (1965) "Research on Variables Affecting Filtration".J. Sanitary Eng. Div. American Soc. Civil Engs.(Aug), 1.

Kawamura, S. (1975) "Design and Operation of High Rate Filters, Parts 1-3." J. American Waterworks Assn. **67**, (10), 535, (11), 653 and (12), 705

Mohanka, S.S. (1971) "Multi Layer Filter Design",Water and Water Eng., **4,**143

O'Melia, C.R. and Ali,W. (1978) "The role of retained particles in Deep Bed Filtration", Prog. Water Treatment, **10,** 167

Stevenson, D.G. (1986) Solid /Liquid Separation Equipment Scale Up, Ch.3, Ed. Purchas,D.B. and Wakeman,R.J., Uplands Press, London.

Stevenson, D.G. (1997) "Flow and Filtration through Granular Media - The Effect of grain and Particle Size Dispersion", Water Research, **31**, 310.

Tien, C. (1989) Granular Filtration of Aerosols and Hydrosols, Butterworth, London.

Further Reading.

Water Quality and Treatment (4th Ed.) Ed. F.W. Pontius, 1990, American Water Works Association, Mc Graw Hill.

Water Treatment, Principles and Design, James Montgomery Consulting Engineers, (1985) Wiley.

CHAPTER 19.

CONDITIONING OF THE FEED SUSPENSION

19.1 Filter Loading

A granular media filter traps and holds suspended solids within the voids or pores between the grains. The process is generally a batchwise one and even with the so called continuous filters, each element of sand absorbs suspended solids and then undergoes cleaning cyclically. The pores within the media have a limited volume in the range of 33-50% of the bed volume which is very small compared with the volume being treated. Thus granular media filtration is primarily a process for low solids concentrations. The loading is a term describing the weight (or more strictly the volume) of solids that can be applied before backwashing is necessary. The designer is interested in the volume of water that can be treated before washing is necessary and hence the concentration of the suspended solids must be known.

As will be discussed later, typically three bed volumes of water are used to wash a filter. Thus if pores of 40% voidage are entirely filled with solids, which they never can be, the wash water will contain no more than 1/3 of 40% ie, 13% of solids by volume. In practice, there are channels through the deposit. These occupy part of the void volume. The 'front' over which the concentration of suspended particles falls from the initial to the final value contains voids that range from the clogged value to the clean bed figure. The average volume will be about half way between the two.

The lower part of the bed must remain almost clean otherwise breakthrough will occur. The volume of the solids in practice is unlikely to exceed more than about 20% of the bed volume, thus with a filtration rate of say 10 m/h and a run length of say not less than 24 hours on a manual plant some 240 m3 of water will be filtered per square metre of bed. It follows therefore that the volumetric concentration in the feed cannot exceed about 0.1% otherwise the wash consumption will become excessive. Granular media filtration is therefore primarily a polishing process and if the feed contains high concentrations of solids primary treatment will be required. Fortunately, settlement and flotation are complementary in that they are able to handle a wide range of suspended solids concentrations in a truly continuous manner and produce a consistent residue which falls in a range which is easily handled by filtration. Figures are sometimes quoted for the solids loading of filters. This is of course dependent on the density of the deposit.

19.2 Particle Filterability and Coagulation

A point made elsewhere which is not generally appreciated is that granular media filters only filter out suspended solids that are filterable by such means. This statement is not nonsense. Many suspensions are not retained within granular beds. Stable colloidal suspensions will pass straight through. An example would be diluted emulsion paint or milk. The surface chemistry must be altered first. It is therefore unreasonable to demand guarantees if the designer has no control over the incoming solids or the pretreatment process, if any exists. The coagulation pretreatment is as vital to the performance of a filter as the correct choice of grain size and depth, and backwash procedure, point stressed by Cleasby et al. (1992). Coagulation is a science in its own right and is covered by several specialist books eg. Bratby (1980), Amirtharajah, Clark and Trussell (1991), and Stevenson (1980).

Colloidal particles with diameters of only a few microns (μm) or less are stabilised by surface charges, usually negative, known as the zeta potential. Such charges reflect the predominantly acidic nature of surfaces in the environment. Organic matter tends to degrade to materials with carboxyl groups on the surface. Soluble organic species readily absorb on the surface and the less polar parts of the molecule tend to be oriented towards the solid surface. The hydrophilic polar groups pointing outwards cause mutual repulsion of the particles. Clays contain silica and silicates. Again any tendency to ionisation produces silicic acid groups at the surface. The most likely naturally occurring materials that differ from the above pattern will be oxides such as Fe_2O_3 and Al_2O_3. However, the omni-present traces of organic matter absorb onto such surfaces and render them negative.

Likewise practically all filter media is naturally negative. This is certainly true of sand, anthracite, activated carbon and many plastics. However, even if they were initially positive, the first mono-layer of solids would change the nature of that media.

In order to obtain effective filtration, it is necessary to overcome the repulsive negative charges sufficiently to enable the particles to break through the energy barrier and come under the influence of Van der Waals attraction. In theory it is possible to achieve such destabilisation by the addition of polyvalent ions, usually cations. The concentration of ions to achieve such destabilisation is governed by the Schultz Hardy rule which states that there is about a fifty fold reduction in the ionic concentration required for destabilisation for each additional charge on the destabilising ion. In practice no tri- or quadri-valent cations exist in solution under the generally prevailing conditions (pH and ionic composition) in water treatment. Calcium and magnesium certainly assists destabilisation, but are not usually present at a sufficient concentration to achieve coagulation on their own, although these ions are certainly beneficial.

Coagulation by inorganic salts (iron or aluminium) is usually the result of enmeshment by the voluminous mass of hydroxide floc (sweep coagulation). This floc is initially positively charged although it tends to revert to a negative condition after a few days, possibly hours in some cases. Thus the negative colloids are effectively precipitated by positive particles. In some cases, eg. coloured waters, the colour compounds are precipitated directly by the aluminium or iron salts (soap for example, would precipitate in this way). Indeed many organic acids have insoluble aluminium or iron salts.

Measurements and mathematical modelling both indicate that the particle size involved in capture within granular beds, lies in the range below 10-15 μm. Differences of opinion exist about the need for separate distinct (paddle) flocculation stage prior to filtration (flocculation here is defined as slow speed stirring to induce particle collision and the growth of aggregates). Flocculation prior to filtration has been fairly common in the USA, but not elsewhere. Certainly the equations presented in Chapter 17 would indicate that flocculation is unnecessary but may be a factor that influences the precise value of the filtration co-efficient. With the correct size and depth of media any thing can be filtered. Chowdhury, Amy and Balls (1993) have concluded that submicron particles must be incorporated into floc at the pretreatment stage if they are to be removed effectively in filtration. Particle size data seen by the author suggests that considerable numbers of micron size and submicron particles penetrate filters. O'Melia and Ali (1978) suggest that very small particles lead to high headlosses in a filter and that flocculation lowers this as well as improving the filtration coefficient. On this basis it would appear that the absence of flocculation can be offset by the correct media configuration but coagulation as distinct from flocculation is exceedingly important.

19.3 Flocculation

The theory and practice of flocculation is discussed in detail in Chapter 4. It seems likely that for filtration contact time rather than energy expenditure is more important especially at low temperatures where the rates of hydrolysis are low and the attainment of low soluble residues of coagulants may require more time than is available in the volume above a filter bed.

Overall the extensive use of filtration without prior mechanical flocculation confirms that the latter has little relevance to granular media filtration, although Bernhardt and Schell (1982) stress the importance of energy input in the case of the Wahnbach design. On the other hand, the contact time provided by a flocculation system can often be beneficial in allowing the precipitation reaction, destabilisation and perikinetic processes to go to completion. Indeed, vigorous mixing such as used at Wahnbach in Germany provides a means of controlling floc size and hence the depth of penetration. More vigorous shearing may allow deeper penetration and a longer filter run in specific circumstance and where the bed depth is chosen accordingly.

19.4 Polymer Destabilisation

Water soluble polyelectrolyte polymers used as coagulants act as a surface active species that absorb onto the surfaces of particles and media. They impart a positive charge neutralising the existing negatives charges. As a result, particles that collide are able to adhere rather than bouncing off each other. In this case, overdosing can cause a complete changeover to a positive charge with consequent re-stabilisation. Such polymers consists of high molecular weight chains with attached ionic groups which bind on to surfaces. Such molecules can be compared with a string with suction cups attached. Separate particles are lassoed and linked by the polymer chains to form aggregates.

Polyelectrolytes are used either to strengthen existing floc or in their own right as coagulants. However, if such polymers are added too soon after coagulants, before much flocculation has occurred the adhesive ionic groups will be neutralised by the very large number of sub micron particles. In this case, it is necessary to allow perikinetic and some orthokinetic flocculation to reduce the number of particles and allow their size to increase. Thus the objective would seem to be the achievement of a chain with equal strength links consisting of intermediate size but tough floc particles linked by the hydrocarbon chains.

Polymers are also effective at destabilising suspended particles and allowing them to be filtered out efficiently without the use of inorganic coagulants. However such polymers are often unable to precipitate colour or similar organic molecules. If the colouring matter is in true solution some inorganic coagulant is usually necessary, (Graham et al. 1992), although the quantity can be reduced by the combined use of the two varieties. In this case it is often possible to mix them beforehand. Liquid quaternary or polyamine polymers tend to be used for the latter purpose and powdered polyacrylamides or starch based materials for floc toughening, although the latter are effective as coagulants for suspended particles on their own.

Chemical pretreatment procedures for use immediately before filters are best evaluated using a small filter as discussed in Chapter 18. A preliminary indication of treatment conditions can be obtained using the bench top jar test procedure described by Stevenson in Purchas and Wakeman's book (1986). Settlement tests are not relevant to direct filtration. More usually the flocculated mixture is filtered through a No.1 Whatman filter paper or through a cotton wool plug before analysing the filtrate

19.5 Mixing

Confusion between the roles of mixing and flocculation exists . The neutralisation of inorganic coagulant salts and precipitation is a rapid but not instantaneous process. Comments have been made by Edzwald (1992) which indicate that one can have post precipitation if insufficient time is allowed between dosing of the coagulant and the separation process. The comments were made in the context of dissolved air flotation but they would obviously apply to direct filtration also. The particle size produced in the precipitation reaction can be varied over a wide range by changing the design of the reaction vessel. This subject is discussed elsewhere (Stevenson (1964) and (1986)). In water treatment the precipitated particles must have the maximum surface area for absorption, hence neutralisation must be very rapid and confined to the smallest volume.

Secondly, the product of reaction must be dispersed intimately and rapidly with the raw water in order that the hydroxide particles or polymer molecules may attach themselves to the maximum number of colloidal particles before they start to agglomerate with themselves. There is a risk that some particles may collect several hydroxide particles and others none at all. The need for rapid mixing is easily demonstrated using the standard jar test procedure. If a given dose of coagulant is added to only half of a water sample, mixed for a few seconds and then the second half added, the product will be significantly worse than if all is mixed in at one time. On the other hand, if half the coagulant is added, mixed and the other half then added, then there should be little noticeable difference or maybe an improvement in the final quality.

Mixing conditions tend to be specific. Coagulation with polyelectrolytes on their own is reputed to require a higher energy for best performance than with inorganic salts. In all cases mixing must be rapid. Achieving rapid mixing on a small scale laboratory rig is relatively simple. It can be very difficult on a large scale. Ideally mixing should be complete in a fraction of a second. Unfortunately, the time taken increases with the path length. To produce a given power input per unit of throughput the time taken to achieve a given standard of mixing, with a turbine or propeller, increases with the size of the vessel or throughput. To maintain a constant mixing time, the power input rises with the throughput raised to an exponent of 5/3. At large sizes, the resultant fluid shear becomes excessive and can itself cause re-stabilisation or homogenisation.

Direct dosing into pipelines also suffers from the same problem because the angle of the diffusion cone downstream of the point of dosing is constant and hence the mixing time increases linearly with diameter. Static mixing devices are also available for pipes and channels. There is a temptation to over-design and incur an excessive headloss. In some situations, these can become encrusted with a hard deposit of coagulant solid and frequent cleaning may be necessary. Some types collect weed and other debris and designs that could act as a strainer should be avoided.

A popular and very effective method of mixing used on large plants is the simple weir where the water is spread out into a thin ribbon and the chemical added from a perforated sparge pipe or gutter into the cascading water. Mixing occurs in the ensuing turbulence. The energy input is virtually independent of the flow.

Where polyelectrolytes are added subsequently an energy input is again required to give rapid mixing, otherwise the chemical may be wasted. Again part of it may produce a multiple coating on some particles while other particles remain uncoated.

Energy inputs for mixing, expressed as shear rates, are quoted in the literature as 500 to 1500 s-1 with times up to 10 to 15 seconds. These should be treated with caution as they relate to stirred tanks and do not take account of the geometry of the system. Neither would 15 seconds be regarded by many as an adequately short mixing time. A weir might have a fall of 300-500 mm and the retention even in the receiving pool of only 2-3 seconds but it would usually give an excellent performance.

Excessive shear can reduce the filterability of a suspension, particularly where it occurs after the coagulation step. Velocities in pipes should be no more than would be used in open channels. Fittings producing a high headloss should be avoided and the water should not fall more than a fraction of one metre into filters. In one example a 2m fall into an empty filter without level control significantly reduced the filtrate quality. Coagulated water should not be pumped unless a polyelectrolyte is available to restore the damage, after the pump. Pressure reducing valves should not be incorporated in this area either.

19.6 Prevention of Surface Filtration

The mechanisms by which filtration occurs including surface filtration have been described in Chapter 17. Mathematical modelling shows that particles typically above 15-20 μm are readily captured in the depth of only a few grains. Particles larger than the pores are retained at the surface and tend to produce a filter cake. It has been mentioned how algae for example can rest on the surface to form a pre-coat which will then retain micron sized floc particles and these effectively seal the filter. Thus it is important that the granular media filters be kept free of material that blinds the surface. Filamentous algae known as `filter blockers' must be removed either by settlement, microscreening or indeed usually more effectively by dissolved air flotation. Large fast settling particles are preferably removed by sedimentation. However, it is possible to modify the filter by providing a large grain material as the capping layer on the Wahnbach plant, which has already been referred to, where 3 to 5 mm low density activated carbon was used. Such a coarse material allows a fair degree of penetration into the bed and avoids much of the difficulty.

19.7 Encrustation

Calcium carbonate and sometimes manganese dioxide can coat the media by direct transfer from the solution. Calcium carbonate can arise from supersaturation in reservoirs particularly during algal blooms when the pH is elevated above the saturation level. Precipitation softening also usually leaves a residual supersaturation in the final water with respect to calcium carbonate. This must be neutralised either by blending with unsoftened water to reduce the pH to below pHs or by acidification with a mineral acid or carbon dioxide.

A fairly common situation which can cause calcium carbonate to deposit on a filter media is aeration of ground water. The carbon dioxide is stripped off to raise the pH thereby making the water less aggressive (corrosive) but also to encourage precipitation of iron. The efficiency of aeration must be controlled to avoid supersaturation. If the quantity of carbon dioxide is not excessive, air can be injected under pressure to assist oxidation, this avoids stripping of carbon dioxide. The pH is then controlled if necessary by dosing an alkali.

If calcium carbonate is allowed to precipitate on the media unchecked, the grains can grow to the point where they are no longer fluidised by the available facilities and eventually the bed can become cemented.

Iron removal filters often rely upon the catalytic action of manganese dioxide which is either included with the original charge or material which has deposited on the grains. Precipitation of calcium carbonate inhibits this catalytic effect and reduces the efficiency of the filter.

Sequestering agents such as sodium metaphosphate are not recommended because they are in fact dispersing agents. Not only will they inhibit precipitation of calcium carbonate but they may inhibit deposition of the particles to be filtered out. Examples of where this has occurred can be quoted.

19.8 References

Amirtharajah, A., Clark, M.M. and Trussell, R.R. (1991) Mixing in Coagulation and Flocculation, American Water Works Research Foundation Report, Denver.

Bernhardt, H and Schell, H. (1982) Energy Input Controlled Direct Filtration to Control Eutrophication", J. American Water Works Assn. 74, 261

Bratby, J.(1982) Coagulation and Flocculation, Uplands Press, London

Chowdhury, Z.K., Amy, G.L. and Balls, R.C. (1993) "Incorporation of Submicron Particles into Floc in Water Treatment", J. Environmental Eng. American Soc. Civil Eng. 119,202

Cleasby, J. Sindt, G.L., Watson, D. and Bauman, E.R. (1992), Design and Operation Procedures for the Optimisation of High Rate Filtration Processes, American Waterworks Association Reserach Foundation Report. ISBN 0-89867-604-5

Edzwald, J.K. and Walsh, J.P. (1992), "Dissolved Air Flotation:Laboratory and Pilot Plant Investigations", American Waterworks Association Research Foundation, A.W.W.A., Denver.

Graham, N.J.D., Brandao, C.C.S. and Luckham, P.F. (1992) "Evaluating the Removal of Color from Water using Direct Filtration and Dual Coagulants", J. American Waterworks Assn. **84**, (5) 105

O'Melia, C.R. and Ali, W. (1978) "The Role of Retained Particles in Deep Bed Filtration", Prog. Water Treatment,10, 167

Stevenson. D.G. (1964) "Aspects of the Design and Operation of Continuous Chemical Precipitators", Trans. Inst. Chem. Eng. 42, T316

Stevenson, D.G.(1982) "Coagulation and Flocculation", Developments in Water Treatment, Ed.Lewis, W.M.,Ch.3, Applied Science Publishers, London

Stevenson D.G. (1986) Solid/Liquid Separation Equipment Scale-up, Ch.3. Uplands Press, London

CHAPTER 20.

BACKWASHING

20.1 Introduction

Unless a filter is properly backwashed, it will work possibly only once. Conditions for backwashing granular bed filters are more critical than those for filtration. Unfortunately this is not always appreciated. Indeed the very first filters constructed in the early 19th Century failed for this very reason. A full review of backwashing was undertaken by Cleasby and Baumann for the US EPA (1977). This highlighted the weakness of washing with water only. Any reader proposing to study backwashing in detail should examine this reference.

Reversal of flow in a packed bed may resuspend floc particles impacted on the leading edges of the grains but the floc may then proceed to filter on the reverse side of the upstream grains. Floc that has been in contact with a grain for a period of time will tend to adhere more strongly, possibly aided by subsequent bacterial action (slime growth) on the grain surface. It is normally considered good practice to wash rapid gravity filters at least once every 72 hours in temperate locations or every 48 hours in tropical areas, otherwise the granular bed may consolidate into massive chunks known as mudballs analogous to clods of soil. It is also conceivable that the dead zones of the bed in between wormholes (see Chapter 17) could go septic, ie. the biological matter filtered out may decompose under anaerobic conditions. This could cause taste problems.

Where biological action is not controlled an active biofilm develops. This occurs in primary filters upstream of slow sand filters which contain much nutritious organic material which has not been coagulated by chemical treatment. The adherence of the solids is often exacerbated by supersaturation of the water by calcium carbonate caused by a high pH resulting from algal activity in the water source.

A similar situation mentioned in the previous chapter can arise on filters following precipitation softening unless the residual pH is lowered with acid or carbon dioxide prior to filtration. In extreme cases, such encrustation of calcium carbonate can change the effective size of the grain and interfere with washing or even cause cementation of the bed.

20.2 Backwashing with Water Alone

The reference point in washing with water alone or preceded by an air scour is the fluidisation threshold (Fig. 20.1) which is the point at which the hydraulic

306

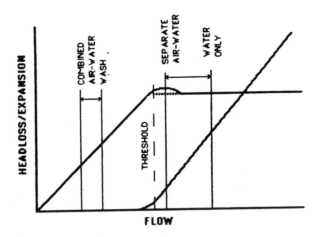

Fig. 20.1 Fluidisation behaviour and wash regions.

gradient through the bed matches the submerged weight of the media in its rest state. At higher rates the bed expands to reduce the hydraulic gradient to maintain the same headloss. This reference point has also been used by Amirtharajah (1984) in empirical equations to predict conditions for washing with simultaneous air and water.

The hydraulic gradient in a bed of porous media was first resolved by Kozeny (1927) for streamline flow and later extended by Carman (1937) for transitional flow to give the equations that are widely derived in textbooks such as Coulson and Richardson's Chemical Engineering (1991), of which the first edition was dated 1955. It is surprising that Leva (1959) and Wen and Yu (1966) did not consider this work and blithely ignored the effect of voidage. These equations are discussed in detail in Chapter 16.

The submerged weight of the bed exerts a limiting pressure loss defined by:-

$$\frac{\Delta p}{h} = (1-e)\ (\rho_s - \rho_L)\ g$$

In order to use the Carman Equation a characteristic size for the grains of media is required. Practical filter media are not single size materials but contain grains

Fig. 20.2 Calculated
fluidisation thresholds.
(40% voidage)

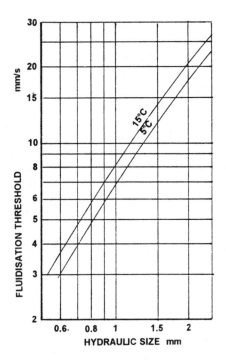

in a size range typically
in a range of 1.5:1 to
3:1. The literature
quotes various proposals
and little justification
for the grain size which
represents the mixture
for use in these
equations. No single
size taken from grading
plots can characterise an
arbitrary mixture. In
view of the basis behind
the Carman and Kozeny
equations the correct
size would appear to be
the area mean size,
which in the British
Water standard (1993) is
designated as the
'hydraulic size'. The
specific area of a sphere is 6/D, hence the hydraulic size is derived by summing the
contribution of each sieve fraction to the area. (Divide the weight of the fraction by
the retained sieve size, convert the resultant sum back to a size by taking the
reciprocal, and then add 10%, which is half the incremental step between the standard
sieves normally used, to produce a centreline size rather than the retained size.) This
gives the characteristic size that would have the same specific area as the disperse
sample. This parameter provides data on the pressure or headloss of the clean media
in filtration.

It should be noted that the fluidisation threshold rises with the square of the
hydraulic size ie, a 10% change in size causes a 21% change in the backwash rate.
Also the threshold varies with $e^3/(1-e)$ and hence a change from say 40% voids to
41% voids would increase the backwash rate by 9.5%. The voidage of sands from

different sources varies with the particle shape, size range or distribution. Materials that have been used have voidages in the packed or vibrated condition ranging from 33% up to 50%. Sources of sand widely used in Britain have a packed voidage of 38%. It is not unknown for plants to be refurbished with new sand from a different source and then fail to operate satisfactorily because of an inadequate wash rate with a new high voidage sand. Indeed it is possible that this is the reason why the myth that sand should be rounded arose. Angular sands with a higher voidage allow more floc to collect and hence provide a better quality for a longer time, but the wash rate must be selected accordingly.

The voidage also varies with the previous history of the bed. In the working condition after fluidisation the voidage of Leighton Buzzard sand has been measured at 43%. If washed in that condition without a preliminary air scour then fluidisation will commence at that voidage. On the other hand if an air scour is applied first the bed compacts to 38% voidage and expansion will start at a lower threshold. Thus voidage cannot be taken for granted. Anthracite has a packed voidage of about 42% and displays a similar variation with previous history. It is somewhat alarming to note that Amirtharajah (1984) published a correlation for the backwash threshold that did not include a term for voidage and also to see the correlations of Wen and Yu, and Leva being accepted in publications such as the AWWA Research Foundation (1991). Humby (1994) has recently confirmed with 1.0-2.0mm sand that the Wen and Yu equation is seriously inaccurate for some media.

All backwashing processes rely upon an upward flow of water under conditions that allow movement of grains. Although there is often a dispute over the amount of movement required, all the surfaces of the grains must be exposed to a shear stress if the adherent floc is to be moved. The average shear stress is given by the following equation:-

$$R' = \frac{\Delta p}{l} \cdot \frac{e}{S(1-e)}$$

where $S = 6/D$ in the case of spheres.
Combining this with the earlier equation an upper limit for the shear stress averaged over the surface of the grains is:-

$$R_{max} = \frac{e_i}{S} \cdot (\rho_s - \rho_L) g = \frac{e_i D}{6} \cdot (\rho_s - \rho_L) g$$

where e_i is the unexpanded voidage. (For single grains $R_{max} = D(\rho_s - \rho_L)g/6$).

The maximum shear stress thus rises linearly with grain size and hence larger grains are easier to clean but of course need a higher wash rate, rising with the square of the grain size. Media with a higher voidage offers more capacity for solids in filtration. The statement which appears in some older textbooks that one should use rounded media appears to have no foundation. It may be that plants designed for rounded media have been unable to provide an adequate wash rate for angular media and hence the media has been blamed. Anthracite, a very angular material has long been accepted as very effective, and a recent paper, mentioned elsewhere, claims that crushed quartz was very effective.

It follows from the above equations that the headloss through a granular bed will rise almost linearly with flow until the threshold is reached (Fig.20.2). Above this the bed will start to expand so that the extra pressure loss is offset by an increase in voidage. Indeed, the above equation may be used to predict expansion rates. However, the Kozeny constant tends to change and the Richardson and Zaki equation is more often used:-

$$u = u_o.e^n$$

where n ~4.6

For practical sands with a 2:1 size range n may be as large as 7. Several workers have studied the values of n. Gunasingham et al (1979) found values as follows:-

$$
\begin{aligned}
\text{Sand} &\quad n = 4.991 Re^{-0.054} \\
\text{Ballotini} &\quad n = 3.94 Re^{-0.100} \\
\text{Anthracite} &\quad n = 7.11 Re^{-0.148} \\
\text{Polystyrene} &\quad n = 6.98 Re^{-0.170}
\end{aligned}
$$

From these, one would conclude that n is a function of particle shape and angularity. Unfortunately these workers did not characterise the shape of their material nor attempt to correlate n with sphericity (aspect ratio), or size range.

The free fall velocity of particles u_o which is needed in the Richardson/Zaki equation is difficult to calculate for practical non spherical particles because of the difficulty of characterising the shape. Direct measurement of expansion thresholds and expansion curves is probably a far quicker and more reliable approach and has the merit of providing data on the bulk mixture. It is essential however that such measurements be made in columns with a diameter greater than 50 particle diameters (see Chapter 30 for the procedure).

Amirtharajah (1978) has developed a theory for the most efficient expansion during backwashing. The optimum is claimed to be that at which the hydrodynamic stress is at a maximum and this occurs at the point where $e = (n-1)/n$ The optimum is however extremely shallow and varies from 0.3 units at 40% voidage to 0.35 units at the so called optimum. The stress falls to 0.3 units again at about 200% expansion. This optimum expansion corresponds to 5.2 times the fluidisation threshold, using Amirtharajah's example. Thus a negligible increase in scouring efficiency is achieved at a very great cost. In practice an expansion of only 10-40% is normal ie. voidages of 50-58%. Amirtharajah notes that particle collisions play a very small part in the cleaning process. The assumption also that hydrodynamic stress is important might be questioned in the light of the above equation defining R' (the surface shear stress) in terms of the headloss which is <u>independent</u> of expansion.

The expansion E of a granular bed during fluidisation is derived from the voidage by the following equation:-

$$e = \frac{e_i + E}{1 + E}$$

$$E = \frac{e - e_i}{1 - e}$$

where e_i is the unexpanded voidage.
The average shear rate in the water is defined by the Camp Stein equation:-

$$G = \sqrt{\left(\frac{P}{V \cdot \mu}\right)}$$

The power dissipation corresponds to the product of velocity and pressure loss, the fluid volume per unit volume of bed is e. Hence substituting:-

$$G = \sqrt{\frac{u(\rho_s - \rho_L)(1 - e)g}{e\mu}}$$

It is certainly necessary to expand the bed beyond the fluidisation threshold into a region where the grains are able to move about. This occurs at about 5-10% expansion. Very high expansion rates can produce other problems such as disruption of packing layers and in the absence of air scour, mudballing, and may be counter productive.

Most workers have sought to achieve uniform fluidisation. Much data is based on small laboratory columns with a diameter considerable less than the height of the media. These stabilise the fluidised bed. On a large scale circulation patterns similar to convection can set in with rising and falling zones. The behaviour of such zones within a cylindrical fluidised bed has been discussed in the chapter on fluidisation by Coulson et al. (1991) but the behaviour in larger beds corresponding to sand filter does not appear to have been studied. This circulation tends to reduce the resistance of the bed by allowing incipient spouts to form. Indeed it is possible that an appropriately designed bed with high and low velocity areas deliberately designed to cause recirculation of the media and the resultant abrasion could be more effective than uniform fluidisation.

Several papers have been published on the prediction of bed expansion at upflow rates in excess of the fluidisation threshold. While this is a satisfying exercise from the academic point of view it is in fact easier to measure these direct by experiment rather than to obtain the necessary size, shape and density data with the necessary accuracy for calculation. Indeed a better estimate of effective grain size can probably be obtained from the fluidisation behaviour rather than the reverse. With porous grains direct measurement is virtually essential. The effective densities of such materials must be derived from the headloss across the expanded bed.

20.3. Stability of Clean Fluidised Beds

With a bed of any granular material, whose linear dimensions are comparable or larger than the depth, over a floor with low headloss a phenomenon known as spouting or boiling can set in when fluidised with water alone. In this situation columns of water can spout through the bed causing eruptions at the surface. In severe cases media can be thrown aside and fluidisation fail. Less severe situations can occur where fluidisation is still complete but uneven.

Consider an incipient spout in a large bed. Over the spout any expansion caused by the excess flow will lead to spillage over into the surrounding bed so that the surface is substantially level. Instead of an expanding bed there will be a constant level and falling concentration and therefore falling back pressure within the spout, which at zero floor headloss can lead to a runaway situation. Any headloss in the floor (which in the case of a nozzle floor will be turbulent with the loss being a function of the square of the flow) will throttle the flow going into the spout and tend to restore stability.

The Richardson and Zaki equation can be rewritten:-

$$e = (\frac{u}{u_o})^{\frac{1}{n}}$$

The floor and media losses ΔH_F and ΔH_M are respectively:-

$$\Delta H_F = K_F u^2$$

Where K_F describes the headloss characteristics of the floor.
Thus:-

$$\Delta H_M = h \frac{(\rho_s - \rho_L)}{\rho_L} \cdot (1 - (u/u_o)^{\frac{1}{n}})$$

$$\Delta H_T = \Delta H_M + \Delta H_F$$

Having set the conditions in the main part of the bed and determined ΔH_T one can explore the effect of a small increase in the voidage in the spout. It will be found that there is a value for K_F below which a runaway situation can occur leading to a stable spout within the bed. For example, taking a deep bed of sand with a fluidisation threshold of 4.5mm/s, a voidage of 42% and a density of 2650kg/m³ it will be found that a floor with a headloss of 50mm at 5mm/s will produce serious boiling whereas a headloss of 100mm at the same flow will be stable. Distribution may be affected not only but the actual nozzles or floor orifices but by the packing layers as well. The latter may undo the work of the distribution orifices. Flow may be swept up from the surrounding nozzles into a spout (Fig.21.6). Baylis (1958) has discussed the boiling of sand beds in some detail and stated that he had never seen a bed that did not boil to the extent of causing the gravel to move. His filters were provided with a deep 6-layer support system which would destroy much of the positive distribution provided by the lateral floor. It should be noted that American practice uses high wash rates with water alone, which places more severe demands on the gravel.

The headloss required to maintain good distribution in the bed is additional to the headloss required for good floor matrix distribution (in the absence of media). A bad floor design will tend to trigger spouting.

A porous plate floor relying on laminar flow to produce the necessary headloss will have to produce an even higher headloss to give even fluidisation.

It follows that the headloss through a floor should be adjusted according to the density of the media and also the bed depth. A granular carbon bed for example will require only about a third of the headloss of that of a corresponding sand filter. High losses are not detrimental apart from the additional pumping power.

The above theoretical argument appears to hold good in practice in conventional gravity filters. In small filters, the walls provide a stabilising effect. Once the linear dimensions exceed the bed depth, boiling within a badly designed filter can become severe. With deep beds, the above basis may be conservative because the surrounding media will tend to slump into the boiling column. The situation is worst when backwashing a dirty bed when floc deposits can cement the media, and the flow in a badly designed filter may bypass clogged regions. Usually a considerable margin of safety is allowed with the distribution headloss to cover dirty bed situation.

The above comments relate to clean media. If a filter bed is cohesive and effectively clogged by dirt then slabs of bound media will tend to rise until cracks form. Such cracks will permit the upwash flow to escape. However if the headloss though the floor is low then the slabs may rest on the floor undisturbed with all the flow escaping through the cracks round the edges. To ensure flow all over the floor under the worst case (a fully clogged bed)) the headloss through the floor must exceed the dead submerged weight of the clogged media. (The situation is comparable to an air or water bearing supporting a weight). Only under these conditions would a flow continue to emerge from under the slabs and give any chance of the slab disintegrating. Thus to be safe the headloss generated by the backwash flow (water alone) through a floor intended for sand should be equal to the depth of the bed (ie. with media of 40% voids>) Air scouring helps to break up such slabs but the above rule will be safe under the worst conditions.

20.4   Stratification in Single Media Beds

Fluidised beds, such as filters during backwash, are dynamic systems and involve diffusional migration and circulation of particles in three dimensions. This is aided by incipient spouts and vertical currents produced by random variations in the solids concentration but not to the extent that occurs in air fluidised beds where bubbles or voids occur. Also the finite spacing of filter nozzles can cause local circulation or convection patterns. Such movements can only be beneficial as far as cleaning of the bed is concerned. Al Dibouni and Garside (1979) have found that stratification of particles of different sizes within fluidised beds occurs only when the fluidisation threshold of the size fractions differs by a factor of two or more, and in a conventional bed with a 2:1 size range (ie. between the 5 and 95 percentiles) there is only a small amount of segregation.

Fine particles below the 5 percentile limit will of course migrate to the surface and if left there such fine particles would create a layer that blocks rapidly during filtration. It is therefore normal practice to skim fines from newly installed beds after a few backwashes.

Combined air scour and water washing mixes the media and even fine particles will remain dispersed within the bed. As discussed elsewhere fine particles are entrained by the air bubbles and if the wash is accompanied by overspilling of the washout weir or launders such fine particles will be slowly lost from the system. The hydraulic size of the media presented in the main clogging zone to the incoming flow during filtration will be slightly coarser following the combined wash. A high rate fluidising rinse will tend to cause restratification and such rinses can be counterproductive. Much of the air left in a bed after air scour has been found to redissolve in the course of a run, particularly in filters which have a depth of water over the media in excess of the clogging head.

20.5 Stratification in Dual and Multi Media Beds

If anthracite or any lighter material is used in conjunction with sand, the anthracite behaves as if it were floating on a medium with a density corresponding to that of the sand/water mixture. Low density anthracite has a density of 1400 kg/m^3 ie, 400 kg/m^3 greater than water. Sand has a density of 1650 kg/m^3 greater than water. The sand/water mixture with say 40% voids before fluidisation has a density of 0.6 x 1650 = 990 kg/m^3 close to the fluidisation threshold. Massive anthracite will float on a sand bed at marginal fluidisation. However, if the sand is expanded by upward flow to a density of less than 400 kg/m^3 ie, 30% solids or 70% voids, the anthracite will sink.

It is unusual for sand expansions to exceed this value. One can have unexpanded anthracite floating on an expanded bed of sand. Particles settling at a slower rate will of course tend to float on top of the faster settling particles and this also ensures that fines come to the surface providing mixing is not too violent (see below). Larger particles of a given density or even of lower density settle faster than small particles and hence aggregates of sand with floc (mudballs) can sink through the fluidised bed of sand, even though the mudballs may have a slightly lower density than the sand itself (but higher than the fluidised bed). This follows from the Stokes equation:-

$$u = \frac{D^2 (\rho_s - \rho_l) g}{18\mu}$$

If this is applied to a particle settling within the fluidised bed the viscosity will not be that of the water but of the fluidised mass (see Coulson et al. 1991).

In normal downflow filtration, the floc deposit collects mainly within the top few centimetres and any consolidation of the floc/media may cause aggregates to break into chunks like pack ice when the backwash flow commences, but in contrast to ice such chunks can sink into the clean media below. Ultimately this can lead to a mudball rubble on the top of the floor or on top of the packing layers. Normal backwash conditions tend to be incapable of breaking down such agglomerates even at high fluidisation of the residual clean material. In such a situation the bed may have to be dug out and relaid.

Most operators like to see a clean interface between the layers and will penalise contractors who do not achieve this. There are however, fairly strong views from researchers that the interface should be graded. To quote Cleasby and Sejkora (1975), "European practice as reported by Ives and Gregory has been to discourage interfacial intermixing. However other researchers and case studies in Europe have shown that filters perform quite adequately when intermixing occurred either by accident or design. The results of this study show substantially better filtrate quality for the mixed interface media rather than the sharp interface media". "The mixed interface media produced higher (head)losses than the sharp interface media for all suspensions due to the greater suspended solids removal."

In practice a sharp or diffuse interface will depend on the width of the size cut and the relative submerged densities. To achieve a sharp interface the ratio of the largest to smallest sizes must be less than the square root of the ratio of the submerged densities.

An interesting situation occurs with garnet in triple media filters. Several so called garnets exist and their density varies. Some commercially available materials have densities as low as 3800kg/m^3 others at 4100kg/m^3. The submerged densities of beds of these at 43% voidage will be 1590 and 1760kg/m^3 respectively. Sand grains have a submerged density of 1650kg/m^3 and thus would be expected to float on the latter but sink into the former. As the expansion increases the bed density will fall and sand grains are expected to sink into both grades eventually. If the flow to a fluidised bed is shut off quickly the settled bed reflects the mixing in the expanded bed. If the flow is turned down slowly the bed has time to re-order itself to the situation that occurs at marginal fluidisation.

This buoyancy effect however operates in tandem with the settlement rate and with the correct size cut reasonably sharp stratification can be obtained with both grades. The size cut however has to be somewhat closer than the usual 2:1 range. The intrinsic diffusion of grains in the fluid bed discussed in the previous section tends to blur the interface and on large beds the convection-like circulation patterns probably render the stratification less sharp than in laboratory columns.

20.6   Air Scour

This term is used in the water industry to describe the use of air either separately or combined with water to assist cleaning of the media during the backwash cycle. Air may be distributed across the filter bed by a set of separate distribution pipes or through a dual purpose underdrain system.

Separately applied air scour has long been used in the United Kingdom to overcome the mudballing conditions as described above. The bed is not fluidised during this process although the top few centimetres of sand may be disturbed. There is certainly no fundamental basis for selecting air scour rates, only experience. Typical values range from 6.5 to about 20 mm/s. The action of this air is primarily to break up any consolidation of the surface floc layer. Indeed if the top of the bed has approached a fully blocked condition, the air will tend to lift it and disperse the clogged material. However, seriously clogged and consolidated sands may actually divert the air leaving a 'water lily leaf' appearance with excess air emerging around the dead zones. Separate air scour is undoubtedly very effective when used in conjunction with the correct low expansion rate of washing, and continues to be used particularly for dual media, triple media and granular carbon filtration, in spite of the criticisms of Cleasby (1977).

Undoubtedly the most effective moment of a separate air scour is the point at which it starts up, when the air displaces water from the underdrain and from the lower layers of the media. There is a short moment of combined air and water agitation when the air is propelled by the cocurrent water. Once this phase has passed, the air does little further good and there seems to be little point in pursuing extended air scour on its own. The air merely flows through established capillaries in the bed. One exception to this is the application of air scour to fine media and low density media where the buoyancy of the air is capable of causing movement within the bed without the additional aid of water. This can be observed with granular activated carbon or finer sizes of anthracite eg. 0.85-1.7mm where churning of the mixture can occur.

At the end of air scour much of the air remains within the bed and on commencement of backwash, a second brief combined air scour occurs. Although it is more effective to have a set of starts followed by a pause rather than one prolonged scour, it is seldom practicable because of a limitation on the number of times that pumps and blowers can be started in quick succession, or valves operated. The failure of the air scour on a filter designed for it usually leads to a rapid failure of the backwashing efficiency.

No filter designed for separate air scour should ever be washed with combined air and water. Serious disruption of the bed, loss of media and possibly penetration of the underdrain may occur. For this reason, pumps and blowers are usually interlocked on such plants.

Columns of air rise within the granular bed by natural buoyancy. There is no distribution headloss requirement equivalent to that for water. However with fine media the bubble pressure of the pores within the media due to surface tension can affect the behaviour.

This pressure is given by the formula:-

$$\Delta H = \frac{4\gamma}{D_p \rho g}$$

The apparent pore size of a granular bed is derived by comparing the Poiseuille equation with the Carman/Kozeny equation:-

$$D_p = D \sqrt{\frac{0.177 e^3}{(1-e)^2}}$$

Hence for sand with a hydraulic size sand of 0.7mm, with 40% voids the equivalent pore diameter $D_p = 0.0124$ millimetres. The significance of this is discussed in Chapter 21.

20.7   Simultaneous Air Scour and Water Wash

Although long used in Continental Europe washing with combined air and water is now becoming popular elsewhere in the world. Its greater efficiency is no longer disputed. Hitherto the specification of the necessary water and air flows appears to have been based on empirical observations. An attempt to put the process on a more scientific basis has been made by Amirtharajah et al. (1984) and their original work has been updated in the AWWA Research Foundation report (1991). Amirtharajah has identified a "collapse pulse" mechanism where small voids are created within the media as the air passes through. These voids collapse as the air migrates to the next cell. The overall effect produces abrasion between the grains with negligible bed expansion, indeed often with a bed contraction from the post fluidised condition. Equations of the following type have been developed by Amirtharajah to predict the threshold for "collapse pulsing" and water passing through various media:-

$$A Q_a^2 + \% \left(\frac{V}{V_{mf}} \right) = B$$

(In this equation Q_a is in m/min, whereas in original Imperial units were given. v_{mf} is the minimum fluidising rate or threshold.) In the 1984 work V_{mf} is based on a correlation of Leva (1959) but in the later work he reverted to that of Wen and Yu (1966). However both of these as already noted ignore the voidage of the media. It is assumed that they apply only to some conventional rounded sand. Furthermore the Leva correlation is based on the 60 percentile size D_{60}, while the Wen and Yu one is based on the 90 percentile size D_{90}. For a commercial sand with a 2:1 size range with the usual log-normal size distribution the hydraulic size, already discussed, is close to D_{40}, but this is only a coincidence. In this context it may be noted that the combined air and water wash provides good mixing of the media and even the smaller gravel layers as used in many filters intended for washing with water or sequential air and water can mix with the working media. The working media is then homogeneous and not stratified. The hydraulic size characterises the mixture. Thus there is little reason to take the 90 percentile which might otherwise be argued to cover the less readily fluidised fractions of a stratified bed.

Stevenson (1995) has found that, with sand, Amirtharajah's data can be predicted fairly accurately if a more fundamental approach is taken, based directly on the Carman and Kozeny Equations, by making the assumption that the air and water are effectively moving in slug flow at the same interstitial velocities, as they would certainly if true capillaries were involved. Indeed if the air cells moving up through the bed spread out they will tend to form a series of local membranes which will be forced upwards by the water, The effective water velocity in such a model will be the sum of the air and water velocities $(u_a + u_w)$. (These are approach velocities as used in the Kozeny Equation. The interstitial velocities will be $(u_a+u_w)/e$). The hydraulic gradient is readily calculated from the Carman or Kozeny Equations. The air will offer negligible resistance and the overall hydraulic resistance must therefore be reduced by the ratio of the water to the combined flow, ie.by $u_w/(u_a+u_w)$.

It also follows that the voidage within the bed will be occupied by air and water in a ratio corresponding to their flow rates. As with water on its own the hydraulic gradient at the fluidisation threshold where movement starts balances the submerged weight of the media but in this case the buoyancy of the air must be deducted from the submerged weight of the media as follows:-

$$\frac{\lambda u_w}{(u_a+u_w)} = g[\,(\rho_s-\rho_L)\,(1-e) - (\frac{u_a}{u_w+u_a})\cdot\rho_L]$$

In this case λ is the hydraulic gradient calculated for water at the combined air and water velocity. This equation must be solved by using the Carman equation to calculate λ and increasing the water flow until the two sides balance. The hydraulic diameter of the media must be used in the Carman equation.

To be precise account should be taken of the absolute pressure at the relevant level in the bed, usually at the base. Even with 1.0m depth below the water level, plus the hydraulic gradient through the bed (possibly adding a further 0.5m to the effective head) the effective volume of the air may be reduced by 15%. In Amirtharajah's experimental work it appears that the base of the beds was 1.6m below the water level and the data relate only to that condition. As would be expected the published data shows that more air is required for deeper beds.

Comparisons have been made for media of various sizes, using an assumed 2:1 size range between the 5 and 95 percentiles, which is common (uniformity coefficient = 1.3), and log-normal distribution. For these specific examples both the D_{90} and hydraulic sizes were derived. On the one hand for the Amirtharajah approach V_{mf} was calculated by the Wen and Yu equation using D_{90}, and the conditions for collapse pulsing obtained from the latest equation of Amirtharajah, where A=8.5 and B= 43.5. (These constants are those derived by Amirtharajah et al, from experimental data where the correlation coefficient was only 0.62, based on 18 experimental points.) Amirtharajah's work covers D_{60} sizes of 0.62mm, 0.86mm and 1.54mm.

On the other hand the Carman based equation used the hydraulic sizes. The assumption was made that the voidage in the Carman case was 38%, which would be representative of most media used in Britain and presumably in the USA, but to demonstrate the effect calculations were also made for 40% voidage. Amirtharajah regrettably gives no information on the voidage of his materials, however normal rounded filter media compacts consistently to a voidage of 38%, although after fluidisation it may have a voidage of 43%. The Kozeny (Carman) constant is not determined with any great accuracy and the data in only given to two significant figures. The calculations assume 750mm depth of media covered by 750mm of water to simulate the experimental conditions used by Amirtharajah et al.

The results shown in Table 1 show a surprising agreement both in relation to the effect of variations with air flow and also to the absolute values, which seem to confirm the assumption that the air and water do effectively pass through in virtual slug flow. They also demonstrate the sensitivity of the water rates to the voidage of the media. The accuracy of agreement is within the scatter of Amirtharajah's experimental data.

It therefore appears that the Carman Kozeny approach enables wash conditions for media of other voidages to be extruded from first principles. The modified hydraulic gradient shown in Table 1 is the equating group in each half of the above equation. This is a useful parameter (bed headloss) which is not available from the Amirtharajah approach. The value for 0.85-1.7mm media corresponds with observations made by the author for these conditions.

Table 1   Backwash Rates with Simultaneous Air and Water

(All calculations are based on the water viscosity at 10°C)

Sand Grade	Air m/h	Amirtharajah	Stevenson 38% voids	Stevenson 40% voids
mm	m/h	mm/s	mm/s	mm/s
0.5-1.0	60	1.9	1.7(0.68)	1.9(0.64)
	40	2.1	1.8(0.70)	2.0(0.65)
	20	2.3	2.0(0.76)	2.3(0.71)
0.85-1.7	60	4.7	4.4(0.76)	5.1(0.69)
	40	5.3	4.9(0.78)	5.7(0.74)
	20	5.7	5.6(0.85)	6.7(0.82)
1.18-2.36	60	7.4	7.9(0.78)	9.3(0.74)
	40	8.4	8.9(0.82)	10.4(0.78)
	20	9.0	10.1(0.92)	11.9(0.85)

The figures in parenthesis are hydraulic gradients (m/m). (The constants in the equations are probably only accurate to ±2%, and voidages cannot be measured with any greater precision.)

The situation with light media such as anthracite and granular carbon is less clear. Firstly it is more difficult to estimate the effective voidage of the media to which Wen and Yu's data relates and one might query the validity of the equation with media with a voidage significantly higher than that of sand. The buoyancy of the air is considerable relative to the submerged density of these materials. For example the submerged specific gravity of a bed of British anthracite (sg. 1.45) with a voidage of 42% is only 0.23. Movement is predicted at very much lower rates. Indeed with the more commonly used carbons and finer grades of low density anthracite, eg. 0.85-1.7mm of sg. 1.45 churning movement has been observed in a 900mm wide 180mm thick transparent sided tank at air rates of about 20m/h, with no water flow. Predictions based on the above approach indicated that movement occurs at lower water rates than indicated by Amirtharajah's equation. It is possible that wall friction in the relatively small columns used in his work may have affected the outcome with such materials. Certainly if the lower density materials permit large air voids to form then the assumed mechanism may no longer be valid.

Predictions are however of less practical use in this case. Some commercially available carbons show rapid attrition (Humby 1994, 1996) and the combined wash procedure is not appropriate. In practice the wash conditions in stratified media will be governed by the requirement of the sand layer and the behaviour of the anthracite will be incidental. Nevertheless there is scope for further work to explore the behaviour in more detail, to define the density thresholds where the above relationships no longer apply. From pilot plant observations it appears that the size of the column has a considerable effect.

Various practical considerations such as the time taken to complete a wash and the cost of the installation will affect the choice of air and water rates. Lower water rates allow longer laterals to be used. The main disadvantage of the combined air and water wash apart from floor design complications is the higher peak power required compared with separate air and water. With the combined wash air has to be delivered at a similar pressure to the water as it is subject to the same headloss through the media.

The main benefit, apart from the higher efficiency, of the combined wash is the reduction in the water flow required for coarser materials. As an example 0.85-1.7mm sand with a separate air scour will need a wash rate of at least 11mm/s whereas the combined wash might only requires 4-5mm/s of water with perhaps 16.7mm/s of air. The action of combined air and water appears to be akin to raking, in that the grains are forced by the collapse pulse process to move over each other as the bubbles pass through the matrix. Locally there are probably zones where the residual dead weight of media presses down on the layers below whilst movement is taking place. In contrast to fluidisation with plain water in which the grains are buoyed up by the hydraulic shear stress which is uniformly spread around the gain surface. The residual load with combined air and water is applied at the grain to grain contact points, which move around. However, this abrasive action limits the process to tough materials and friable materials such as activated carbon can abrade seriously under such conditions.

Interesting endoscope studies have been made by Fitzpatrick (1990) who has observed the motion of individual grains and the displacement of floc both with water alone and simultaneously with air.

20.8 Surface Scouring

This term has nothing to do with air scour as discussed above. It involves jetting the media near the surface of the bed with high pressure water during the backwash. It is used, particularly in the USA often with rotating distributors. Such jets cause collision between the grains and thus scouring of the floc from the grains. The designs are empirical and are uncommon in the United Kingdom and in Europe, indeed they would be difficult to install in the majority of low level side weir designs because there would be insufficient depth.

20.9 Raking

At one time, it was common practice to fit rakes into filters. There were many in Britain particularly on 2.4 metre diameter vertical pressure filters which were installed in large numbers for treating upland waters in direct filtration plants. Such rakes were mounted on a shaft passing through the top of the filter vessel or located centrally in the case of circular open filters. The tines consisted of rods passing down into most of the depth of the media, and were rotated slowly during the backwash. Indeed raking appears to have been the first procedure used to assist upward flow water washing before fluidisation was fully understood. Such installations often used water power and line shafting to drive the rakes which would were undoubtedly effective with low expansion fluidisation, or even below the fluidisation threshold. On small filters or in primitive environments raking is still a sound approach. On small filters such rakes can be operated manually.

The maintenance of the mechanical components and the replacement of many older works with larger modern units has led to the demise of raking but it still constitutes an effective method of assisting the backwash, and is appropriate for many smaller manual installations.

20.10 Fluid Displacement

Fresh floc can be regarded as having little adhesion to the grains. It is rapidly flushed away from the surface. Any set of conditions imparting sufficient shear to overcome the bond is likely to be quick acting. A slower response may be expected where abrasion is required to scour chemically or biologically bound material from the surface, and any aggregates such as mudballs have to be broken down. The slowest process in backwashing would seem to be displacement particularly from the supernatant water above the bed. Any material not backwashed will go into supply, otherwise the media will progressively foul up. Any floc not flushed away will either be re-filtered or will pass forward. Overall it would appear that the washing must displace 99-99.5% of the material filtered out although there may be a permanent residue which forms a coating over the grains of media.

Displacement comprises two steps, firstly displacement from the granular bed and secondly from the supernatant water layer. The dispersion of flow through the bed is discussed in Chapter 16. A sharp front of clean water entering a bed filled with suspended solids will not displace the solids in a sharp piston-like manner. The front will diffuse and blur. For a typical bed depth from around 700 particle diameters, the dispersion to 1% of the slug concentration is +/-0.27 about the retention time within the bed. 99% of the material should be displaced within 1.27 bed volumes. In practice about three bed volumes are employed in filter washing. As discussed in Chapter 22 the reason for this discrepancy seems to be associated mainly with the depth and mixing profile in the supernatant water.

Practical experience and theory indicate that displacement is related to the *volume* of backwash passed and not to the *time*. A given number of bed volumes plus a factor for the supernatant water configuration is required and *higher* backwash rates enable the time to be *reduced*. It was the realisation that the depth of the supernatant water influenced the quantity of wash water required that led to low side weir designs in Britain. If the backwash rate in a given filter is marginal a higher rate may actually reduce the volume required. High rates on the other hand lead to more expensive pumps, pipework and valves, and also an increased danger of losing media, a feature which is discussed in Chapter 22.

In conclusion it would appear that there is still room for further information on the basic processes involved in backwashing and the most cost effective process employing the minimum amount of wash water and the minimum power. There is also a lack of quantitative information on the behaviour of air in submerged granular beds.

Amirtharajah (1985) has examined the transition from backwashing to filtration. It is well known that the first filtrate is of poorer quality than the main run and filters are said to mature. This effect is explained in Chapter 17. At the end of the backwash the base of the bed will be filled with clean backwash water and any residual particles of dirt will have been carried to the top of the bed. Thus when the filter returns to service the first water to emerge is the clean wash water. Residual solids from the wash emerge next and then fresh filtrate with new solids which have not be adequately filtered. This produces a twin peak which Amirtharajah has studied in detail. The duration is only a few minutes and the effect has little impact on the running of the filter. The benefits or otherwise of slow start are discussed in Chapter 23.

20.11 References

Al-Dibouni, M.R. and Garside, J. (1979) "Particle Mixing and Classification in Liquid Fluidised Beds", Trans. Inst. Chem. Eng. **57,** 94

Amirtharajah, A. (1978) "Optimum Back Washing of Sand Filters", J. Environmental Eng. **104**, 17

Amirtharajah, A. (1984) "Fundamentals and Theory of Air Scour",J. Environmental Eng. **110**, 573

Amirtharajah, A., (1985) "The Interface between Filter Backwashing and Filtration", Water Research, **19,** 581

Amirtharajah, A., and Hewitt, S.R., (1984) "Air Dynamics through Filter Media during Air Scour",J. Environmental Eng. **110**, 591

Anon. (1993) Standard for the Specification, Approval and Testing of Granular Filtering Materials, British Water, London

Baylis, J.R. (1958) (Title nto known) J. American Waterworks Assn. **50**, 126

Carman, P.C. (1937), "Flow through granular beds", Trans. Inst. Chem. Eng. **15**, 150

Cleasby, J.L., Aboleda, J., Burns, D.E. Prendiville, P.W. and Savage, E.S. (1977) "Backwashing of Granular Filters", J. American Water Works Assn. **69,** 115

Cleasby, J.L. and Baumann, E.R. (1977) "Backwashing of Filters used in Wastewater Filtration", EPA Report 600/2-72-016

Cleasby, J.L. and Sejkora, G.D.(1975) "Effect of Media Intermixing on Dual Media Filtration", J. Environmental Eng. **101,** 503

Coulson J.R., Richardson J.K., Backhurst, J.T. and Harper, J.R. (1990) Chemical Engineering. Vol.II.,Pergamon

Gunasingham, K., Lekkas, T.D., Fox, G.T.J. and Graham, N.J.D.(1978) "Predicting the Expansion of Granular Filter Beds",Filtration and Separation, (Nov/Dec), 619

Fitzpatrick, C.S.B.(1990) "Detachment of Deposits by Fluid Shear during Filter Backwashing," Water Supply (Int. Water Supply Assn. Conf. Jönköping) **8**, 177

Humby, M.S. (1994) "The Friability of Filter Media", M.Sc. Thesis, School of Water Sciences, Cranfield, UK.

Humby, M.S., (1996) "Attrition of Granular Filter Media during washing with Combined air and Water", Water Research, **30**,291

Kozeny. J. (1927) "Uber Kappilare Leitung des Wassers im Boden", Sitzb.Akad. Wiss. Wein, Math.-naturw. Kl **136**, (Abt,IIa) 271-136

Levenspiel, O. (1972) Chemical Reaction Engineering, Wiley, p275

Leva, M. (1959), Fluidisation, Mc Graw Hill. N.Y.

Stevenson, D. G. (1995) "Process Conditions for the Backwashing of filters with Simultaneous Air and Water", Water Research, **29**, (11), 2594.

Wen, C.Y. and Yu, Y.H. (1966) "Mechanics of Fluidisation", Chem, Eng. Prog. Symp. Series. No.62, American Inst. Chem. Eng.

CHAPTER 21.

FILTER FLOORS

21.1 Introduction

All granular media filters must of course contain media and this must be supported or held down by a floor.

A filter floor must:-

a) Support the granular media ie. carry its dead weight both submerged and drained down.

b) Withstand the maximum filtering headloss in addition to a).

c) Allow percolation of the filtrate uniformly across the bed and provide collection to a single outlet pipe.

d) Distribute backwash water uniformly across the entire bed with a defined accuracy and control the uniformity of fluidisation.

e) Distribute, where applicable, the scouring air either simultaneously with the water or separately according to the design.

f) Withstand the upthrust created by the pressure losses inherent in d) and e) at the maximum peak rate and degree of clogging of nozzles etc.

g) Withstand any abrasion or be designed not to encourage abrasion.

h) Be constructed in a material that is non corrodible under the most aggressive conditions of use and be acceptable to any regulatory bodies.

The filter floor should be regarded as separate from the structural floor or containing vessel, although the floor design does influence the latter. For the most part filter floors comprise either a suspended concrete slab or a steel plate, these having a space or plenum below, or a set of parallel pipes known as laterals which connect into a header pipe or duct. Both types of floor may be provided with orifices fitted with strainers which may be part of a nozzle assembly, or they may merely have orifices covered with coarse gravel. Some newer floors have laterals which comprise slotted grids eg. wedge wire.

Much has been written about granular media filtration as a process. Floor design tends to be proprietary and far less information is available, yet the success or failure of a practical filter depends upon a good understanding of the design principles. It is here that most of the cost lies and also here that 'cowboy' designs can fall down. The skill is in achieving a sound design at lowest cost.

21.2   Filtrate Collection

The primary purpose of the filter floor is to prevent the media passing forward while allowing the filtrate to do so. In almost all practical cases, the floor design will be governed by backwash considerations. The headloss in forward (downward) flow will be small because filtration rates are usually lower than backwash rates. During filtration the headloss through the media dominates the situation. For example, a traditional filter washing at say 6 mm/s with a headloss in the nozzle of about 700 mm would have a headloss of only 80mm at a filtering rate of 2 mm/s, (7.5m/hr). The clean bed headloss in a conventional filter is often around 300 mm. It is common practice to run filters to 2 m headloss. Thus the underdrain design will have little influence during forward flow, except that the spacing and exposed area of the strainer must be sufficient not to cause local headlosses around each nozzle as the flow converges.

If the filter floor is not level as is the case with some types of granular media filter designed for carbon transfer, the flow rate through the shallower parts of the bed will tend to be higher (ie, it will vary inversely with depth). However, if such filters are intercepting floc, then the shallower zones will tend to clog more rapidly causing the local flow rate to even out. If such filters are used purely to absorb soluble material then the uneven flow will continue and premature exhaustion of the shallower parts of the bed may be expected.

Overall therefore if the design is satisfactory for backwashing it will certainly be satisfactory for filtration.

21.3   Backwash Distribution

The floor must distribute the water fed in from below during backwashing in a even manner. The importance of good distribution is illustrated by inserting a hose in a bucket of fine sand. The sand will remain settled but a boiling spout will be observed above the end of the hose unless a high velocity jet is used. Conditions for stable fluidisation are discussed at the end of this chapter. The headloss through the bed with a water flow above the fluidisation threshold is constant and the bed has no regulating effect on the local flow rate. It will be shown that a headloss at least one tenth of that through the bed, and preferably more, is needed purely to distribute the water evenly through the media. A greater headloss may be required to ensure an acceptably even distribution through the floor assembly. Furthermore, the headloss must not generate a vertical jet that punches a hole through the media. If the working media is supported on packing layers ie. graded gravel layers, these may have a greater influence than the floor itself. One can have a good floor design but maldistribution may occur as a result of a poor choice of packing material which may allow water to migrate horizontally within the gravel into spouts within the media.

The floor headloss must meet the requirements for good distribution within the two dimensional matrix of the floor. Velocities under a <u>suspended</u> floor are low and it is easy to obtain good distribution. Floors involving parallel lateral pipes produce higher velocities and particularly if they carry nozzle stems where friction and the Bernoulli effect can be significant. There are strict limits on the lengths that may be used if good distribution is to be maintained. Typically a flow range of 5% across the floor may be demanded.

The whole of the underdrain is usually submerged. Small errors in the level are usually not critical. The surface of the media after washing will be level. However if there is a significant slope in the floor then the headloss through the media will not be constant and the backwash rate will tend to be higher in the shallower parts of the bed. In this case an additional pressure loss must be incorporated in the nozzles to balance the flow.

Interestingly, combined air and water washing conditions are normally well below the fluidisation threshold and the positive flow/headloss characteristics of the media assist good distribution, in contrast to a fully fluidised bed. A much lower nozzle loss can be used. In the fluidised bed the headloss is either constant or if spouting is occurring it can actually fall locally with increasing flow (see Section 21.13).

21.4 Air Distribution

If either separate or combined air scour (ie. the use of air either before or simultaneously with the backwash flow) are employed to assist washing, the air must also be distributed evenly. The efficiency of washing is less sensitive to the air flow, hence the accuracy of distribution can probably be somewhat lower than that for water. However, it is easier to measure the local air flow using a bell jar type collector placed above a section of the bed with a variable orifice flowmeter to measure the airflow from the section. Users may therefore tend to be more critical in their demands, whereas the accuracy of water distribution can only be inferred from local measurements of the headloss across the floor, making the assumption that all the nozzles are identical.

Air is distributed commonly by one of two methods. Firstly with plain perforated pipe laterals, orifices are located on the crown of the laterals (Fig.21.4). These must produce a sufficient head to provide an adequate cross section for the air flow along the lateral as discussed later. Secondly a purpose designed nozzle with a metering orifice may be used. This regulates the airflow from the nozzle but requires a level water interface in the lateral or under floor plenum, assuming that the nozzles themselves are all level.

If air is distributed into a submerged granular bed with no water flow, bubbles force their way up through the capillaries in the media once the maximum pore pressure has been exceeded. The hydrostatic pressure at the nozzle corresponds to the depth below the free water surface. The resistance to flow of the air is small and less than the buoyancy of the air. For simultaneous air and water, the air emerges into the media which is also subjected to the back pressure of the water flow though the media and errors in water distribution will affect the airflow. Conversely, at least one floor design uses virtually no headloss for water distribution and relies almost entirely upon positive distribution of the air, the latter providing an air lift mechanism to ensure an even water distribution. A zone of low water flow would receive a higher proportion of air, thus increasing the local buoyancy and increasing the flow.

Water distribution systems remain full of water in service. Air distribution systems will tend to refill after the air has been switched off. The designer must ensure that the air displaced from the under drain emerges uniformly and does not cause problems of local disruption. In some designs a considerable volume of air may be trapped below the floor or in distribution ducts. Commencement of a backwash while such air is present can cause a short period of combined air scour and backwash which may be localised. In a floor not designed for the combined wash this may lead to an expensive disruption of packing layers and even penetration of media into the under drain. Floors should therefore be self venting.

21.5   Manifold Theory

A manifold is a pipe or duct into which a set of other pipes, ducts or orifices connect. One can have collecting manifolds or distributing manifolds, depending on the direction of flow. The flow into and out of the branches is not uniform even with identical branches.

In the present context, the Bernoulli equation states that the sum of the potential energy and kinetic energy in a given (in this case incompressible) flow stream is constant.

$$\frac{p}{\rho} + \frac{u^2}{2} = Constant$$

where u is the velocity, p the pressure and ρ the fluid density. (For a full discussion see a text book such as Coulson and Richardson (1990).) Hence the effective pressure in a flowing stream is reduced by the `velocity head', $u^2/2g$. Thus if one considers a manifold pipe with branches, (Fig.21.1), the effective pressure driving the fluid into each branch is lower at the entry end by an amount corresponding to the velocity head.

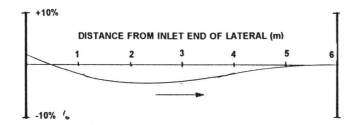

Fig. 21.1 Discharge from a manifold with significant friction.

On the other hand, pipe friction operates in the reverse direction and increases the pressure at the upstream end. The frictional headloss is given by the Colebrook/White equation which is used extensively in the water industry together with an equivalent, more easily managed, version given by Miller (1990), which is reproduced below.

$$f=\frac{0.25}{[\log_{10}(\frac{K}{3.7D}+\frac{5.74}{Re^{0.5}})]^2}$$

where f is the friction factor, K is the roughness factor (m), D the pipe diameter, and Re the Reynolds Number.
The latter is defined by the equation:-

$$Re=\frac{uD\rho}{\mu}$$

u being the velocity, μ the viscosity and ρ the fluid density.
The headloss ΔH in a pipe of length L is given by:-

$$\Delta H=f.\frac{Lu^2}{2g.D}$$

Simple programmes can be written to analyze the driving pressure along a lateral pipe and hence the flow through each successive branch starting at the dead end and working back. The flows are summed and if the total does not match the actual flow entering the pipe, the individual discharges are corrected proportionally. It may be assumed that with turbulent flow throughout the duct or pipe, all the flows in the system remain proportional to each other regardless of the absolute flow. A typical flow pattern across a manifold forms an asymmetric U shape (Fig.21.1). A 5% flow range is an acceptable target which makes the error in the floor less than the intrinsic uncertainties about the media (actual hydraulic size and voidage).

In the above, it is assumed that the main headloss element in the branch (eg. the control orifice) is not exposed to the tangential manifold flow but is within a nozzle. In the case of a plain perforated pipe where the discharge orifices are swept by the flow, the oblique angle of discharge reduces the effective area and hence the flow emerging from that orifice. This situation is discussed by Miller (1990), who provides curves for the discharge coefficients. Care is needed should any re-entrant stems have orifices on the upstream side that could act as pitot tubes. These will pass a higher flow. For a quick evaluation a useful equation for short manifolds where the friction is low states that a 5% flow range can be achieved if the velocity u_{max} (in m/s) at the entry end of the manifold is less than:-

$$u_{max} < 1.4\sqrt{\Delta H}$$

where ΔH is the headloss produced by flow out of the branches (m). The constant includes the gravitational constant and only applies with the units given above.

21.6   Two Dimensional Manifolds

Most large filters have a header duct or tube with a herring bone or half herring bone arrangement of laterals branching from it. (Fig.21.2.) The same method of calculation is employed in designing the header and the laterals are then treated as branches. The headloss in the branch includes entry, friction and discharge losses are treated as a composite branch loss. If the individual laterals are allowed to produce a 5% range, and the headers also produce a similar level of maldistribution, then the total error across a filter   would correspond to a 10% range. The upwash velocity must be chosen so that areas with the minimum flow still receive an adequate wash rate. However, even a 10% range is equivalent to a 5% change in hydraulic size of the media and such variations can occur on a batch to batch basis. Standard sieves used for grading media come in size steps of approximately 20%. Thus there is little point in striving for a distribution accuracy much greater than the above figures.

Floor designs lend themselves to computer calculation. This speeds the process, increases accuracy and removes most of the tedium. Such programmes are easily written but those that exist remain the proprietary property of consultants and contractors.

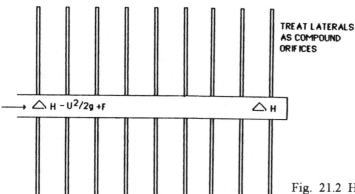

TREAT LATERALS
AS COMPOUND
ORIFICES

$\triangle H - U^2/2g + F$ $\triangle H$

Fig. 21.2 Header/lateral - 2 dimensional manifold.

21.7 Manifolds for Air

The same design procedure may be used with air, providing that the change in pressure is small and compressibility effects can be ignored. It is essential with lateral floors in which there is an air/water interface during the air scour, that a sufficient cross section is available for the air to pass without waves forming and also to meet the velocity requirements outlined earlier (Fig 21.3).

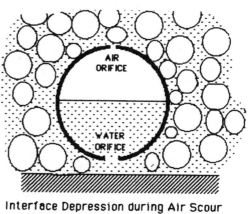

AIR ORIFICE

WATER ORIFICE

Interface Depression during Air Scour

Fig. 21.3 Nozzleless perforated pipe lateral floor.

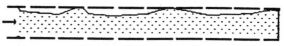

Waves with Inadequate Depression

A point easily overlooked with air is the adiabatic heating that occurs on compression coupled with frictional heating in the blower which with deeper filters can produce temperature rises that can soften some materials of construction which may be in contact with hot air on both sides. There is less of a problem if one side of surface is always in contact with water.

21.8   Packing Layers (Support Gravel)

Filter floors must either have fine slot or mesh strainers to keep the filter media out of the underdrain or they must have graded gravel to prevent percolation. A hybrid version is the no fines concrete or porous resin bonded aggregate. Sintered materials either ceramic or plastic also come under this category.

Nozzles with strainers having slots fine enough to retain the working media can block up if the backwash water is not entirely free of grit. Such strainers however provide a positive distribution into the base of the media during backwash. During filtration, there is some convergence of flow to each strainer. If these are small the high localised flow can increase the headloss at this point.

Packing layers of graded gravel enable fairly coarse slots or orifices to be used eg, up to 12 mm in some designs. Normally the maximum pebble size of each successive layer is less than the smallest fraction in the one below. The maximum permissible size of step between adjacent layers is that which prevents percolation of the upper layer into the lower layer. A 4:1 difference is usually safe and even wider steps are sometimes used. (The average pore diameter is discussed in Chapter 20. For 40% and 50% voidage the effective pore size is 0.177 and 0.29 times the hydraulic diameter of the media). With angular media a closer size step may be needed. It is fairly easy to check for percolation using a small laboratory column.

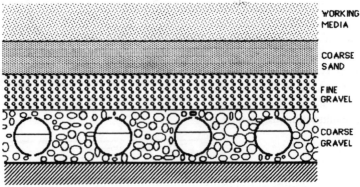

Fig. 21.4 Graded packing (support) layers for an exposed lateral floor.

Fig. 21.5 The hydraulic gradient through gravel, 40% voids at 5°C.

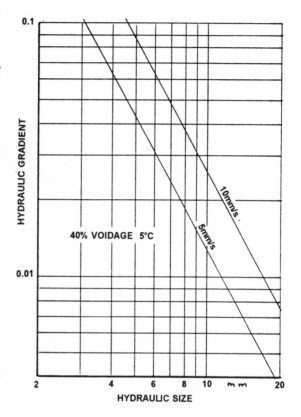

The depth of each layer is a matter of practicality in ensuring that a consistent layer over the entire floor can be achieved. 50 mm layers are sometimes specified, but on larger filter 100 mm layers are common. The first layer over strainers or laterals is often deeper to ensure adequate cover (Fig.21.4).

The headloss through the packing layers is small (Fig.21.5). This in fact is a weakness in that this permits horizontal short circuiting at the nozzle level. Although the nozzles provide positive control of the flow into the base of the bed, an incipient boil may lead to the flow from several nozzles converging to a single spout through the packing layer (Fig.21.6).

The packing gravel must remain stationary during backwash. If it disrupts the filter is ruined and must be relaid. Backwashing will not regrade the material once it has mixed. An attempt to do so may only make matters worse. The top layer of gravel, particularly at higher bed expansions may be close to its fluidisation threshold which is a dangerous situation and accounts for some of the problems described in American literature where wash rates are higher than in British and European practice. Any over

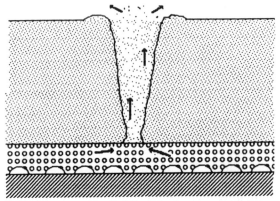

Fig. 21.6 Spouting in an unstable fluidised bed.

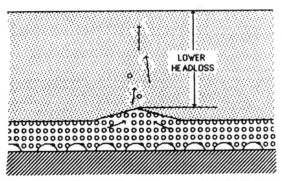

Fig. 21.7 Spouting and incipient disruption caused by uneven packing layers.

running of a filter which allows penetration of floc into packing layers will increase the danger of disruption (cf. upflow filters, discussed in Chapter 25.)

Surface irregularities in the packing layers, such as footmarks, particularly in the top packing layer are most undesirable. A hump in the interface reduces the depth of working media at this particular point (Fig.21.7) so that the resistance to flow through that part of the bed is less. This enables flow in the packing layer to converge towards the summit of any dimples and the fluidisation threshold of the finer gravel can be exceeded locally, allowing it to be carried up into what rapidly becomes a spout. This can progress to the point where coarse media can be seen at the surface of the filter. There is little that can be done to resolve a serious case other than to install new packing media. The situation is more critical with high velocity washing.

The behaviour with air is interesting. If the difference in size of adjacent layers is too large or again if there is any undulation of the interface, the difference in the bubble pressure for pore penetration will cause the air to run up the sloping interface and emerge from the summit of any humps. This concentration of air and water at the summit of any undulations, particularly during the transition at the beginning or the end of air scour, can be doubly damaging. Smaller steps in size make the bed more tolerant and for this reason wide differences in size are usually avoided.

Washing with simultaneous air and water rapidly disrupts conventional packing layer configurations which are designed for separate air scour and water washing. For this reason, blowers and upwash pumps are usually interlocked and such interlocking should not be overridden.

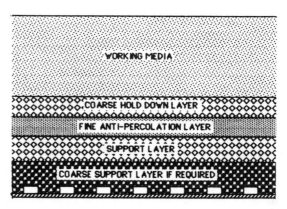

Fig. 21.8 Reverse graded packing layers.

One method of overcoming packing disruption, used in the USA on filters with high wash rates and also for combined air water and washing is the reverse graded configuration (Fig.21.8) where a finer gravel layer capable of preventing percolation of the working media is held down by a one or two coarser layers above. The low headloss and dead weight of the coarse pebbles prevents the 'jam' layer in the middle from moving. Such a configuration however, gives a poorer fluidisation pattern with water alone because of the horizontal short circuiting in the upper coarse layers. With combined air and water the plumes of air bubbles provide even distribution and the buoyancy of the air ensures that it continues to rise vertically.

There is of course a limit to the headloss that can be applied to the centre of a 'jam' layer and if abused it will lift and disrupt. Severe clogging, for example, resulting from an overdose of polyelectrolyte will test this concept to the limit, and it is understood that disruption can still occur.

Lastly with some nozzle designs, packing layers can be useful in keeping the working layer away from the slots of the nozzles and in so doing prevent abrasion. A good design however, should keep the emerging velocity low enough to prevent sandblasting.

Overall, packing layers are considered by many users to be an essential feature of filters. They reduce the danger of nozzle blockage but they have many disadvantages and even the reverse graded packing can disrupt if abused. Many designers like to avoid using them altogether. On the other hand packing layers have the advantage of being replaceable whereas a porous matrix such as no-fines concrete cannot be cleaned easily should it become blocked.

21.9   Filter Nozzles

This term is applied to the components which retain the media and also provide the means for good distribution of water and air (where used). Nozzles may be installed in lateral pipes or in suspended floors.

A nozzle has three main components, a strainer to exclude the media or packing gravel, a water control orifice and a stem with an orifice(s) to control the airflow.   Many makes of nozzle rely upon the open area of the strainer slots to provide the controlling headloss for good distribution. However, it is expensive to modify slots and an independent orifice within the assembly is often preferred. Strainer slots can become partially obscured with media. This will alter the characteristics. High velocities through such slots can also cause abrasion of the nozzle material. If the control orifice is within the body of the nozzle and the slots have a larger cross sectional area, the behaviour of the nozzles will be less affected by partial obscuration by the media in contact with the strainer.

Strainer slot sizes vary considerably from about 0.2mm to around 10mm, the latter being used in upflow filters. Slots of 3mm for example are used to exclude 6.7-13.2mm packing material. Finer slots are used where the strainer has to exclude the working media as in the case of combined air scour and water backwash. Normally the slot width is no larger than 0.5-0.7 of the smallest fraction of the media in contact with it. Very great care must be taken to avoid grit particles entering with the backwash water if slots less than 1.0 mm width are to be used. If filtrate is always used for backwashing there is usually little difficulty. With very fine slots suspended floors are recommended as they allow grit particles to settle out whereas in lateral floors the grit can be sucked up into the strainers.

Some floor suppliers promote nozzle-less floors but these rely either on packing layers or porous compositions which would seem equally or more likely to block and more difficult to clear. Floor blockage where filtrate is used for backwashing is very rare. Disruption of support gravel is more common.

The function of the stem of filter nozzle is to control the distribution of air. If air is not used stems are unnecessary. They are usually provided with a top orifice (Fig.21.9) and lower orifices or slots. As air is applied to the floor whether suspended

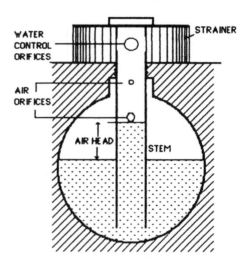

Fig. 21.9 Components
of a filter nozzle.

or of the lateral type, the air collects at the crown and depresses the water to form a continuous interface giving clear access to all the stems. If the air velocity is too high, waves can form causing an intermittent discharge. In severe cases waves can act as pistons and become propelled by the air and cause damage. The upper hole is regarded as a pilot hole to trigger the flow of air and the headloss caused by the airflow causes the initial depression of the water interface on the outside. Once the lower holes or slots are exposed, the larger area available to the air stabilises the water level.

One can have a situation particularly with lateral floors where too much water can be extracted during start up, exposing too large an area of orifice in the stem. This may allow water to flow down some nozzles allowing back filling of the plenum from above. On starting up the air scour the air advances long the lateral in a sharp wave, sometimes deeper than the final depression. The air applied is of course that required by the whole length whereas initially it is able to discharge only through the nozzles that have been exposed. Under certain situations, it is possible to have a stable situation (Fig.21.10) with some nozzles allowing backfilling while others airlift water

340

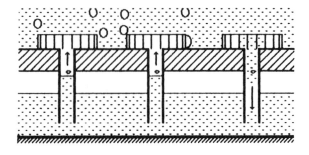

Fig. 21.10 Back syphoning in filter nozzles.

from the under drain into the bed above. This is an undesirable feature and can be avoided by correct design. Nozzles should allow back filling if over evacuation occurs but they should not allow on going recycling, with water in some continuing to flow downwards while others provide an air lift.

Orifices in the stem of each nozzle provide more accurate distribution if there are errors in levelling but slots allow a wider range of flows to be accommodated in a given stem. The situation is comparable to the difference between orifices and weirs for dividing a water flow.

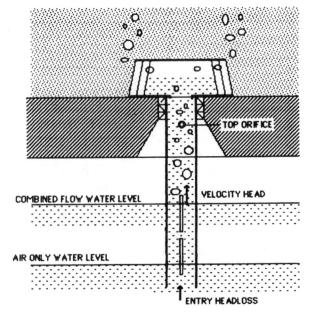

Fig.21.11 Features of a Nozzle for Combined Air and Water.

If air and water flow simultaneously the water entry loss at the bottom of the stem plus the velocity head of the water rising up the stem both increase the driving head forcing air into the stem. This causes the water level within the plenum or lateral to rise (Fig. 21.11). However, by correct selection of pilot hole size, slot sizes and stem bore the level can be maintained within an acceptable band to maintain good distribution under both conditions. The water level in the plenum is maintained by the incoming flow. It is essential that all nozzles within given laterals or plenum bays are level with each other if accurate distribution is to be achieved. Some makes of nozzle incorporate a means of easy adjustment to offset errors in the concrete structure.

The top pilot hole in a nozzle allows venting of air from the underdrain after the air scour has ceased.

Various designers have their own views on the spacing between nozzles. Often nozzle densities such as the number per square metre are quoted in specifications. This parameter on its own has little meaning because the diameter of the strainer and the direction of the emerging water also influences the effective coverage provided by each nozzle. Some designs of strainer are over 100 mm in diameter and are able to deliver the water at a radius of almost half the pitch between adjacent nozzles. One US design described in the patent literature takes this approach to an extreme with a second smaller strainer ring mounted on top of a large pancake strainer. It could be argued that the optimum configuration would use strainers with a diameter of half the nozzle pitch (Fig.21.12). Small nozzles require a high density particularly if they give an upward flow. High nozzle densities on the other hand interfere with reinforcement bars within concrete slabs. These bars must be covered with a minimum thickness of concrete in order to remain protected against corrosion.

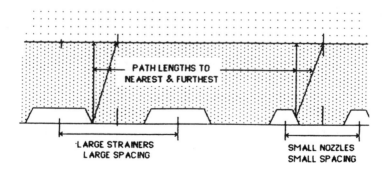

Fig. 21.12 Effect of nozzle size on nozzle density.

Filter nozzles are sometimes installed in lateral pipes embedded in gravel packing. However, they are mostly used to provide a smooth floor in which only the strainer domes protrude above the concrete or steel plate. In this case the whole of the granular material including the packing layer is exposed to the wash flow including the air and any danger of growth of organisms in stagnant regions is minimised.

21.10   Parallel Pipe Lateral Floors

One common long established type of pipe lateral floor is the type where the pipes are cast into concrete (Fig.21.13). This floor has either a side collecting duct or a central underfloor duct. The latter version has header pipes to distribute the air into the laterals whereas the former version operates during air scour with a continuous air water interface throughout the floor and air enters via the collecting duct. Such a floor is not intended for combined air scour and water wash but it can be fitted with a separate air lateral above the packing layers if a change in washing technique is required. Filters of this type have been constructed up to sizes of 150-200m^2.

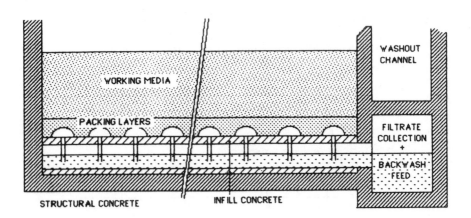

Fig. 21.13 A lateral floor set in concrete.

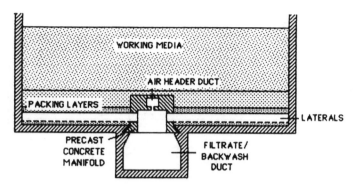

Fig. 21.14 An older type of perforated lateral floor.

The simplest floor concept is probably the simple header/lateral assembly (Fig. 21.14) for water alone with perforations on the under side to avoid upward jets penetrating into the fluidised bed. Graded packing is placed around and above the pipe to prevent percolation of the working media and to provide a level base for the latter. The size of the water orifices is designed to give satisfactory distribution both within the manifold and also into the fluidised bed, (the latter aspect is discussed in more detail in Section 21.13). The spacing of the laterals in typical cases ranges from 150-300mm. This will in turn depend to some extent upon the depth of the packing layers which distribute the water from the laterals uniformly into the base of the working media. A wide spacing however increases the cost of the packing layer and makes greater demands upon it but reduces the cost of the laterals.

If air scour is required with perforated laterals, holes must be provided both on the top and the bottom (Fig.21.3). The top holes are calculated to give a headloss corresponding to 1/2 to 3/4 of the lateral diameter at the design airflow. This ensures that the air/water interface within the lateral is deep enough to provide a sufficient cross section for the air to pass along the entire length of the pipe without forming waves. The water which is displaced at the start of the air scour and passes down through the water orifices at the bottom. Air is admitted to the laterals via metering orifices, from a separate air header above or from a collecting duct below (Fig.21.15). Such a design usually requires a fairly deep layer of packing gravel which adds to the cost. In some examples, strainers are fitted over the orifices to avoid using packing layers. One advantage of the perforated lateral design is that fairly long laterals can be used because internal friction is low. However it has been criticised because the

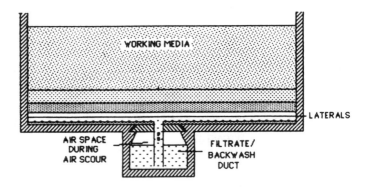

Fig. 21.15 A centre fed pereforated lateral floor.

packing layers below the lateral are out of the line of flow, neither are they air scoured. This has been claimed to allow the growth of organisms.

It is difficult to use this floor for combined air and water washing, except in short lengths, and almost impossible to design a lateral to distribute both separate and combined air and water washing, except in very short lengths. The headlosses for the air differ very considerably in the two cases and it is not possible to achieve one size that suits both duties. However the Leopold floor discussed below illustrates one way round this problem. Combined air and water is normally applied either from plenum floors where the velocities are very low, or with floors employing separate laterals for the air and water respectively. These do not have an air/water interface within them whilst in operation. Separate air laterals are often located above any packing layers to avoid disruption of the latter.

Many proprietary variations on the above theme exist. Some use strips of perforated or wedge wire mesh as the strainer. However flow is controlled by orifices set beneath the mesh in the same way as discussed above.

21.11 Suspended Floors

This term comes from the building industry and implies a floor which is supported between dwarf walls or on columns, and which has a space or plenum below. Several varieties of suspended floor are available. They mostly following a common theme (Fig.21.16). Lateral floors tend to use tubes 100 mm. diameter or less and entry for inspection is impossible. However, once access to the plenum becomes possible then means are usually requested. Steel filters tend to be the exception but large filters often have a plenum between 600-1000 mm deep with an access hatch. Such floors are constructed either from precast slabs or with a monolithic concrete slab cast on permanent shuttering panels. Both versions are laid on dwarf walls. The pre-cast slabs which have grooves around the edge for sealant are bolted down and the joints sealed with grout. In the monolithic version, reinforcement bars are linked to the structural reinforcement to produce a very strong design. It avoids the considerable amount of grouting and the potential points of leakage of the slab design.

The large plenum of typical suspended floors produces low velocities. Such floors are widely used for combined air and water washing but are equally suitable for separate air and water washing and combinations

Nozzles in such floors are usually inserted in bushes which are cast into the concrete. The pilot hole already referred to is set within the bush to enable residual air under the slab to escape at the end of the air scour. Means of dissipating the energy of the incoming flow from the upwash main may be rquired and a side collecting/feed duct is commonly used for this purpose. This serves as a secondary manifold with air entering at a high level and water at a low level respectively.

The main disadvantage of the suspended floors compared with lateral floors is the additional depth of excavation and concrete walling. If a nozzle should be broken which is usually rare, a lateral floor can be flushed through to the collecting duct whereas a plenum floor requires access for the removal of media. Nozzle breakage is most commonly attributed to careless use of shovels by labourers although designs which rely on solvent or ultrasonic welding may be less reliable than those which use mechanical assembly. With such large numbers (perhaps 100,000 in a large plant) absolute perfection in this respect is essential. It should be noted that more air is trapped in the continuous air space that forms under a suspended floor during the air scour and this displaces much more water than from a lateral floor. This has implications in the design of the top end.

Steel floors used in smaller filters and pressure filters follow the same principle and nozzle operation is identical. One problem with suspended steel floors is corrosion protection. Often corrosion has been accepted and an extra thickness allowed. For this reason, many pressure filters use lateral floors set on top of a mass concrete infill so that all of the internal surface is accessible and can be repainted after removal of the media. The concrete is non-corrosive with respect to steel.

It is interesting to note that perforated pipe lateral floors cannot be pressure tested although header blocks on floors have been known to blow up. Pressure testing of suspended floors appears to be common and for this purpose blanking plugs for each bush are supplied. These also prevent ingress of construction debris before the fitting of nozzles. Nozzle type lateral floors can be pressure tested but seldom are. In some cases stand pipes are included in the wash main to limit the pressure applied to the floor.

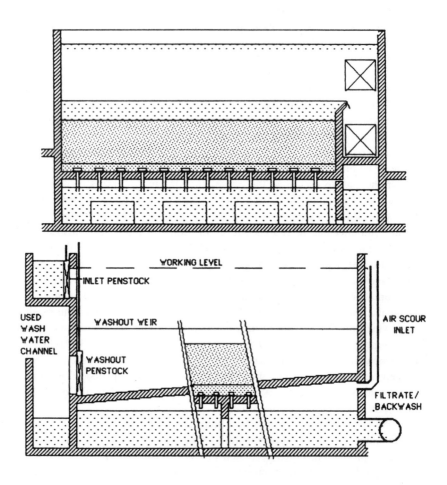

Fig. 21.16 Typical arrangement for a suspended filter floor.

One particular advantage of the suspended floor is the assurance against blockage that is provided by the plenum. It can act as a settlement basin so that any solids entering with the backwash water are removed and do not enter fine slot strainers from below. However by the same token a suspended floor would be undesirable in an upflow filter fed with settleable solids.

A disadvantage of the suspended floor which is not always recognised is the considerable volume of the plenum which increases the volume that must be run to waste to avoid the initial lower quality filtrate produce after washing from going into supply. It also interferes with the procedure of injecting polyelectrolyte into the filter in the final stages of backwash, which is sometimes useful in minimising the above early penetration.

21.12 Other Designs

Whereas the above are most common, there are other interesting designs of which the following are representative.

21.12.1 No-Fines Concrete (Fig. 21.17)

This and the following design are related although very different in appearance. 'No-fines concrete' is produced by omitting or reducing the sand in the sand-gravel-cement mixture normally used. It contains pores like gravel however the pebbles are cemented together but with less strength than normal concrete. The no-fines concrete floor uses slabs of this material to provide a permanent packing layer over a plenum with pre-cast or standard dwarf walls for support. The headloss for water distribution appears to be low ie less than the dead weight of the slabs and hence low velocities in a deep plenum are essential to good distribution. The air is

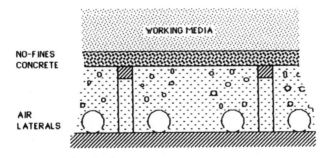

Fig. 21.17 A "no-fines" porous concrete floor.

348

distributed by perforated lateral pipes laid under the no-fines concrete slabs. Such pipes are able to provide accurate distribution of the air and the columns of air rising through the slab and through the media carry to water uniformly up through the media. This type of floor is used extensively in France with combined air and water washing. In this situation the working media remains below the fluidisation threshold and has a positive influence on distribution. There is little upthrust and the no-fines floor slabs behave like packing layers. Thus such a floor is automatically self venting at the end of the air scour. With water backwash on its own it is not likely to give very even fluidisation. Should blockage occur the no-fines concrete will be difficult to clean and expensive to replace.

21.12.2 Tunnel Block Floor (Fig.21.18)

This related floor which was developed in Germany which has been installed in the USA and to some extent in the UK, operates on a similar principle to the above but uses precast concrete tunnel shaped blocks laid in rows on the structural floor. These blocks accommodate air lateral pipes laid underneath and reverse graded packing (ie. coarse-fine-coarse) is placed on top. These blocks which are 150-200mm high are designed to allow both air and water to pass from below. The water is distributed from an under floor duct but because of the relatively small section of the tunnels laterals, the reach from the central duct to the side of the filter is less than most other floor designs. Filters therefore tend to be long and thin, not the shape for a minimum wall area. In a more recent development the tunnel blocks are formed as light weight moulded plastic containers which are filled with concrete on site before laying.

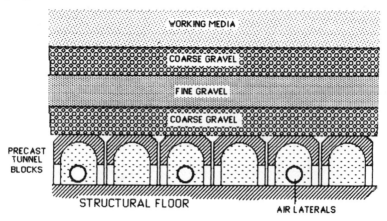

Fig. 21.18 Tunnel block floor.

Like no-fines concrete floors, this system is most appropriate for combined air scour and water backwashing and the water rate tends to be below Amirtharajah's optimum with air rates correspondingly higher in order to reduce the amount of maldistribution inherent in the design. It would appear to be a specific design for smaller filters for combined air and water washing of relatively coarse media or where gravel is cheap.

21.12.3 Leopold Floor (Fig.21.19)

This is a hybrid that has been used extensively in the USA, originally for high rate water washing but has been adapted for air scour and combined air/water washing. It could be described as both a suspended or a lateral floor. In the latest version it comprises a large moulded lateral box (originally in clay but now in plastic) with a primary manifold that distributes into a pair of secondary ducts. The secondary ducts have a upper face which forms the perforated filter floor. The main headloss for distribution is at the orifices between the primary ducts and secondary ducts. The first version had perforations across the virtually continuous exposed surface and support gravel was laid on top. The later version has 25mm sintered granular polyethylene slabs screwed to the plenum boxes. The intention is to provide a continuous floor equivalent to a sintered porous plate or no fines concrete floor but with the necessary headloss characteristics for good distribution and fluidisation under all methods of operation.

The primary duct has holes at more than one level and is able to operate in the same way as the perforated laterals of Fig.21.14, except that the media is kept back from the orifices.

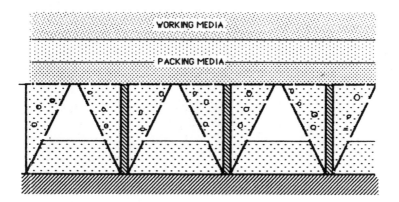

Fig. 21.19 Leopold moulded lateral floor,

The manufacturers claim that the floor can be used for combined air and water washing. The combined duct is rather deeper than the conventional round lateral and by having orifices at several levels, it is possible to alter the area available for discharge of air and so offset the effect of the headloss caused by the water flow. This helps to control the depth of the air channel.

21.12.4 The Twin Duct Lateral Floor (Fig.21.20)

This is another patented version which is related to the separate air lateral design but uses separate air ducts incorporated within a single extrusion instead of using completely separate pipes. This particular design retains all the merits of completely separate distribution systems but provides saving in the cost of materials and installation. The bush assembly connects the twin ducts to the media via strainers which are set in a concrete floor which appears from above to be a suspended floor. Such a design eliminates the extra concrete and excavation of suspended floors as well as providing even greater flexibility for varying the air and water rates independently. Support gravel is optional.

The above examples illustrate the basic principles used in filter floors. There are many minor variations but these tend to fall into the categories discussed above.

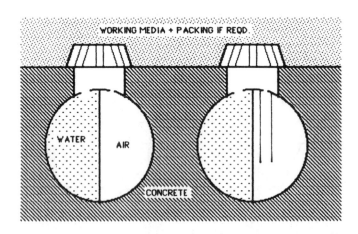

Fig. 21.20 Twin duct lateral floor.

21.13 The Effect of Floor Headloss on the Stability of Fluidised Beds

This subject has been discussed in detail in the previous chapter. To summarise, the stability or uniformity of fluidisation depends on the headloss through the floor even in suspended floors where the flow distribution based on manifold theory might be perfect. The resistance to flow through a spout in a fluidised bed has a negative coefficient against velocity so that without a positive coefficient in the floor all the flow would escape through the spout and the rest of the bed would settle to an unfluidised condition. The mathematical argument given provides one basis for predicting the required floor loss, but the situation is complex. Real spouts are not sharp columns but blend into the surrounding media. Also the proportion of the bed involved in spouts varies and also one cannot differentiate between spouting arising from spontaneous bed instability and spouting triggered by even small errors in distribution within the floor.

Clogged beds require a higher headloss to ensure flow distribution into the whole area below the clogging. This becomes more important if beds are maltreated. A safe rule is to provide a headloss equal to the loss through the fluidised media. If such a loss involves no particular penalty then it there is little point in going for a lower, less safe, figure. It may be possible to use much lower headloss when the bed is first commissioned.

The air scour plays a major role in breaking up consolidated areas and will help to prevent the worst case scenario occurring. With combined air and water the columns of air are able to punch their way through the bed and carry the water by an air lift action. Lower headlosses may then be used with impunity, as in the case of the tunnel block floor described above.

21.14 References

Baylis, J.R. (1958) "Nature and Effects of Filter Backwashing",J American Water Works Assn. **50,** 126

Miller, D.S. (1990) Internal Flow Systems, 2nd. Ed. B.H.R.Information Services, Cranfield, U.K.

CHAPTER 22.

TOP SIDE DESIGN

22.1   Introduction

The design of a filter above media level is perhaps less critical than that of a floor. A bad floor can quickly lead to inadequate washing and irreversible clogging. However a bad top side design can lead to a serious loss of media or inefficient flushing of the solids during backwashing.

22.2 Feeding

The hydraulic resistance of the granular bed during filtration automatically provides good flow distribution and variations in the underdrain resistance have little effect on the flow pattern.   Likewise there is no need to provide any form of distributor at the top of the filter providing high velocity jets strong enough to gouge the sand are absent. Thus in a pressure filter with a top inlet a target plate is normally fixed across any central downward inlet to disperse the incoming stream. With light materials such as ion exchange resin more elaborate arrangements may be necessary, as discussed in Chapter 14.

Gouging of the media is a particular consideration in large gravity filters but the problem is usually restricted to the refilling stage after backwashing when water cascading from a height may gouge a hole, down to the underdrains in a severe case. Deflector shelves are provided in such instances. A good arrangement is to place the inlet above the washout channel so that the energy is dispersed away from the granular media. The flow may then be diffused and move gently over into the zone above the media.

An operational solution for this type of problem is to refill the filter partially by extending the backwash flow with the washout valve or penstock closed. Extra water used for backwashing in this way is not lost, but is merely refiltered.

Refilling of gravity filters from the inlet can also cause problems if not controlled. For example, if the inlet is opened fully it is sometimes possible to starve the remaining duty filters. This can cause hunting of the flow which in turn can lead to breakthrough of turbidity as well as making extreme demands on the chlorine residual control system downstream as the filtrate flow fluctuates.

Problems of this type are fairly easily overcome providing that they are anticipated.

22.3 Washout Cills and Launders

The used water welling up from the bed during backwash is decanted off either into overhead troughs or over a side weir into a channel constructed within the body of the filter. The cill of the side weir is well below the rim of the filter. The choice between side weirs and overhead launders is an area of considerable controversy but the reasons can perhaps be explained. American practice has favoured high level wash launders whereas British and European practice has favoured side weirs.

French (1981) has considered the flow pattern converging towards wash launders and has concluded that for a uniform draw, round bottom launder troughs should be no further apart than three times the height above the media surface, the trough width should be 0.25 times the centre to centre spacing and the depth 0.2 times the same parameter. (Fig.22.1). This may be valid for a horizontal flow settlement tank where the floc has a negligible density but it ignores the self levelling effect of fluidised beds of real filter media, and the fact that the combination of a properly designed floor and a uniform bed depth provides a very accurate distribution of the flow up to the surface of the media. The velocities rising towards a high level trough will never generate heads comparable to the regulating headloss from below and the density of the media ensures that the fluid bed surface is for all practical purposes level.

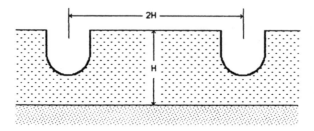

Fig. 22.1 Launder spacing (after French).

The reason for the use of high level washout launders in the USA appears to have followed from the use of high wash rates and bed expansions often employed there. High level troughs ensure that the washwater rises vertically but the displaced floc particles will only be removed if they settle more slowly than the vertical rise rate of the water. On washing, floc particles in the media rise to the surface of the media.

Residual floc that is too large or heavy to rise to the launders will form a fluidised layer above the media at velocities far below the free settling velocity. The fluidisation threshold is about 0.01 to 0.05 of the free fall velocity of a particle, depending on the voidage of the settled bed and a similar factor may be expected for floc. Thus the solids washed out of the media are mobile horizontally across the surface of the media at much lower velocities than their free fall rate and they can more easily be removed with side weirs. Residual particles are carried towards the weir by drag of the cross flow. Also the Bernoulli effect tends to lower the effective pressure as the flow accelerates towards the weir. The surface of a fluidised bed will therefore tend to slope up towards the weir (see next section). Velocities near a weir often approach 1 m/s. A highly expanded sand bed will start to be carried over if the free board is too small. Low level side weirs are more compatible with low expansion washing preceded by an air scour and high level launders are more appropriate for high expansion washing.

22.4 Displacement of Suspended Solids

Earlier, it was indicated that displacement of dirt from the bed requires not much more than the interstitial bed volume because conditions within the bed are similar to plug flow. The situation above the bed is rather different with high level launders. There is likely to be a fair degree of back mixing because of differential densities. The floc will be denser than the washwater and density convection can be expected. (The system is similar to fluidised bed floc blanket clarifiers where the concentration tends to be uniform or to increase with depth into the blanket.) One therefore has the equivalent of a fully back mixed reactor where the concentration decays exponentially with time. The decline of concentration with volume passed is illustrated by the following table:-

Table 22.1

Displacement	1	Residue	37%
(units of	2		13.5
supernatant	3		5.0
volume)	4		1.8
	5		0.7

Thus for 99% removal of floc, 4.6 volumes are required and for 99.5%, 5.3 volumes. Because of practical considerations and the launder spacings proposed by French, the volume above the media beds with high level launders is often similar the bed depth and hence a wash of five bed volumes or more appears to be necessary.

A low level side weir with a much shallower supernatant layer might have a cill height of only 150 mm above the sand level. A wash rate of 5 mm/s over a 5 metre wide bed will produce a crest height of 60 mm, thus 5.3 times the supernatant volume would only be equivalent to 1.1 metres or about 2 bed volumes. Adding a further bed volume to displace the dirt from the bed gives 3 bed volumes, a fairly common quantity in Britain. The velocity approaching the weir in the above example would be about 0.12 m/s and is not likely to carry sand overboard.

22.5 Cross Flow

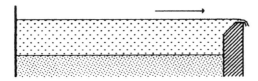

Fig. 22.2 A common side weir profile for separate air/water washing.

When side weirs are used to decant the wash water the supernatant water must flow across the surface of the media bed to the weir, accelerating from zero at the back wall to a reasonable velocity as it approaches the weir. Concern is sometimes expressed about the distance of travel from the back wall of the filter to the washout weir (Fig.22.2). In fact observations suggest that within reason, a longer draw is more efficient because the cross flow velocity over most of the bed is higher. Problem areas in filters are more usually located at the dead end by the back wall. Obviously there is a limit but a 10 metre draw may be found in many filters using traditional fine media and a separate air plus water wash. The velocity approaching the weir is given by the equation:-

$$u_{max} = u_b \cdot \frac{W}{h}$$

where u_b is the backwash rate (m/s), W the width of the filter and h the depth of water over the media (ie. the freeboard to the cill plus the crest height of the water spilling over the cill). (u_{max} is used here to describe the maximum value and is not to be confused with other uses of this term, all of which signify the maximum in varying situations.) The profile of the weir is important. High level launders have traditionally had rounded bottoms. It is doubtful whether there is any significance in this with sand media although if the density of the sand is ignored, there is an argument for rounded bottoms. Side weirs are more critical and an upstream chamfer is found to be beneficial in promoting a smooth convergence to the crest. A sharp edge sometimes appears to encourage the formation of a frontal vortex that picks up media under the upstream face. There must always be a clearance between the expanded media and the cill crest.

The effect of the cross flow is more pronounced with anthracite. The frictional drag on the anthracite can be calculated using the equivalent of the Colebrook/White equation . The hydraulic diameter of the supernatant layer is the equivalent of 4 times the depth. Water emerging from the bed in areas away from the weir must accelerate the down stream mass causing a relative pressure across the bed of u^2/g (not $u^2/2g$). Consider an example of a 10 metre long bed of sand plus anthracite, the latter having a density of 1500 kg/m^3, expanded to say 60% voidage. The submerged density of the fluidised bed will be 200 kg/m^3. Let the freeboard between the expanded bed and the weir be 200 mm. The crest height of at a backwash rate of say 10 mm/s (36 metres per hour) would be 146 millimetres ie. a total depth above the media of 346 mm. The velocity across the bed towards the weir will be 0.286 metres per second.

The velocity head $u^2/2g$ is 8.5 mm water gauge which at 200 kg/m^3 will elevate the interface by 42 millimetres. This deflection will further reduce the depth. On recalculation it appears that the actual elevation will be nearer 60 mm. Thus the proposed free board may well be inadequate. The surface profile is actually parabolic with a elevation relative to the level at the back wall varying with the square of the distance from the back wall.

If in the above example the media is not cleaned properly and becomes coated with a slime layer, the effective grain density will fall. Thus if the diameter is doubled by a slime layer, the volume will be 8 times larger and the effective density will fall to 1062 kg/m^3. A 60% expanded fluidised bed will have a submerged density of only 25 kg/m^3. The 8.5 mm momentum head will correspond to a deflection of 314 mm. Considerable quantities of such coated media would be lost from the system. The side weir system is therefore best suited to clean media in filters of not exceptional length. Even in the latter case, a 5 m. wide bed will provide a deflection of only 85 mm. which with an appropriate free board can be tolerated.

22.6   Air Scour

The weir design becomes far more critical when air is applied to the bed because media is entrained in the wake of air bubbles and carried to the surface. Thus even with a separate air scour light media, such as anthracite, is carried upwards to the surface from whence it settles back. When air is applied to a filter there is always some displacement of water from the underdrain and if the water level is already level with the washout cill some media can be lost. The chamfer on the upstream face of a washout weir helps to allow the media to settle back into the filter and keeps the bubble path a little way back from the edge of the cill. A flat topped weir collects a little media at each air scour and is washed overboard when the backwash starts. A little at each wash amounts to a significant amount after several hundred washes. Sometimes, filters are washed with the washout channel flooded to prevent overspilling of the weir when the air scour starts. If the weir is drowned there will be no cross flow during displacement of water from the under drain.

When air and water are applied simultaneously, a much wider front face chamfer must be provided on the washout weirs. The width is given by the equation:-

$$W=F\left(\frac{Q_w}{u_a}+\frac{Q_w}{u_s}\right)$$

where Q_w is the flow (m^3/m length of weir), (equal to $u_b.L$), u_a is the air bubble rise rate (m/s), typically 0.3 m/s, u_s (m/s) is the settlement rate of the smallest fraction of the media and F a safety factor, say 1.25-1.5. Thus for example with a 4m wide filter, washing at 5 mm/s the horizontal width of the chamfer would need to be 300mm multiplied by the safety factor for 0.5-1.0mm sand.

This equation which is derived on the same basis as the Hazen equation for settlement assumes that the air bubbles rise, release the grains of media which then fall back to the chamfer and slide back into the bed. Because of the size of the chamfer in practical situations, the crest of the weir is typically about 500 mm above the media with combined air and water washing.

To accommodate the resultant weir width, some designers place the chamfer centrally over the weir wall with a return chamfer on the front face (Fig.22.3). This has the disadvantage of diverting air rising from below and produces an extra air curtain at the toe of the chamfer which in turn reduces the disentrainment efficiency. A similar situation exists with high level washout launders if used on filters with combined air scour and backwash. The air rising from below spills out each side. It is difficult to accommodate the necessary chamfers on such launders and several designs for separate deflector plates (Fig.22.4) have been promoted e.g. Louboutin et

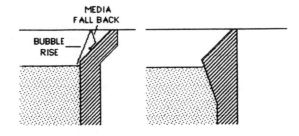

Fig. 22.3 Side weir profiles for combined air water washing.

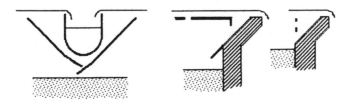

Fig. 22.4 Deflector plates used to minimise the loss of media.

al. (1986), O'Brien (1993), and Trainor (1984). Certainly, side weirs have been used most extensively with non expanding washes and have proved very satisfactory on plants designed by many contractors and consultants. The disentrainment problem of course becomes easier as the media size increases and particularly when the settlement rate exceeds the rise rate of the air bubble.

It is often said that combined air scour and backwash is not possible with anthracite or dual media filters. Certainly the chamfer in the simple side weir design becomes rather large but examples now exist. Again the situation improves as the media becomes coarser. One solution for anthracite and dual media is the so-called ,rising wash' where the supernatant water depth is usually greater than 1 metre. The filter is allowed to drain down nearly to the level of the media or indeed into it and the air scour initiated. The backwash is then brought on and the level allowed to rise in the chamber without overspilling. The air is then shut off early enough to allow venting of the underdrain and displacement of air from the bed before the water overtops the weir or launders. The duration of the combined wash is of course limited in this case and there is an advantage in aiming at a lower wash rate and a higher air rate when using Amirtharajah's equation (Chapter 21) in order to prolong the wash. This procedure is a sound one but has the disadvantage of using additional wash water. A higher backwash rate is needed to carry the floc up to the troughs although with dual media filters a regrading rate will in any case be required to separate the anthracite from the sand after the wash, during which the two will mix.

22.7   Surface Flush

Many open gravity filters are equipped with a perforated trough mounted on the back wall of the filter opposite the washout weir. Unfiltered water can be introduced into this trough towards the end of the backwash and this water augments the cross flow. It can be allowed to continue after the upwash flow has ceased. This procedure was promoted originally as a further development of low level side weir filters as a means of reducing wash water consumption. The intention was to reduce the wash duration to the point where the displacement from the bed was complete and then to flush the floc from the surface using raw water (or settled water). In theory, this was a sound approach and which should reduce the wash volume to not more than about one bed volume. The flush was achieved by arranging for the inlet penstock to discharge into a trough manifold (Fig.22.5) which was flooded when the filter was in service. This also neatly overcame the problem of gouging of the media during refilling of the filter.

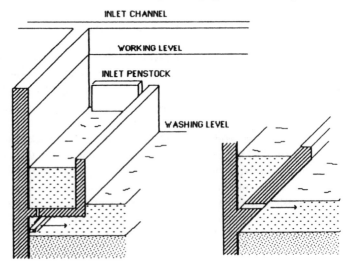

Fig. 22.5 Two types of surface flush trough for side weir filters.

In practice, operators and commissioning engineers tended to let the backwash run until the media could be seen through the clear backwash water which made the surface flush pointless. In this arrangement, the cross flow during the surface flush is probably about only 25% of the backwash flow. This is rather marginal for conveying particles resting on the unfluidised bed, and being unfluidised the mobility of the solids is greatly reduced.

Some filters washed with combined air and water also have this type of surface flush but, with the somewhat higher weir level in such filters, the cross flow velocity is even lower. There is nevertheless a theoretical saving in that some the displacement volume can be made up of raw water rather than filtrate if the cross flow is started early enough. There has been a tendency in recent years for this feature to be scrutinised and abandoned as not being particularly cost effective.

22.8   Surface Scour (Surface Wash)

This confusing term, already mentioned in Chapter 21, describes a rather different concept to the above, namely static or rotating reaction propelled arms carrying high velocity jets mounted close to the surface of the filter media. They are popular in the USA where they have been employed to assist cleaning of media washed with water alone. They do not appear to have been used in filters with air scour which is becoming increasingly popular in the USA. Some details are given by Cleasby (1990).

22.9   Pressure Filters

Much of the foregoing relates specifically to open gravity filters. Many filters, particularly smaller ones are enclosed and contained in a pressure vessel. They may operate at pressures ranging from close to atmospheric pressure to several hundred metres head. Some of the above considerations still apply to such filters. These may be horizontal or vertical (this description applying to the axis of the cylinder). Horizontal filters have lengths corresponding to the reach from the back wall to the weir of many gravity filters eg, 6 metres. The height available in a horizontal filter makes rising washes (Section 22.6) rather difficult. On a vertical filter washed with combined air and water the washing arrangement must still include a disentrainment system if simultaneous overspill is intended. In many vertical filters a top connection is used both for the inlet and washout. Larger filters often have troughs across the diameter, similar to open filters.

22.10   Launder Design

Launders (or decanting troughs) must be able to carry away the water spilling into them without flooding or backing up. Side weirs usually discharge into a concrete or steel channel which often forms the upper deck of the filter collecting channel (Fig.22.6). The cross section will often be governed by the latter. High level washout launders fabricated in steel, GRP or concrete on the other hand are purpose-made and will be designed more precisely. Launder design is an exercise in critical flow, and a full treatment of the subject is beyond the scope of this book. Readers are referred to books on open channel flow such as Chow (1973).

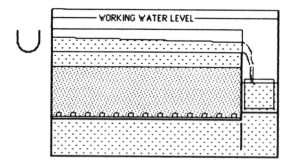

Fig.22.6 High level washout launder configuraion .

Various shapes of trough may be used. (Fig. 22.7) It makes little difference whether such launders receive a continuous feed along their length or at one point. The depth of flow for a given width is governed by the critical depth which is that which provides a minimum total energy situation. For the present purpose the total energy per unit volume is given by :-

$$E=h\rho g+\frac{u^2\rho}{2}$$

where u = velocity, ρ = fluid density, h = depth/height of flow.

For any flowing system:-

$$u=Q/A$$

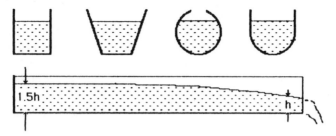

Fig. 22.7 Washout or decanting launder sections.

Hence for any specific shape of cross section, one may write a computer programme to consider a set of thin depth increments and calculate the area for the cross section in question ie, rectangular, trapezoid or circular. From this cross section the velocity for a given flow may be calculated and thence the total energy per unit volume ie. depth plus velocity head. Starting at a shallow depth the energy sum will fall as the velocity falls. The point where the energy sum levels out and starts to rise is the critical depth. The driving head is 0.5 of the critical depth, ie. the dead end depth is 1.5 times the critical depth (not 1.732 times the critical depth as stated in some texts).

For a level rectangular trough:-

$$u = \frac{Q_L}{h}$$

where Q_L= flow per unit width, and h the depth of flow.
Substituting and differentiating :-

$$E = h\rho g + \frac{Q_L^2 \rho}{2h^2}$$

$$\frac{\delta E}{\delta h} = \rho g - \frac{Q_L^2 \rho}{h^3}$$

For critical flow this differential term is zero. Thus:-

$$h = \left(\frac{Q_L^2}{g}\right)^{1/3}$$

Q_L is in m³/s/per m width of trough, h is in m.

Computed data for circular launders is included in Fig. 22.8. The data will be safe for other sections which can enclose the diameters indicated, eg. squares or rounded bottoms. For more precise information specific calculations are necessary. The level in the receiving channel must be below the critical depth in the decanting launders. Launders usually have a level invert.

In designing a trough it must be remembered that the main load on a high level launder is the buoyancy when empty ie. the upthrust.

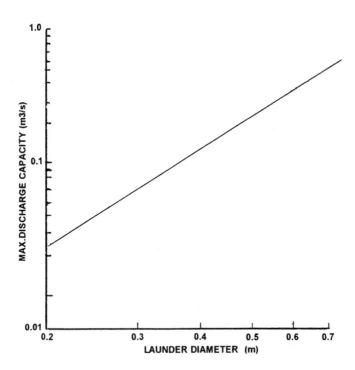

Fig. 22.8 Capacities of round collection pipes in trough type flow.

The crest height over the top is that of a simple weir where the following simple equation given by Coulson and Richardson is adequate:-

$$h = \left[\frac{Q_w}{2/3 . C_d . \sqrt{2g}} \right]^{2/3}$$

Where $C_d \sim 0.6$ and Q_w = flow per unit width of weir.

Where high level launders are used, the total length is considerable and hence the crest height is small. Accurate levelling is therefore essential unless the depth is deliberately increased by castellation.

A filter that has been air scoured may well carry a head of foam. For this reason, the surface should preferably spill over a weir without obstruction. A low level submerged washout dsischarge or submerged orifices would not be acceptable under such circumstances.

22.11   Syphons

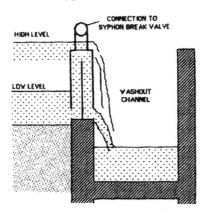

Fig. 22.9 The Paterson dual level syphon weir.

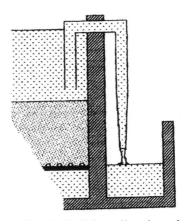

Fig. 22.10 Side wall weir syphon.

In some designs of filter weirs have been provided with syphons to allow the draw off level to be adjusted eg, Fig. 22.9. These syphons enable the level to be drawn down to a point close to the media and also allow a degree of backing up later in the wash to permit overspilling to remove floating foam and scum. The object of this arrangement was to reduce the volume of water used for displacement of supernatant floc. A somewhat different use of syphons is illustrated in Fig.22.10

where washout weirs are replaced entirely by syphons. This arrangement is used in one particular context for combined air scour and water backwashing where the air is applied while the filter is filling and then cut off before overspilling commences. Priming of the syphons is encouraged by a higher rate of backwash following the air scour. The process is repeated several times until a sufficient volume has passed.

Washout syphons are not common but are popular with certain designers.

22.12 Water Depth

Depth of the supernatant layer of water varies with the general philosophy of the designer. If the water within a filter contains dissolved air above saturation at the prevailing hydrostatic pressure, then such air is likely to nucleate on any surfaces present and therefore grow within the pores of the media. In so doing they will cause clogging, described as air blinding. The hydrostatic pressure in a filter is depicted in Fig.22.11. It rises linearly with depth into the supernatant layer and continues with a different slope into the bed as the resistance to flow counteracts some of the hydrostatic head. As the filter clogs, the floc and the headloss are concentrated near the surface of the bed. If the supernatant water is shallow and the clogging headloss high, the hydrostatic pressure can become sub-atmospheric, described as a 'negative head.' A safe design for most situations is to ensure that the water depth is always greater than the clogging head ie. the total headloss less the clean bed headloss at the same rate.

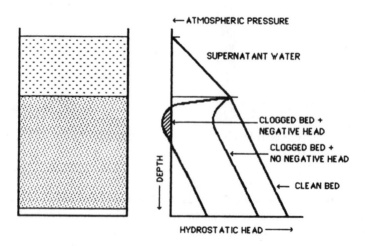

Fig. 22.11 Hydrostatic profiles in a working filter.

In practice, this approach is often very conservative particularly in coarse media and with anthracite. With coarser media the floc does not deposit in a shallow layer, but is distributed into the bed. Computer modelling can be used to predict the point at which the hydrostatic pressure falls to the atmospheric level which can be well down in the bed. Less supernatant water is then required to maintain safe conditions.

With very coarse media, the local clogging gradient can be close to unity in which case, no water cover may be needed apart from that required to feed the filter. The above assumes that the water is not super saturated. This may not always be the situation with reservoir waters during algal blooms or in river waters with vigorous weed growth. In such situations, cascade aeration (de-aeration) may be necessary to avoid problems. In some applications such as tertiary filtration of sewage, the less soluble gases such as oxygen are probably well below the saturation level, not likely to cause a problem.

Another approach has been to limit filtration rates and deliberately operate filters to a low headloss to avoid negative head conditions with a relatively shallow water depth eg. 500 mm. The penalty will be a shorter filter run, but this is often an acceptable solution as it offers a simpler design without washout penstocks. Such shallow water filters are operated with the water level below the washout weir and no washout penstock is required. On backwashing, the filter is allowed to spill over into an otherwise dry washout channel as illustrated in Fig.24.4.

22.13 References

Chow, V.T. (1973) Open Channel Hydraulics, McGraw Hill

Cleasby, J.L., (1990) Water Quality and Treatment, p517, American Water Works Assn. Denver.

French, J.A. (1981), "Flow Approaching Wash Water Troughs", J. Environmental Eng. **107,** 359

Louboutin, R. Durot,J and Gibaud, P. (1986) "Filter Bed Device", British Patent 2,174,013

O'Brien, M.J. (1993) "Baffle Assembly for Air and Water Backwash of Media Filter", US Patent 5,207,905

Trainor, A.I. (1984) "Filter backwash baffle", US Patent 4,479,880

CHAPTER 23

OPERATION AND CONTROL OF MULTIFILTER INSTALLATIONS

23.1 Introduction

Granular filters are seldom installed singly. They must be cleaned by backwashing and unless the plant flow is reduced the remaining duty filters must carry the diverted flow when one is being washed. Furthermore the resistance to flow through a filter increases as it collects solids. If a clean filter is placed in service alongside a clogged filter the flow through the former may be so high that solids break through. Thus means of regulation are required. These vary according to the design and whether the plant is an open gravity one or enclosed and under pressure.

23.2 Operation of Pressure Filters

Pressure filters have already been mentioned in the previous chapter, Section 22.9. There is no fundamental difference between open gravity filters and pressure filters except that the latter may be allowed to run up to higher headlosses and cannot overflow. A high loss gravity filter would need a very deep structure.

Very small filters tend to be pressure units. The size is limited by the dimensions of readily available pressure shells. Small ones are usually 'vertical' (ie. a vertical axis) with many examples of 2.4m diameter and some up to 5m. However transport becomes difficult at this size. Large pressure filters are more usually horizontal, again 2.4m being a common diameter. The length is often 6m or so but this is more variable. Pressure filters are convenient when treatment under pressure avoids a pumping stage. One example is where an upland supply is to be treated at low level and forwarded to an elevated service reservoir.

A practice sometimes encountered, particularly on small transportable plants but occasionally on larger ones, is the use of an open clarifier followed by a high pressure pump delivering to a service reservoir via a pressure filter. Generally this is bad practice because the floc is often reduced to a partially unfilterable colloid. If such a practice appears attractive then pilot trials are strongly recommended as it may be difficult to meet guarantees. Polyelectrolyte dosed after the pump is sometimes useful in restoring the filterability.

23.3 Flow Balancing

If a battery of pressure filters is to be washed in sequence with the plant off line so that all the filters come back into service at the same time in the clean condition it is merely necessary to ensure that the effective dynamic hydraulic resistance across the parallel paths is equal (Fig.23.1). In this ladder situation the Bernoulli effect in the feed manifold and the momentum effects in the collecting manifold act together to increase the flow through the filter nearest the outlet. With the feed and discharge at the same end the effects are to some extent compensated but a calculation on the matrix is essential. If there is an imbalance then the filter taking the largest flow will tend to clog more quickly and compensate for the inequality providing that it is not serious. The headloss through the clogged filters is likely to be several times that through the clean bed and in practice a certain amount of imbalance can be accepted. In any case the solution is to add a balancing orifice in the outlet branch (not in the section common to the backwash) to swamp Bernoulli and momentum effects.

Typical pipe velocities are around 1.5-2m/s so that the velocity head will be around 203 mm at 2.0m/s and the collection momentum head 406mm. The net effect, ignoring friction effects in this example, ie. with short pipes, will be either 203mm or 609mm depending on the arrangement. To keep the difference in flow to within 5% the turbulent loss would have to be 2.0m or 6.0m respectively ie. $1/(1.05^2-1)$,(assuming that the clean bed loss is low by comparison.

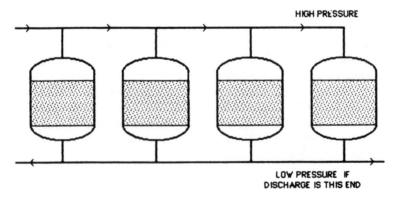

HIGH PRESSURE

LOW PRESSURE IF
DISCHARGE IS THIS END

Fig. 23.1 A group of pressure filters in parallel.

If the filters are not washed while the plant is off line but taken out in sequence, washed and returned in a clean condition then the one washed most recently will carry a far higher flow. The situation is worst when they are washed, probably manually, in a bunch at a particular time of day. If breakthrough is to be avoided either the system must include a significant turbulent loss to balance the flows or an automatic active control system will be required to measure the flow from each filter and control the valves . If the filters are washed at equi-spaced intervals this produces a declining rate mode of operation. The degree of decline can however be controlled, if necessary at a low value.

Ignoring manifold errors the losses between the inlet and outlet manifolds may be listed as follows:-

Inlet = ΔH_i Outlet = ΔH_o Media = ΔH_m Dirt = ΔH_d

Total = ΔH_T All these losses are based on the mean or design flow.

Let the permitted flow variation as a ratio about the mean be from 1+R to 1-R. Thus balancing the clean and dirty situations:-

$$(1+R)^2 (\Delta H_i + \Delta H_o) + (1+R) \Delta H_m = H_T$$

'Clean'

$$\Delta H_T = (1-R)^2 (\Delta H_i + \Delta H_o) + (1-R) (\Delta H_m + \Delta H_d)$$

'Dirty' (The values of the above headlosses are those at the design flow rate.) Thus for example if ΔH_i=0.3m, ΔH_m=0.3m, ΔH_d=2.0m, and R=0.2, ie. ±20% flow range, then by substitution in the above equation ΔH_o= 1.55m. This could be exchanged with the inlet loss. An orifice must be included to make up the difference between this figure and the existing valve + pipe losses. If preferred a value for R of 0.1 might be used to maintain a more constant flow.

The above situation is that of a filter which has been washed and returned to service while the remainder are dirty as with bunched washing. The same calculation will provide a design for declining rate operation if the filters are washed at equal time intervals.

To be more precise ΔH_d in the above calculations should be only (N-1)/N times the final dirt loss because the cleanest, most recently washed filter, comes back into service when the most clogged filter of the remainder still has one wash interval to run. Declining rate operation is discussed further below.

23.4 Washing of Pressure Filters

The washing of pressure filters introduces a number of practical problems. On some works the water is derived from the outlet manifold without using any storage. Thus for example four filters operating at 10m/hr can furnish wash water for one filter being washed at 40m/hr. However in a pressure installation the discharge from the filter being washed is likely to be at atmospheric pressure (Fig.23.2) whereas in service the filters are probably delivering to a considerable head. Thus unless the design includes some form of restriction an excessive flow with accompanying breakthrough of turbidity from the duty filters may occur. The situation is further complicated if air scour is required, because with the latter the filter being washed will have to refill before the full backwash flow is applied. The changes in flow through the plant as the wash sequence proceeds and the back pressure changes are likely to cause surge pressures in the supply and delivery pipelines and need appropriate expert attention if damage to the lines or pressure vessels is to be avoided.

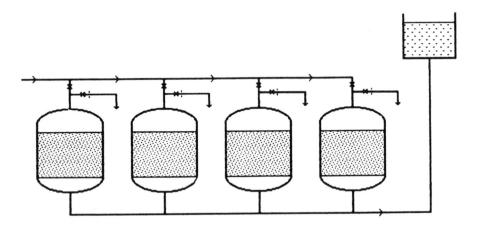

Fig. 23.2 Washing of filters from the outlet line pressure.

While the media in a gravity filter is open to view, seldom can the surface of the media in a pressure filter be inspected and users are less likely to be critical. Indeed one occasionally hears reports of neglected plants where it appears that the media had never been installed or if it had most of it has been lost.

23.5 Flow Changes in Multiple Filter Installations

In all multi filter plants whether pressure or gravity, when one filter is taken out of service while the plant is in operation the remaining filters usually have to take on the extra load. In an equal flow system this extra corresponds to the ratio $N/(N-1)$, ie 33% for 4 filters and 50% with three. Such surges can cause turbidity break through particularly on the filter next due for washing. To prevent such an occurrence the design must provide an extra margin of safety between the headloss and breakthrough end points (for example by providing a deeper bed). Alternatively more filters may be used, for example a minimum of 6 (an extra 17%), to keep the surge down to an acceptable level. Another partial solution is to adopt declining rate operation as described below so that the surge is reduced.

On small plants it may be possible simply to reduce the through put while filters are being washed to avoid excessive flows.

23.6 Gravity Filters

As already indicated, these involve open tanks. The water usually flows into and from the filters in open channels. One immediate difference in designing an open gravity filter installation is that levels have to be controlled and flooding is possible. Filters can also 'dry bed' ie. the water level can fall below the surface of the sand so part of the bed is out of service. If the level subsequently rises the bed will probably be 'air blinded' and only partly active. The are several methods of control in common use.

23.6.1 Weir Splitting and Weir Level Control (Fig.23.3)

This is the simplest arrangement and uses a set of level and equal size weirs discharging from the inlet channel into the filters. The flow over the weir is independent of the down stream level. However the weir flow is very sensitive to variations in the upstream level. Bernoulli effects and friction in the feed channel affect the accuracy of flow splitting. Friction is usually small but in some cases may need to be taken into account.

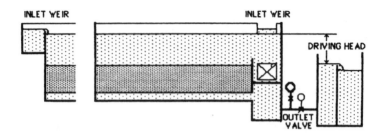

Fig. 23.3 A weir controlled gravity filter.

The discharge flow over a weir is defined by the following (approximate) equation:-

$$Q = \frac{2}{3} \cdot C_d B \sqrt{2g} \cdot h^{1.5}$$

where C_d is the weir discharge coefficient, B the breadth of the weir and h the crest height, ie. the height of the upstream water level above the weir cill. Differentiating:-

$$\delta Q = \frac{2}{3} \cdot C_d B \sqrt{2g} \cdot 1.5 h^{0.5} \cdot \delta h$$

Dividing the above equations by each other:-

$$\frac{\delta Q}{Q} = \frac{1.5 \delta h}{h}$$

The level error δh corresponds to the velocity head $u^2/2g$ (less friction). Thus:-

$$\frac{\delta Q}{Q} = \frac{1.5 u^2}{2gh}$$

Thus for example with an inlet end channel velocity of 0.6m/s and a weir crest height of 0.1m the flow range will be 27%. This error is not linear but parabolic along the line of filters. Such an error can be largely eliminated by tapering the inlet channel (possibly stepwise). It is recommended that a computer programme be drafted and used, taking account of frictional effects as well. Normally an accuracy of say ±5%, ie. a 10% range might be deemed acceptable in terms of filtrate quality but such errors will have some effect on the running time of individual filters.

The latter can also be influenced by segregation of suspended solids in the inlet channel. For example the first or the last in a line may tend to receive more than its fair share of the solids even if the flow is perfectly divided. Errors in flow can be counteracted by changing weir lengths or height but the correction is then only valid for one flow rate.

In theory orifices would be equally suitable for splitting the inlet flow providing that they discharged above the downstream water level. They are somewhat less sensitive to errors in the upstream water level but because the loss varies with the square of the flow rather than the 2/3 exponent for weirs such an option would be limited to a small flow range unless very deep inlet channels were used.

Downstream of filters an outlet weir box or a bell mouth above the water level in the filtered water channel is commonly used as a means of isolating the filters and outlet valves for maintenance when necessary. If these are placed at a level just above that of the surface of the media the latter will remain flooded even when no flow is entering the filter. To provide the necessary driving head for filtration the filter chamber must be deep enough to accommodate the media, dirt and outlet system headlosses. With such a rising level system the water entering a clean filter has to cascade down possibly some 2m which can adversely affect the floc and consequently the filtrate quality. Nevertheless the system is simple and avoids mechanical or electrical control systems with their maintenance problems. No slow start (Section 23.9) is possible apart from manual control of the inlet penstock, or diversion to waste.

23.6.2 Float Control (Fig.23.4)

Most of the objections to the above system are overcome by using a float or syphon outlet control system. Weirs are again used to split the inlet flow but the water level within the filter is controlled over a relatively narrow band. A float is linked mechanically to the outlet valve so that the valve opens as the float rises and more importantly the outlet valve closes before the level has fallen very far. Syphons are used to perform a similar action by using an arrangement as shown in Fig.23.5. A small float operated air valve admits air into the syphon as the level falls thereby reducing the pull of the syphon. In this case a separate isolating valve for back

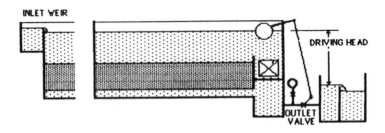

Fig. 23.4 Float control of a gravity filter.

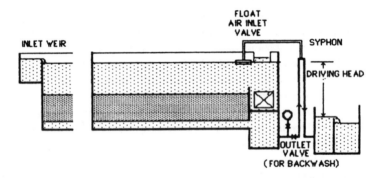

Fig. 23.5 Syphon control of a gravity filter.

washing is necessary whereas with the float outlet control the valve usually performs both functions. The same basic principle is sometimes used in electrical and pneumatic control systems where the outlet valve is controlled from the water level within the filter concerned.

With these means of control the water level in the filter does not vary nearly as much as in the previous case and the depth of the chamber can be reduced considerably. It need only be as deep as the media plus the 'clogging head' even if 'negative head' conditions are to be avoided. Also the water no longer cascades from a great height when a clean filter is returned to service ie. after refilling. The main headloss in a clean filter occurs in the outlet valve or syphon rather than at the inlet cascade, and the suspended solids being filtered escape damage. If the level control system is fairly precise submerged inlets may be used, providing that they are identical. A common inlet channel level and equal filter levels will guarantee an equal flow into each filter.

23.6.3 Declining Rate Operation

If a set of filters is served by an inlet channel with submerged orifices (valves or penstocks) rather than weirs then the hydraulic situation will be more like that in a set of pressure filters. The level or pressure in each filter will interact with that in the inlet channel or manifold and the flow will decline as the filters clog. In a plant working at constant rate the flows in the filters between each successive wash will tend to converge as the faster ones tend to clog more rapidly. The overall headloss increases. When a filter is washed a step change occurs and the cleanest filter becomes the fastest while the remainder move one step down a stairway profile (Fig.23.6).

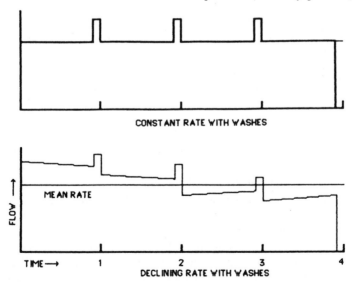

Fig. 23.6 Filter flow profiles on a constant flow plant with 4 filters.

It is possible to use the calculation procedure given earlier for pressure filters to calculate the size of a fixed orifice to control the maximum flow and flow range. Some installations have been built without such calculations and this has given declining rate operation an unjustified bad reputation. The penalty of a fixed orifice is the presence of the residual loss when the bed is clogged although the square law relationship reduces the impact somewhat. Cleasby (1981) and Hilmoe and Cleasby (1986) have vigorously promoted such a method of control particularly for plants which receive poor maintenance. Cleasby (1993) has reviewed declining rate operation more recently and quotes examples where 33% more water was obtained per filter run and with an average of 20% reduction in turbidity.

With any form of outlet control using with a float or an electrical equivalent the system can be designed both to control the maximum flow and also the remove any unwanted throttling action as the filter clogs. The designer can stipulate the flow variation required, even to the point of making the result almost equivalent to constant rate operation. Using the earlier nomenclature but adding the clean condition valve loss ΔH_v the earlier equation may be re-written:-

$$(1+R)^2 (\Delta H_i + \Delta H_o + \Delta H_v) + (1+R) \Delta H_m = \Delta H_T$$

'Clean'

$$(1-R)^2 (\Delta H_i + \Delta H_o) + (1-R) (\Delta H_m + \Delta H_d) = \Delta H_T$$

'Dirty'

Again these headlosses are those at the design flow rate. From this it is possible to calculate the valve loss (based on average flow) and hence the loss coefficient. (Data for valves is given by Miller(1990)). The valve movement or linkage ratio is designed so that it is fully open at the intended top operating water level (Fig.23.7) and at the indicated throttling position at the clean bed working level. The closed valve position will be proportionally lower down. For example the closed valve position may be 750mm below TWL and the clean bed throttling position may be say half way between. Several examples of such a system, using float control exist.

To exploit declining rate operation the filters must be washed at equal time intervals as mentioned earlier and back washing must be initiated from a datum level in the inlet channel if it is to be automated. If filters are washed in a bunch then the flow will remain equal as the filters clog. The situation is then no different from conventional practice. With declining rate operation the headloss at the inlet is small and hence floc is less likely to be damaged. The filters slow down as they approach

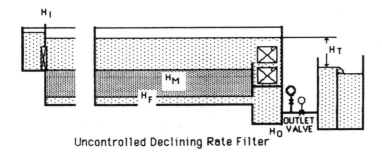

Uncontrolled Declining Rate Filter

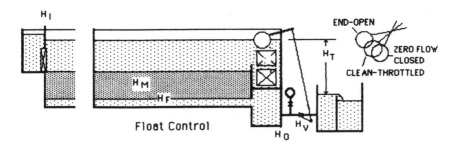

Float Control

Fig. 23.7 Forms of declining rate filter.

washing and turbidity is less likely to break through. The filter being removed from service for washing is the slowest and the additional flow transferred to the duty filters is therefore less than with constant rate operation.

Because the flow declines as the filters become clogged a greater load can be accommodated for given driving head, ie. the runs are longer. This bonus follows from the fact that energy is not being lost by cascading or through a throttled valve in the cleanest filters as occurs in constant rate operation. The main fear expressed is that the cleanest filter is the fastest. This might run counter to the philosophy of slow start which is sometimes used to offset the poorer initial quality produced by a filter. Slow start can of be added to a declining rate system by controlling the opening of the inlet or on an automated plant by over-riding the rate of opening of the outlet. A small amount of additional media will also offset the lower initial efficiency, and other options such as refiltering the first filtrate are still available.

There is a widespread lack of understanding of declining rate operation and specifications frequently (probably unwittingly) exclude it by stating that the flow shall be divided equally between the filters in service. The <u>overall</u> flow averaged over a run <u>is</u> divided equally but the <u>instantaneous</u> flow is not. A reported complaint from one operator was that he could not get all filters to run at the same rate! One unpublished evaluation of a large plant in Scotland confirmed the improved quality. It was found in particular that the background quality was similar for both modes of operation but spikes when filters were taken out for washing were noticeably less with declining rate operation. In this test it was found that the filters were not being run to the full headloss and therefore the flow variations were less than the design values. The comparison might otherwise have been even more dramatic. Cornwell (1984) has reviewed the impact of regulations and concludes that the policies in the USA tend to discriminate against declining rate filtration. Cleasby's review (1993) has already been mentioned.

23.6.4 Active Flow Control

Various means have been employed to control filters using hydraulic, pneumatic or electrical actuation. In some cases the electrical control system is merely the equivalent of one of the mechanical or hydraulic procedures already described. However there is a more flexible system that has been employed for vary many years although the hardware used differs widely. In an open gravity plant the most important point is to keep the water contained and not to allow it to overflow. Thus a level signal is transmitted from the inlet channel (Fig.23.8).

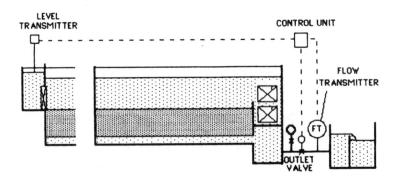

Fig. 23.8 Active filter control.

The flow enters each filter via a submerged orifice, valve or penstock. All filters have some means for measuring the flow and the overall function of the control system is to adjust the outlet valves to bring the flows into line at a value that maintains the target level in the inlet channel. Any increase in flow will initially cause the level in the inlet channel to rise, the resultant signal will open the outlet valves and allow the filters to run faster. As the filters clog the valves are opened gradually to offset the higher resistance. The signals can be processed for example to allow the most recently washed filter to run up to speed on a defined ramp before switching over to the level signal and the flow signals from a set of filters can be biased in sequence of washing to produce a declining pattern if so desired. The rate of change may also be controlled.

The earliest form of this control philosophy used hydraulic techniques and pneumatic devices for signal transmission with pneumatic or hydraulic valve operation. Electrical systems have taken over in many cases but not entirely. A hazard of such a control system is 'hunting' ie. constant over correction which produces a cyclic flow pattern which encourages breakthrough of floc. With electrical valve actuation this also shortens the life of the actuators. Filters, in contrast to many other systems that require control, operate for very extended periods but conditions change very slowly so that correction for clogging for example is not required very frequently. With modern data logging systems it is possible to follow valve positions and monitor the conditions of filters, and take action in advance of possible events. For example if a plant is running at low flow a situation can be encountered whereby all the filters are clogged to a point where they can handle the existing flow but would be incapable of handling the design flow should it be demanded. Instrumentation can anticipate such a situation and take the necessary action in advance.

23.7 Sequencing

On automatic plants the operation of valves and pumps during filter washing is controlled by a unit generally known as a wash sequencer. This may be a separate stand alone box or an integral part of a wider system.

The resistance to flow through granular media filters increases approximately linearly with the volume of filtrate passed or more precisely with the solids collected, providing that the rate is constant. Filters in a set do not necessarily behave identically because as mentioned previously there will be slight differences in flow and sometimes in the solids concentration entering the set.

With flooded inlets as in the case with a declining rate system or with set of pressure filters there is an element of self compensation. With weir flow division or with an active control system errors will continue uncorrected and as a result each filter will tend to run for a slightly different time. One feature that is not infrequently included in the sequencer is a queuing system that enables not only the initiation of the wash programme on a particular filter to be delayed until the previous one is complete but if more than one has signalled that the wash is due, eg. by virtue of headloss or turbidity breakthrough, then the washes are initiated in the order in which the signals were received. Queuing systems have been in fairly common use for some years. With declining rate operation or with uncontrolled pressure filters the headloss can be regarded as a parameter common to the set and it may be argued that there is little point in a queuing system. The filters should be washed in numerical sequence triggered by a common headloss threshold.

23.8 Low Flow Operation

When a set of granular media filters is operated at flow rates well below their design rate they are able to accumulate considerably more solids but these may well be shed if the flow is brought up to the design figure. Also as already mentioned, it is possible to find that the plant is unable to handle the design flow due to headloss limitations in this situation. It is wise to take filters out of service if low flow rates are likely to continue for substantial proportions of a filter run or longer. If the duty filters continue to run at a rate close to the design value they cannot be overloaded and when the plant flow rises spare filters can be brought into service. This may be tedious for manual operation but presents little problem with an automatic plant. This procedure also avoids the rush of filters to join the queue when the flow is increased.

Where this situation continues beyond the next wash the filter last washed will be held on standby until required or until the next filter is due for washing. There could be several filters on standby on large plants at low flow. In this situation the standby filters should rotate so that all are kept serviceable. It is unwise to leave a filter idle for long periods. The residual organic matter in the floc can become septic (putrid), causing taste problems when the filter is returned to service, algae can grow in the supernatant water and valves become unserviceable. Where filters have to remain unused for very long periods they should be drained down and left dry.

If filters are to be withdrawn from service short periods for the above reasons other than after washing the cleanest one should be selected. Filters with substantial dirt loads should always be washed otherwise the dirt may putrefy and/or dislodge when the filter is returned to service. This is one recommendation of the Badenoch Report (1990).

23.9 Slow Start, Flow Changes and Particle Breakthrough

As discussed in Chapters 17 and 18 freshly washed filters are less efficient than filters that have been in service for a while. The initial deposit acts as a collector towards particles arriving later and the filters mature. Also from Amirtharajah's work is it known that residues of the backwash also contribute to the initial turbidity and particle count. To minimise the effect on the final water quality is has been a relatively common practice on larger plants for the flow rate to the filters to be raised slowly to the working level over a period of 20-30min. rather than immediately. This procedure is known as a slow start. An alternative procedure used on some plants, particularly in the USA is to run the first filtrate to waste.

It is certainly true that slow start improves the apparent quality of the water from the plant by reducing the excursions in filtered water quality, but to some extent this is the result of reducing the flow while poor quality water is produced so that it is diluted further by the good quality water from the duty filters. There may be no change in the total solids going into supply. Indeed the first Badenoch Report, Anon. (1990) quotes trials on 14 plants in the USA where even running the first filtrate to waste had no perceptible effect on the breakthrough of Cryptosporidium oocysts. Measurements made by the author on a plant in southern England also suggested that the solids going forward were unchanged. The second Badenoch Report (Anon. 1995) comes down more firmly in favour of the slow start and gives some curves for the particle counts against time. Even these are not entirely convincing as the slow start produced higher particle counts later on. Slow start ramp rates vary from 5% per minute down to 1% per minute. Colton et al. (1996) examined particle counts during the ripening period with no slow start, and with 30 and 60 minutes respectively, and expressed the counts as a percentage of the counts emerging during the entire filter run. The 60 minute start was a little better than the 30minute start but both reduced the overall count during the ripening period by about 50%, while the particles in this period contributed about 40-50% of the total penetration. The improvement is not particularly dramatic considering that the numbers arising vary on a logarithmic scale. It is probably best where flexibility is available on modern plant to make this an operator controlled variable so that it can be optimised by experience.

The safest and most positive method of avoiding the initial particle breakthrough is to run the washed filter to waste for a period. As increases in rate are likely to displace further particles the filter should be run at the intended flow rate or even faster until the initial turbidity has subsided. Hitherto such a facility has involved additional hardware to divert the flow. However it is possible to use the upwash main for this purpose by providing a pumped connection from this main over to the filter inlet channel (but not submerged in it). The pump capacity should be based on the design filtration rate of one filter. The upwash main is invariably designed for a higher flow than the output from a single filter. By this means the initial filtrate is refiltered through all of the duty filters and no water is lost from the system.

A further method of minimising the initial breakthrough is to feed some polyelectrolyte into the last stages of backwash. This conditions the bottom of the bed and improves the initial capture. Relatively little has to be added. The beneficial effect tends to wear off in the usual maturation period. The procedure however is limited to lateral type floors with a small under drain volume. With a suspended floor the space below can be more than one bed volume and only part of the added polyelectrolyte would reach the bed.

Breakthrough in loaded filters can occur when the flow rate increases. Flow rates can be increased slowly and would no doubt reduce the danger but this is an area where little work has been done. Rates of increase of only 1% per minute have been suggested. This would greatly restrict operators in meeting fluctuating demand.

Taking filters out of service for washing produces regular surges in the flow through the remaining duty filters, as already discussed. However if the washed filter is held on standby until the next one is due for washing surges in flow can be eliminated. This is not yet a common practice but it has already been reported in the trade press in Britain, as being adopted for a new plant in Yorkshire. Many plants are designed to handle the design throughput with one filter washing and one out nominally for maintenance. This spare filter would be put to good use in such a manner rather than reducing the average flow slightly. The above important aspect of filter operation is frequently overlooked in pilot plant trials, with the result that such trials may give misleading data.

23.10 References

Anon (1990) Cryptosporidium in Drinking Water (Badenoch Report), H.M.S.O., London.

Anon (1995) Cryptosporidium in Drinking Water (Second Badenoch Report), H.M.S.O., London.

Cleasby, J.L. (1981) "Declining Rate Filtration", J. American Water Works Assn. **73**, 484

Cleasby, J.L. (1993) "Review of Declining Rate Filtration", Water Science Tech. **27**,151

Colton, J.F., Hillis, P. and Fitzpatrick, C.S.B. (1996), "Filter Backwash and Start-up Strategies for enhanced Particulate Removal ", Water Research, **30**, 2502

Cornwell, D.A., Bishop, M.M. and Dunn, H.J. (1984) "Declining Rate Filtration: Regulatory Aspects and Operating Results", J. American Water Works Assn.**76,** 55

Hilmoe, D.J. and Cleasby, J.L. (1986),"Comparing Constant Rate and Declining Rate Filtration of a Surface Water", J. American Water Works Assn.**78,** 26

Miller, D.S. (1990), Internal Flow Systems, B.H.R. Information Cranfield, U.K.

CHAPTER 24

FILTER DESIGN

24.1 Introduction

Granular media filters come in two basic varieties, the open or gravity version or the closed or pressure version. The most common gravity filters are downflow with fine media of either 0.5-1.0 or 0.6-1.18mm size (based on standard BS sieves.) In the USA the sand is slightly finer at 0.425-0.85mm ie 0.5mm effective size. Depths usually lie in the range 0.6 to 0.7m excluding any support layers. Traditional filtration rates are around 5-6m/hr. However many more adventurous operators have selected very different configurations of media and higher filtration rates.

Pressure filters as mentioned previously may be contained in vertical or horizontal cylindrical vessels depending on the unit size and also the depth of media. Deep beds are accommodated more easily in vertical shells. The detailed design varies according to the manufacturer (Fig.24.1). Some vertical filters have a plain top inlet and outlet connection, some have an internal rudimentary weir box for wash water collection and diffusion of the inlet flow (to avoid gouging of the media.) Horizontal filters resemble gravity filters and either have an internal washout weir and transverse channel at one end or in the centre, or a high level longitudinal launder. The variations are limited (Fig.24.2).

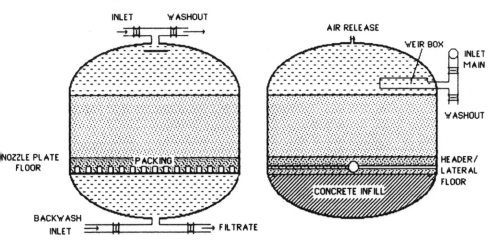

Fig. 24.1 Two varieties of vertical pressure filter.

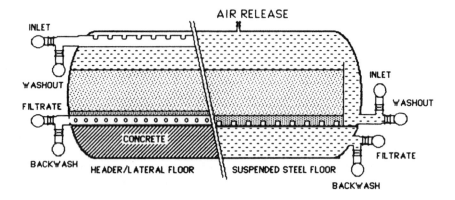

Fig. 24.2 Two types of horizontal pressure filter.

24.2 Optimisation of Gravity Plant

Gravity filters come in a wider variety of designs. Depending on the floor chosen a side washout weir may be at the centre (twin beds), or at the side of the chamber. Alternatively high level launders may be used. The underfloor collecting duct if present, may also be at the centre or at the side, or in one example, at a high level with down comers to the laterals.

The shape of the filters in a set with common walls affects both the civil and mechanical cost (ie the cost of excavation and concrete, plus the cost of the floor, valves and external pipework.) For a given area of floor and number of filters there is an aspect ratio which gives a minimum wall length or volume of concrete. For N filters in a row (Fig.24.3) with a unit area A, the total wall length L_T is:-

$$L_T = (N+1)L + 2NW.$$

where L is the length of each filter.

The area of each filter A=LW, thus $L_T = (N+1)L + 2NA/L$
Differentiating - $\delta L_T/\delta L = N+1 - 2NA/L^2$
At the optimum the slope $\delta L_T/\delta L = 0$

Substituting for A, the optimum value of L/W occurs at a value defined by the following equation:-

$$L/W = 2N/(N+1)$$

Thus:-

N	1	2	3	4	5	∞
L/W	1	1.33	1.5	1.6	1.67	2

The above formula may need to be adjusted according to the design. Thus with a half height longitudinal weir wall:-

$$L_T = (N+1)L + NL/2 + 2NW$$

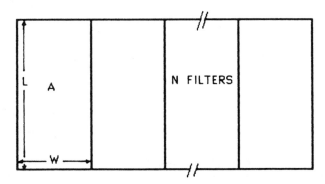

Fig. 24.3 Optimisation of filter shape.

A further consideration is the optimum number of filters in the plant . The surface area of the structure is a function of unit size to an exponent of 2/3. Valves, pipework and pumps are usually considered to vary with the same relationship although often there are step changes as one moves from one range in the manufacturers models to another. Thus fewer larger units usually prove cheaper. As has been discussed elsewhere in this book there are process limitations because of the steps in flow which can be tolerated when filters are taken out of service for washing. Sometimes it is possible to reduce the plant flow or run the raw water to waste to overcome such constraints. With fewer but larger filters clean and dirty washwater tanks will also be larger but these are cheaper structures than filters but nevertheless should be taken into account in optimising costs.

The balance between civil and mechanical cost depends on the tendering procedure and method of evaluation. The life expectancy and amortisation policy (eg. 50 years for concrete and 20 years for steel) also affects the choice.Where the design is based on an optimised mechanical/electrical element there may be a temptation to choose long thin filters with a short filter gallery. This would probably increase the civil engineering cost. A designer working for a civil engineering contractor would probably do otherwise. It is possible to choose between concrete ducts and pipes for upwash mains or inlet manifolds for example, and between steel, GRP (glass reinforced plastic) or concrete channels for washout launders. Concrete comes into its own on large plants but at intermediate sizes steel has made considerable inroads. Some would argue that a turnkey bid or a design that considers both sides will produce the least cost design.

Six filters is a good compromise for medium size plants (eg. 50 Mld). There is a maximum economic size to a filter and large plants may have many filters. One example has 60 each of 150m^2

Most filter installations form part of an essential service and they must be capable of being maintained while the plant continues to operate. Thus the components that are likely to fail at some time such as valves, penstocks, pumps etc. must be capable of being isolated and duplicated. It is usual for gravity filters to have an isolating weir or bell mouth down stream unless the filtered water channel is at a much lower level. Hand stops or means of blanking off filter inlets are needed and all filters need a drain connection. It is common for specifications to insist that the plant should be able to handle the design flow with two filters out of service (one washing and one for maintenance). Suspended floors and collecting ducts on large lateral floors are usually provided with means of access, although this may involve the removal of pipework. All of these are features that affect the design and layout.

24.3 Pressure Filters

One noticeable difference between pressure and gravity filters is the depth of water above the media. In a pressure filter this is usually the incidental outcome of the geometry whether it be a standard dished end on a vertical filter or the curvature of a horizontal filter. In the former the expanded bed will come close to the top of the straight wall (Fig.24.1). In a horizontal filter the working media will be placed equally above and below the axis with care taken to see that the expanded bed and the packing layers do not become cramped by the reduced area nearer the crown and the invert of the shell.

Pressure filters as their name implies operate with an internal pressure often much above the differential pressure and the considerations discussed below do not apply. There may be some instances where an additional delay time is required for reaction or flocculation, in this case an additional vessel may be used. Sometimes an aeration cascade may be incorporated in the head space. In this case the latter may be enlarged to accomodate the necessary fall.

Floors in pressure filters include the steel plate suspended variety with nozzles, proprietary porous compositions, perforated laterals with or without nozzles and indeed other designs used in open gravity filters. A concrete base infill is often used with lateral floors to provide a level base above the dished end in vertical filters or the cylindrical base of horizontal shells (Fig.24.1 & 24.2). Designs are available for separate air scour as well as without air scour and also for combined air and water washing, but in the latter case means of retaining the sand must be included.

The ends of pressure filters are sometimes built into the side wall of the operating gallery to reduce the size of the building . They must have an access manhole and appropriate safety procedures must be observed if a vessel is to be entered. A filter that has been in service will often contain residual biological material that can deplete the oxygen level. Activated carbon filters are an even greater hazard in this respect, but with all such closed vessels forced ventilation during access is essential.

24.4   Gravity Filters

24.4.1 Water Depth

In open gravity filters, depending on the conditions, a wide choice of supernatant water depths is available. As discussed in Chapter 22, it is possible, indeed probable with fine media, for the hydrostatic pressure at a point within the filter to fall below atmospheric pressure (negative head). This will be within the clogging zone just below the surface of the media. If this happens and the water was saturated at atmospheric pressure, air will come out of solution. This air which nucleates on the surface of the grains effectively blocks the pores and greatly increases the resistance to flow. If the depth of water over the media always exceeds the 'clogging head' then this cannot occur, providing that the water is not supersaturated. In practice a margin of safety is advisable or alternatively means such as a desaturation cascade should be provided. This philosophy accounts for the depth of water used on most filters using fine media, which is commonly 1.8-2m at the end of the filter run in rising level filters or with filters with level control.

With coarser media air can escape more readily. In any case the depth of penetration of the floc is deeper while the hydraulic gradient is less, even in the clogged bed. Thus the point at which the minimum pressure occurs is deeper and the depth of water cover required to prevent 'negative head' is much less. This comes out clearly in computer simulations in which the necessary depth for this can be displayed. An extreme case exists with 2-4mm media as used in biological aerated filters where air is deliberately bubbled even through down flow beds while in service and is able to escape upwards continuously.

24.4.2 Shallow Water Filters

With coarse media and also where one can be sure that the water is always undersaturated shallower depths may be used. The depth of water over the media is governed by other considerations such as the accommodation of devices such as floats, inlet weirs, penstocks, diffusers etc. The wash out weir design may also control the water depth. 500mm is a fairly common minimum depth and very many filters with such a depth may be seen in Europe.

It is possible to use such shallow filters even with saturated water if the filtration rate is low and the media of intermediate size as long as long filter runs are not demanded. If such filters are washed daily in the morning then the period of high oxygen saturation, which occurs early in the afternoon when algal activity and rising temperatures combine, will coincide with the time when the headloss is low and negative head unlikely. The filters will be most vulnerable as they reach the limiting head which in this case will be at the end of the night when the water is undersaturated.

A filter with a shallow supernatant water depth is usually provided with a side washout weir (or two) and a dry wash out channel. (Fig.24.4). No washout penstock is needed. The outlet valve is shut and the backwash flow merely causes the filter to

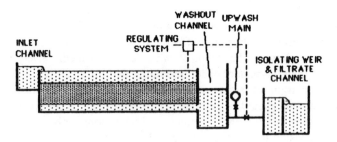

Fig. 24.4 A shallow water gravity filter.

overflow the weir. No preliminary drain down is required and the wash can be started without the associated delay. Likewise when the wash has ended the outlet valve is opened and the filter returns to service without the delay for refilling.

Because the wash period is so short with the shallow water design the incoming raw water is sometimes allowed to continue entering during the wash. For example a 3 bed volume wash on a 900mm bed would consume 2.7 m^3 per m^2 over 540 seconds. Thus with a filtering rate of 7m/h the additional water loss would be 1.05m^3 per m^2. If the filters run for 24 hours the wastage would rise from 1.6% to 2.2%, which many would regard as an acceptable penalty for the resultant simplification. If the water enters via a shallow weir with the cill close to the water level during washing then the flow will be unaffected and also no back flow into the inlet channel can occur. An alternative is a non return flap valve which closes when the level in the filter chamber rises. In some designs this is assisted by a float. The inlet must be designed so that the water does not fall or be directed towards the media otherwise gouging may occur. There must be means of isolation such as a separate handstop or a lock on the non return flap.

Control of filters including level control has been dealt with in Chapter 23. It should be noted that the accuracy of level control is particularly important with the shallow water design because the water level is usually only 50-150mm below the washout cill. An excursion could cause overspilling or loss of residual bed cover.

24.4.3 Deep Water Filters (Fig.24.5)

These usually have an inlet valve or penstock which is closed when the filter is to be washed. Normally the outlet remains open and the filter is allowed to drain down to the level of the wash out weir, close to the media level. Draining down may take 20-30 minutes. This reduces the wastage of water although in emergency situations where filters have to be washed as quickly as possible the water above the bed can be dumped straight to the dirty washwater tank. This must be taken into account in designing the latter as it can almost double the volume to be handled. In the deep water design the washout channel and indeed, high level launders where used, are all flooded while the filter is in service and a washout penstock, not necessary with the shallow design, is required.

After draining down, the wash procedure may commence and the washout valve or penstock will open either before or shortly after the backwash flow has started. There are variations in the detailed procedure which are used according to the choice of media in order to minimise loss. The deep design leads to problems on refilling and unless the layout is correct the media can be disturbed by raw water cascading into the filter. This is discussed further in Chapter 22. In some instances the filter is partially refilled by the backwash pumps to avoid difficulty.

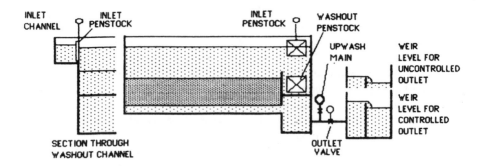

Fig. 24.6 General arrangement of a deep water filter.

Surface flush has been discussed also in Chapter 22.

24.5 Backwash Water Supply

The majority of gravity filters use a pumped backwash system drawing from a clean wash water tank which fills preferentially from the filtered water channel. The refilling of this tank as indicated elsewhere may produce major excursions in the flow passing forward to the disinfection contact tank, which in turn produces problems in controlling the chlorine dose. On large plants with many filters where the minimum plant flow exceeds the filter wash flow it may be possible to pump directly from the filtered water channel and dispense with a separate clean water tank.

Elevated wash water tanks have been used. These are filled by a small pump over a long period and discharge by gravity to wash each filter. In this case the residual head with an almost empty tank must still be sufficient to maintain the wash rate at an acceptable level. Such high level systems are useful where power supplies are limited. A parallel approach is the use of pressure vessels to hold air for an air scour. This again provides a means of avoiding peak power demands. The wider availability of public electricity supplies and the lower relative costs of generators has led to a decline in the use of such systems.

24.6 Single Valve and Syphon Filters

With small plants there is an incentive to avoid the cost of a pumped backwash system. It is possible to backwash the filter from a local storage tank by operating a single on-off valve. (Such single valves are not to be confused with single multi-port valve filters which merely reverse the flow through a conventional pressure filter and do the task of <u>four</u> simple valves. Such a flow reversal is not good practice and can lead to the blockage of the under drain system with solids present in the raw water particularly if surface water is to be treated).

The single valve filter (Fig.24.6) comprises an enclosed low pressure filter combined with a wash water tank. The incoming flow which is pumped direct or possibly derived from a deep or slightly elevated clarifier passes through the filter and into the tank which is kept full, or allowed to fill prior to a backwash.

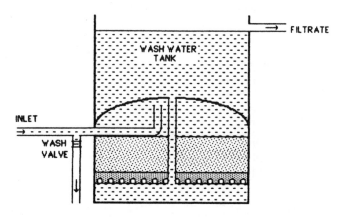

Fig. 24.6 A single valve self washing filter.

A single valve on the inlet allows the feed to run to waste when the filter is to be backwashed and at the same time the water in the tank is able to run back through the filter and out of the same valve. Thus the valve has to pass both the backwash and

396

forward flow but like the shallow water gravity filters already mentioned the forward flow is small by comparison. Again the head available in the tank at the lowest level must be sufficient to maintain the minimum wash flow. The wash outlet pipe is taken down to the lowest possible level to form a syphon and maximise the available head. The wash must of course stop before air is drawn into the filter. The design is simple and effective for washing with water alone, and many examples have been built around the world in steel and some also in concrete. A patented variation on this principle incorporates the filter into the column of a water tower which also includes a stack of rudimentary settling tanks. A single valve actuates the backwash.

Another related design (Fig.24.7) available from more than one supplier replaces the valve with a syphon on the inlet which 'makes' when the head on the inlet pipe exceeds the crest of the syphon. A large bore syphon will only trickle when the water reaches the top and additional features such as an auxiliary ejector are required to make the syphon prime when the critical head is reached. Water in a 12mm pipe, for example, will flow as a slug and can be used with an ejector for this purpose. An air break is also included to halt the flow when the wash tank reaches the lower limit. This is essential, to prevent air disrupting the filter media.

All of the above are simple low maintenance systems. The filtrate flow ceases while the filter is washed. It is possible in theory to add an air scour to the single valve concept but the simplicity of the concept would be destroyed.

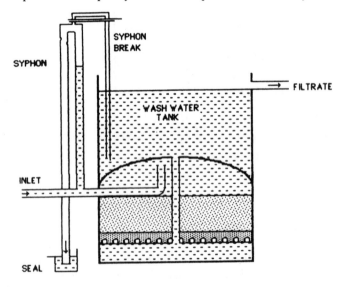

Fig. 24.7 A syphon operated self washing filter.

24.7 The Moore Filter

One South African gravity design by Moore (1966) appears to be unique in retaining dedicated wash tanks in downflow filters (Fig.24.8). In this case each filter is built over its own wash tank which is allowed to refill slowly with a controlled vent releasing the air slowly to maintain the water cover over the media. The wash water is propelled from the appropriate tank by applying the air scour supply, after the air scour itself is complete. The wash water tanks serve as pressure vessels. This system avoids the use of upwash pumps, mains and valves. It is a good example of how costs can be switched from mechanical. to civil engineering construction to suit local economic conditions or the tender evaluation procedure.

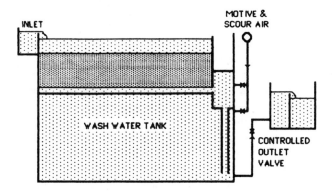

Fig. 24.8 The Moore filter.

24.8 Open Multicell Self Washing Filters

The concepts of washing under gravity and washing with the immediate product of other filters, which is not unusual in pressure filtration, come together in a design (Fig.24.9) that has been used extensively in the USA as well as in many other parts of the world. A very deep filter chamber is used, corresponding to the combined filtering and backwash headloss heights. The inlet may be conventional with a penstock or even a weir which may continue to feed the filter during the wash. In one variation a syphon is used to control the inlet. A filtered water channel is set at an appropriate intermediate height and the level in this is controlled by a common outlet weir.

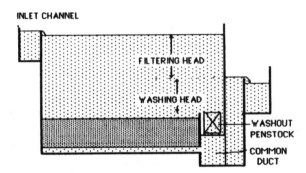

Fig. 24.9 A section though a self washing multicell gravity filter. The installation comprises a set identical to that shown.

The filters may be of any conventional type with wash out launders or side weirs, but the level of these launders or weirs must be lower than the level in the common outlet channel by an amount corresponding to the back wash headloss through the floor and the media. To wash the filter the washout valve or penstock is opened and the contents of the chamber dumped. The wash commences when the washout cill or launders have been uncovered although there is a run up in flow as the water level falls below that in the filtrate channel. This system may have a common under floor plenum running under all the filters in the set although this would make it impossible for one filter to be serviced. However there is not much to service in such a design.

There is an incentive to keep the backwash head losses low in order to reduce the chamber depth and anthracite would have a specific application in such filters. Likewise some compromise on the filtering head is desirable. With typical conditions, eg. a filtering head of 2m, plus a media backwash head of 750mm and a similar floor loss, the wash out cill would be 3.5m below the top working level in the filter. The water involved in this depth has to be dumped before the backwash can reach the design rate. The total wash consumption in this example would be about 6m³ per m², which might be 4-5% of the plant flow. This is not high compared with some figures quoted for conventional plant but more than twice what can be achieved with low level weir designs. However in areas where simplicity is paramount this self washing design has much to commend it.

It would of course lend itself to combined air /water washing with an appropriate floor but not separate air scour because the water cannot be held off. Air blowers would in some cases remove the main justification for the design which, like the single valve filter, is the absence of any power requirement.

24.9 Enclosed Multicell Filters

Another related design of Australian origin (Fig.24.10), is built as a multicell block but covered to form a low pressure system. Like the above it has a common under drain system and an outlet to preserve flooded conditions above the media. The top of each cell is connected to a single proprietary multiport valve which supplies the feed to each cell and connects the cell to be backwashed to a drain. The design avoids the wastage involved in the dumping of water in the above concept.

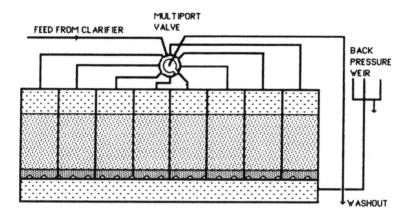

Fig. 24.10 Enclosed self washing multicell filter.

24.10 Reference

Moore, R.P. (1966), "Filter Bed Operating System". British Patent 1,016,843

CHAPTER 25.

UPFLOW FILTRATION

25.1 Introduction

As discussed in Chapter 17 all of the filtration mechanisms operate in any orientation, upwards, downwards or horizontally. In an upflow mode any settling particles will settle on the top of each grain, ie. the down stream side, while particles removed by interception will collect on the underside. Filters exploiting almost all possibilities have been constructed. During the late 1960's and early 1970's upflow filtration using sand gained a popularity which has since waned. An alternative form uses buoyant media and operates like inverted down flow filters. This is being used both for roughing filtration and also as a biofilter.

The upflow sand filter is attractive in theory because a fluidised bed of sand with a wide size range will stratify with coarser grains tending to settle to the bottom. Thus with upflow filtration one should have the benefit of roughing filtration through the equivalent of an anthracite layer on a down flow filter. Sand is of course cheaper than anthracite and the concept appeared to offer the promise of a low cost heavy duty filter.

25.2 Design

The most common type was one developed in the Netherlands (Fig.25.1). This employed a fairly deep (1.5m) moderately coarse sand layer over graded packing layers totalling about 600mm depth. In this particular design a steel grid was used to hold the media down against the upward thrust of the headloss but it tended to interfere with expansion of the media bed when the filter was washed. Other examples do not use this grid. Providing that the depth of the sand exceeds the headloss to which the filters are taken the submerged weight of the media would be expected to resist the upthrust, eg. 1.5m of sand will withstand a total headloss of 1.5m providing that the headloss is concentrated towards the bottom of the bed, as will usually be the case.

The incoming water is distributed through a fairly conventional filter floor system but this has of course to convey the suspended solids present. A plenum floor is not a good choice if settleable solids are present. The nozzles must be able to pass such solids and large orifices are necessary. Such filters can be operated like a battery of pressure filters without positive flow division but most filters in Britain were installed with weir chambers upstream to ensure equal flow splitting.

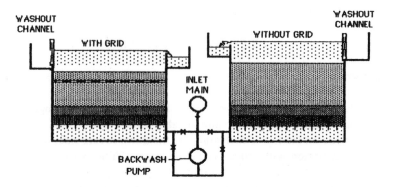

Fig. 25.1 Dense media open upflow filters.

The filtrate is collected from the top of such filters via a weir in open gravity versions and via a plain pipe in pressure upflow filters. The dirty wash water is collected by a weir set at a lower level than the filtrate outlet.

Such filters have been installed for several applications such as potable water filtration both from ground water and surface sources, industrial filtration and most extensively, tertiary filtration of sewage, where the co-current wash procedure is less objectionable.

25.3 Disadvantages

Upflow filters using sand possess some inherent problems as follows:-
1. The wash water proceeds in the same direction as the filtrate and there is no longer any segregation between the clean and dirty side of the media bed. This is a severe limitation where the final filtration of potable water is concerned and would not be acceptable following the Badenoch reports on cryptosporidium. Organisms in the raw water may easily proceed into supply if they are retained in the supernatant volume or on the walls.

2. The under drain carries the raw water with its suspended solids into some form of nozzle or orifice below the sand and careful screening is essential. Experience has shown that in some circumstances nozzles can clog with coagulant deposits and regular remedial action is necessary to prevent this causing damage.

3. Packing layers take part in the filtration process and indeed provide effective prefiltration but it is very difficult to wash these layers. They cannot be fluidised. If they were, the working layer would be flushed out from the filter. Combined air and water washing would mix the packing layers into the working layer and allow the working media to percolate into the under drain. Compromise is therefore necessary, and air scour alone must be relied upon to maintain a thoroughfare through these layers.

4. Any air reaching the media from below as a result of entrainment at the dividing weirs will percolate through the bed and dislodge floc (hiccoughs).

5. The majority of the floc is trapped at the bottom of the bed in the same way as it is at the top of a down flow filter. Thus more wash water is required to flush the floc through the full depth of the bed in the upflow filter compared with a downflow filter.

Advances in conventional down flow filtration, washing problems, the ready availability of anthracite coupled with the above disadvantages have led to a decline in popularity, but there remains a challenge for an ingenious designer to solve some of these problems. It is possible to design a satisfactory upflow filter with coarse media with combined air scour and water washing if the packing layers can be effectively eliminated and the feed is free of large particles. One or two examples exist.

25.4 Biflow Filters (Fig.25.2)

Related to the above upflow filters are the biflow filters which combine a down flow filter on top of an upflow filter in the same vessel. Very high filtration rates based on the area of the vessel can be claimed. The bottom of the upflow filter is as before but the filtrate is collected by centre strainers. The coarse filtering action of the packing layers in the bottom half can be complemented by coarse anthracite above the top half.

There is an obvious comparison with equivalent designs used in split flow ion exchange columns. Care must be taken with the design of the centre strainers so that they do not interfere with expansion on backwashing nor suffer damage from the upthrust if they do. The fact that few such devices have been promoted suggests that the additional complexity does not produce a more competitive device.

An alternative biflow filter which will be obvious after reading the next section is the combination of the conventional downflow filter with a buoyant filter above in a

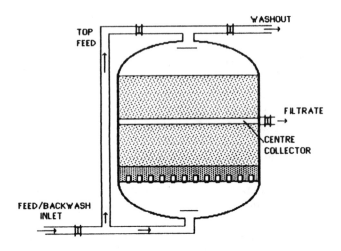

Fig. 25.2 Biflow filter.

single shell. Such a device, described by Makwana (1971), uses low density polyethylene as the upper buoyant section and PVC for the lower section. This combination overcomes the problems associated with upflow filters which are fed through the floor. The particular filter described by Makwana had a retaining screen which would limit the application of such a device.

25.5 Buoyant Media Upflow Filters

According to Hunter (1987) the first patent on such filters was issued in 1960. At least three types of buoyant upflow filter are available in Britain and there may be others. One of the earliest originated from Czechoslovakia and uses expanded polystyrene media (Fig.25.3) with a fairly conventional inverted filter floor. The space above the floor which holds filtrate can serve as a wash water tank so that back washing can be achieved by partially draining down. The feed enters from below and all functions are the same as with a down flow sand filter apart from being inverted. Air scour if applied must be from below. It mixes the media in a turbulent mass after which it must be allowed to restratify. The main advantage in conventional filtration is the very considerable weight reduction which could be beneficial for mobile applications. The media is reported as being adequately robust providing of course that

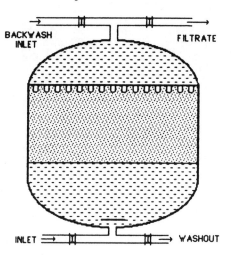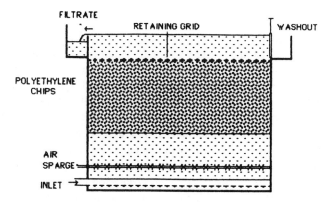

it is not exposed to certain solvents. This type of filter is now being promoted for biological use as discussed in Chapter 27.

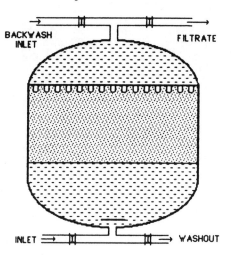

Fig. 25.3 Expanded polystyrene filter.

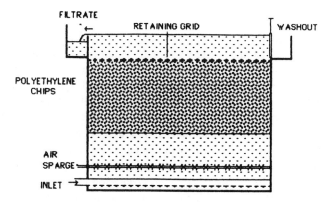

Fig. 25.4 Buoyant media upflow filter, Type 1.

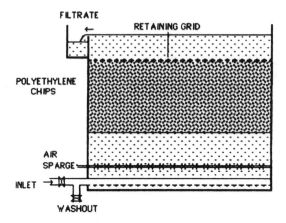

Fig. 25.5 Buoyant media upflow filter, Type 2.

Filters using media less buoyant than expanded polystyrene media are also offered. In one example described by Cantwell et al. (1991) (Fig.25.4) with a media believed to be polyethylene. It is claimed that no back wash is needed. An air sparge mixes the media and removes adherent dirt so that it can be removed by draining down. In this design the buoyant media is held down merely by a screen soffit. Clear water passes through the screen and a sufficient depth is allowed below a weir or launders to retain the necessary wash volume.

The third design (Fig.25.5) is similar to the above except that the dirt is flushed forward through the screen with the continuing filtration flow accompanied by an air scour. In this case the waste water must be diverted via a lower weir as in the case of the upflow sand filter. It is also necessary for the feed to be free of debris that would block the screen. The media appears to be about 3mm in size so that the screen is not unduly fine, nevertheless an upstream screen at least as fine as this will be essential One patent relating to this type (Hsiung, 1985) claims outlet laterals comprising mesh covered pipes or channels comparable with washout launders. The top of the filter is not entirely covered with a screen.

Both of the last two have applications primarily as roughing filters upstream of polishing filters and one is described as an absorption clarifier. In view of their simplicity they clearly have potential for such duties on small plants but in larger sizes the high intrinsic cost of the media is likely to make them unattractive compared with other primary treatment options. In one version which is promoted as a compact clarifier high doses of polyelectrolyte are used with coarse media to achieve filtration rates in excess of 20m/h, but problems have been encountered on the conventional secondary clarifiers as a result of this treatment.

25.6 References

Cantwell, A.D.C., Evans, I.D. and Whittaker, J. (1991) "Filtration Method and Apparatus", PCT Patent. GB 91/00912

Hunter, J.S. (1987),"Recent Developments in Buoyant Media Filtration"' Filtration and Separation, November/December, p399

Hsiung, A.E. (1985),"Process Apparatus for High Rate Upflow Filtration with Buoyant Filter Media", US Pat. 4,547,286, Canadian Pat.810,721

Makwana, J.G. (1971) "Two Way Filtration", Filtration and Separation, (January/February), 90

CHAPTER 26.

CONTINUOUS FILTERS

26.1 Introduction

In most areas of chemical processing where large quantities of material have to be treated continuous operation is preferred, as it is usually more economical and produces a more consistent product. Some steps used in water treatment are truly continuous, for example settlement, flotation and disinfection. It has been assumed hitherto that granular media filters operate batchwise, ie. they run until clogged and are then taken off line and washed before being returned to service again. However it is possible to have truly continuous granular media filtration. The conventional rapid granular media filter which has been built in sizes up to 150-200 m² is a remarkably simple device that can be built in any size below this. Slow sand filters are considerably larger. The low intrinsic cost and simplicity have perhaps acted as a deterrent to the adoption of truly continuous versions.

A similar situation has existed in the parallel field of absorption with ion exchange and activated carbon. 20 years ago there was considerable interest in continuous ion exchange and at a conference (Society of Chemical Industry, 1969) several examples were described. In the event they have not competed with static beds except in certain special applications. The same is partly true of continuous filtration. One difference between ion exchange and filtration is the delicacy of the floc and the ease with which it is dislodged. Materials absorbed on ion exchange resins or activated carbon are held internally and the adsorbate is not removable mechanically.

Continuous filters fall into two classes, multicell and moving bed. The multicell filters are in fact sets of ordinary static sand filters which are washed on a short time cycle with a moving facility. Four examples will be described.

26.2 Multicell Travelling Bridge Filters

The first (Fig.26.1.), which originated in the 1950's, comprises a set of long narrow beds which are divided from each other by thin walls which finish close to the surface of the sand. The sand depth is typically 300mm or 600mm in some cases. Each cell has its own plenum or underdrain which connects to a filtered water channel running alongside the end of these narrow parallel filters. The cells lie in a single tank with a common volume of water covering them all, this volume being the raw water to be filtered. The filtered water overflows a weir which maintains the level in the filtered water channel and also in the body of the filter. The total headloss is low (~300mm).

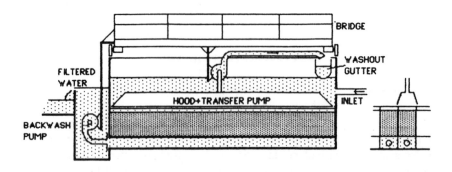

Fig. 26.1 Twin pump travelling bridge filter.

When the headloss exceeds a given limit washing is initiated. The backwash flow is produced by a pump which is submerged in the filtered water channel and suspended from a travelling bridge. The pump delivery is taken to a shoe that slides over or engages with ports at the end of the under floor laterals. The dirty wash water is sucked from the surface of the cell by a hood connected to a second pump which draws a somewhat higher flow than the first pump and transfers the mixture to a launder running the length of the tank.

A more recent development of this design incorporates an air scour facility, and applies this with a second sliding shoe and a second collecting hood above the filter cells.

The second type of multicell filter (Fig.26.2) uses somewhat larger cells with a flat top to the surrounding walls. A common under drain plenum serves the entire set of cells. Again this connects to an external level control weir which by virtue of the low filtering headloss regulates the water level in the common volume above the cells. The backwash water in this design is sucked back from the plenum or under drain using a hood that is lowered onto each cell and seals on the surrounding flat face. A single pump is able both to suck the water and transfer the product to a similar wash out launder. The hood is again carried on a travelling bridge. In contrast to the previous design the external wash water channel does not exist and does not therefore need covering if used for potable water treatment. The production of dirty wash water would also appear to be less because there is no secondary dilution from

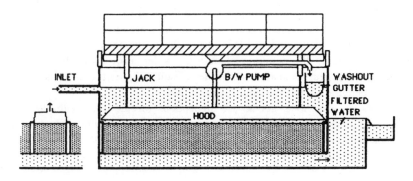

Fig. 26.2 Single pump travelling bridge filter.

the unfiltered water in the tank.

A Chinese version of the latter (Fig.26.3) uses manually propelled trolleys instead of an electrically driven bridge and the wash water is extracted by a large syphon which discharges into a low level dirty wash water channel running alongside the filter. The bottom of the syphon is sealed with residual water within the channel and primed with a small vacuum pump. Full details are not available but it is believed that the jacking down of the hood was achieved by the application of the additional weight of the full syphon acting against balance springs.

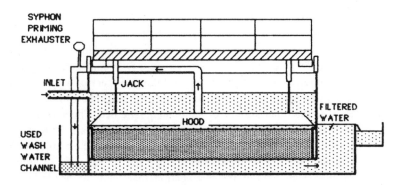

Fig. 26.3 Syphon operated travelling bridge filter.

Yet another variety described by Eimco in manufacturer's publicity material uses a travelling bridge carrying a rectangular caisson, 600mm wide, which is lowered into or forced into the media and comes to rest on the suspended floor. The washwater is sucked through from the common plenum as in the second and third examples above. One would expect some abrasion of the caisson and possible disturbance of the media during penetration but the concept is simple.

Filters as described above are dependent on the serviceability of the mechanical equipment and many users require complete standby units which reduces the attraction somewhat. The main advantages are the low headloss which follows from the simplicity of the hydraulic path, the absence of valves and the ability to wash frequently before the headloss has built up. The fairly continuous low back wash flow simplifies subsequent treatment and disposal. The main disadvantage of such filters apart from the limitations of washing with water alone appear to be the lack of control of flow rate through the cells. Any throttling of the cell outlet will interfere with backwashing. If the headloss is allowed to rise the most recently washed cells will run at a rate that can cause breakthrough. This has been reported as a problem on some installations.

Although the instantaneous wash flow is low the percentage use is reputed to be higher than with conventional filters. Likewise although the peak power demand is much less the total power involved with backwashing is higher than with conventional filters. This type of filter has been used extensively for tertiary treatment of sewage, where the solids are less prone to breakthrough at high flow rates. The ability to apply combined air and water washing would make such filters more attractive.

26.3 Continuous Moving Bed Filters

True continuous filters use a moving sand bed. Two commercial designs have achieved some success. The earliest version originated from Hungary and became known as the Ten-Ten™ filter (Fig.26.4). It has a cross flow configuration with water entering through a louvred or mesh screen around the periphery of a cylindrical 'silo' or though the inclined surface of the sand bed within the silo. The filtrate is collected via another screen on the inner surface of an annular sand bed. The sand is extracted from the conical base of the silo and elevated as a slurry by an air lift to a cleaning system. The cleaned media is deposited back on the surface of the charge within the silo. In a later development a wider size range of sand or possibly an appropriate blend has been chosen to exploit the segregation that tends to occur in disperse mixtures of granular media whereby the coarser fraction tends to fall to the edge of a heap. Thus with this development the equivalent of a multimedia bed is achieved.

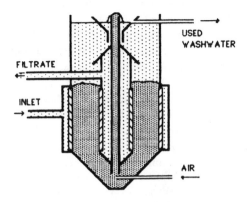

Fig. 26.4 Radial flow moving bed filter.

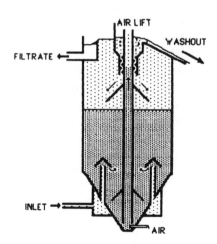

Fig. 26.5 Countercurrent moving bed filter

The Dynasand™ filter (Fig.26.5) is a truly countercurrent device which also uses a charge of sand in a silo. The feed is admitted via downward facing tundishes or inverted troughs located low down in the silo and the water flows upwards through the bed to emerge into the supernatant filtrate layer. Like the above, the dirty sand is extracted and lifted as a slurry by an air lift to a countercurrent washing cascade from whence it falls back onto the top of the sand charge. The design draws much from the early "drifting sand" filter mentioned by Baker (see Chapter. 1.)

Both designs have found application in several fields including especially tertiary filtration of sewage. The sand cleaning efficiency would appear to be good. The main disadvantage of such filters is their limited upscaling potential because in both cases the sand inventory will increase with throughput raised to an exponent of 1.5. The Dynasand version has been developed into a multi hopper system to increase the unit size. The height also increases and such designs often require additional pumping to elevate the feed to the top of the filter, which increases the running cost. It is doubtful whether such filters are appropriate for high quality filtrates because the slight interparticle movement during passage of the sand is likely to dislodge floc. One would expect such continuous units to be more suited to a high load roughing duty.

Continuous moving bed filters are an example where rounded sand may in fact be more desirable and where the virtues of a high voidage angular media cannot be exploited. Angular media will not flow as readily and is more likely to be abraded or to abrade the structure.

Filtration rates in the above designs are understood to remain close to conventional practice although again in principle the continuous moving bed should be capable of very much higher rates. Short runs or short transit times should not present a problem. In both designs the bed depths can be modified to maintain the desired treated water quality. The Dynasand bed depth can be altered by merely adding or removing sand.

26.4 Reference

Society of Chemical Industry (1969), **Ion Exchange in the Process Industries**, Soc. Chem. Ind. London

CHAPTER 27.

BIOLOGICAL APPLICATIONS

27.1 Slow Sand Filters

Whereas there are several aspects of rapid gravity filters, particularly design details, that are seldom discussed in the open literature, slow sand filtration has attracted considerable attention in recent years culminating in an American Water Works Association Manual on the subject (1991). Graham (1988) and Graham and Collins (1996) have acted as editors of books claimed to be a state-of-the-art review, which are still current. Huisman and Wood (1974) also produced a fairly detailed description of these filters for the World Health Organisation. Against this background it would be presumptive to go further than a brief review of the subject for the sake of completeness.

As indicated at the beginning of this book slow sand filters were the first truly successful form of filtration other than infiltration wells. The success was due to the recognition of the principles of mass balance in that solids do not disappear and that which is filtered out must sooner or later be removed from the media.

The slow sand filter requires a water tight static tank (perhaps better described as a lagoon) containing sand with an under drain to collect the filtrate with supporting gravel or some other means of retaining the working media, and an outlet weir to maintain the water level above the media. The incoming flow must of course be divided between duty filters in proportion to their area and the only precaution is the avoidance of gouging of the sand by the incoming water.

Whereas rapid gravity filters operate at 5 to 15 m/h with some examples going further, slow sand filters traditionally operate some 50 times slower and hence the land area occupied accounts for a considerable proportion of their asset value. (A 50Mld plant requires 2 hectares at 0.1m/h). Filters of such an area cannot be washed by the means employed for rapid gravity filters. The design has evolved on the basis that filter runs are normally 6 weeks or longer and any installed wash system would be grossly under utilised. On the other hand there is probably a disincentive to advance filtration rates if the filter run falls below this limit because the sand cleaning facilities would be unable to handle the situation unless there was a complete change in the entire system.

The floor or under drain of the slow sand filter is only required to act as a collection drain and in contrast to rapid filters with a backwash it does not have to withstand any upthrust. There is therefore considerable flexibility in the design. The variations in the resistance from the interface against the working media to the under drain ducts must be small compared with the resistance of the bed at minimum operating depth but a tolerance of say 10% range in flow across the bed would seem to be a reasonable basis for design. Graded packing with a size difference of not greater than 2:1 between the finest fraction of the coarser material and the largest fraction of the finer material will prevent percolation. The depth of packing will depend on the spacing of the under drain laterals and a designer will be able to balance the cost of the packing material against the cost of the under drains. Proprietary flexible field drain systems now available are ideal for low cost slow sand filters up to sizes governed by the available pipe diameters. The under floor provides an exercise in the design of collector manifolds, analogous with the rapid filter under drain. A variety of precast concrete block under drains used in more permanent municipal filters are described by Huisman and Wood (1974). (Fig.27.1)

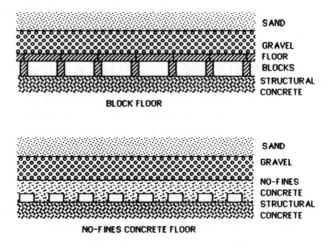

Fig. 27.1 Examples of slow sand filter floors.

It is important that the packing gravel is not disturbed when the working bed is turned over. For this reason Thames Water for example have developed a no-fines concrete under drain which is cast in situ over the structural floor. The casting procedure includes removable forms that produce ducts in the porous concrete which in turn connect with header ducts. The working media can be laid direct on this porous base. Such floors have the advantage that they can take traffic without damage whereas gravel floors can easily be disrupted (by skimming machines).

Slow sand filters have been cleaned, traditionally, by skimming approximately 25mm at a time from the sand bed which initially may be 1-1.5m deep. (The sand is usually of 0.15-0.35mm effective (10 percentile) size). The skimmed sand is cleaned, usually hydraulically with a high pressure jet conveying system and finally stored in a heap or bunker with free drainage. The sand bed is progressively removed until only 300-500mm is left or until there are signs of deterioration in the treated water quality. At this point the residual sand is excavated in strips down to a safe level that will not interfere with the support layer, new or cleaned sand laid in the resultant trenches and the residual sand relaid on top of the fresh material so that it is the first to be removed in subsequent skimming (Fig.27.2). If thin skimming is to be feasible the sand surface must be smoothed over carefully.

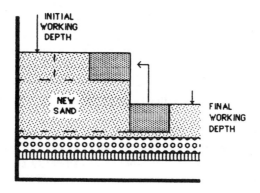

Fig. 27.2 Turnover and relaying of slow sand filters (after Huisman and Wood).

All this is very labour intensive and in industrialised countries mechanical means have been developed to skim and relay the sand, ranging from low pressure tyred vehicles which skim and elevate the sand into accompanying dumper trucks to travelling bridges which skim and convey the sand. Thames Water have adapted dredger technology with a floating skimmer boat which has a floating umbilical hose, which progresses under laser guidance.

The slow sand filter is a bioreactor perhaps more than a filter. The water it receives is not normally dosed with a coagulant, certainly not as it enters the filter, nor is it pretreated by a coagulation process. In the course of say 6 weeks use it will probably pass 100m³/m². The sand is fairly fine and the solids will produce a higher resistance than in rapid gravity filters. Thus the incoming suspended solids concentration must be low. Slow sand filters are more often than not used in conjunction with reservoirs which have a considerable storage time, perhaps of months. This greatly reduces the settleable solids and smooths out fluctuations but if nitrate and more especially phosphate levels exceed certain limits problems occur with blooms of algae which can block the slow sand filters not only with algae but with calcium carbonate. Reservoirs with a long storage time reduce nitrate levels very significantly to the extent that Croll (1993) has reported that in one instance the nitrate standards for a certain river source could be met by blending with reservoir water derived from the same river.

Considerable advances have been made in reservoir management whereby the populations of algae are monitored and thereby the best water for filtration is selected. Mixing of reservoirs to prevent stratification is also used selectively to discourage the growth of algae which tend to proliferate at certain most favoured depths. By such means the water to be treated can be improved considerably.

Primary rapid gravity filters are now commonly used before slow sand filters. Operating at rates of 5-10m/hr usually without coagulation these are able to remove 40 - 70% of the particulate organic matter and thereby extend the slow sand filter runs. Since washing of the latter is an expensive process such prefiltration is well justified.

In Britain, slow sand filters have been built mainly out of doors and exposed to sunlight. This has encouraged the growth of algae even after primary filtration and thick growths of blanket weed (Cladophera) can easily accumulate. For this reason some slow sand filters are now covered with a permanent building or with a floating cover. Nearer the equator they have more often than not been covered.

After all this protection one may ask what there is left for the filters to do. There does not yet seem to be a complete understanding of the process involved but the media certainly provides a substrate for bacteria which thrive on the degradable organic carbon and in so doing add to the biomass on the media. This biomass probably serves as a membrane and filters particles by interception and straining. The biomass layer is described by the German word 'schmutzdecke'. This may take some days to form and mature and a freshly charged filter may have to be run to waste or the filtrate recirculated before the best quality is obtained. The finer media used in slow sand filtration and the vary much lower filtration rates also contribute to the efficiency of filtration even though no caogulants are used. The shear stress at the grain surfaces will be very low. Pathogenic organisms including species such as *Cryptosporidium* and *Giardia lamblia* are removed efficiently by slow sand filtration and this was the immediate benefit of the first slow sand filters in 1829. (These organisms are in fact removed efficiently by well operated rapid filters with coagulation but the recycling of wash water allows the concentration of such organisms to rise until breakthrough inevitably occurs.) One outbreak of cryptosporidiosis discussed by the first Badenoch Report (Anon. 1990) is suspected to be caused by slow sand filters being bypassed.

Physical chemical treatment removes particulate solids and higher molecular weight organic matter efficiently but a residue can pass through to the distribution system to allow biofilms to form which can eventually produce a bacteriologically unacceptable water at the consumers tap. Slow sand filtration reduces the so called assimilable organic carbon (AOC) passing into supply as well as having a disinfecting effect. However in the latter respect a new bacterial flora can appear so that final disinfection is normally still practised. There is increasing interest in biological final treatment as a means of reducing the residual AOC and hence the biofilms which grow in the distribution system in spite of residual concentrations of disinfectant.

A recent development is the introduction of granular activated carbon as a sandwich layer within the slow sand filter (Thames Water) as a means if increasing the removal of traces of organic matter. The 150mm of carbon is placed below the normal 450mm skimming depth, and over 150mm of sand ie. the layer normally turned over. The carbon has an effective size of 0.7mm, and the sand 0.3mm. However when turnover is due the sand and carbon are segregated hydraulically, the sand being cleaned in the usual way and the carbon is sent for regeneration. This innovation has enabled the performance with respect to pesticides to be improved substantially without major capital investment. (Paper 23 in Graham and Collins, 1996).

One major limitation of slow sand filtration is its inability to remove colour. Coagulation is required. This defeats much of the objective of slow sand filtration, which otherwise produces a sludge (from the sand washing plant) which has no added coagualant residues. Odegaard (in Graham and Collins (1996)) has developed a process involving ozone with a bioreactor which provides a solution to this problem.

Slow sand filtration is very appropriate for small communities in developing countries. Efficient filters can be built using locally available materials. They require no chemicals and although chlorination might be beneficial such filters remove most of the pathogenic organisms from surface waters and greatly reduce the exposure of the population. One difficulty however is that it is not often possible to construct reservoirs and such filters must be fed with river water containing possible high levels of suspended solids which rapidly block the slow sand filters. To solve this problem gravel prefilters have been developed. This is a specific example of the exploitation of the settlement mechanism discussed in Chapter 17. Such filters are constructed for upflow and horizontal flow. They are usually cleaned by providing a large drain valve or syphon to empty the bed rapidly. Most of the accumulated solids are released as the air water interface passes down through the gravel. Fuller details are given in a number of papers in the book edited by Graham and Collins. (1996).

27.2 Tertiary Sewage Filters

This is a borderline example of biological filtration. It involves the filtration of residual biomass. The filtrate quality depends on the quality of the solids coming from the upstream processes. Tertiary filtration is generally used as a longstop to catch residual solids that overflow from the final settlement tank. Usually there is no chemical treatment and again it is only possible to filter solids that are intrinsically filterable. (Not all the solids present are in fact filterable by granular media filters. Colloidal material and particles protected by `protective colloids' will pass through the media.) Thus guarantees should be based on filterability tests made on the feed. There is no significant biological action but capture of the residual biosolids depends on the formation of a biofilm on the media which probably adds to the effective roughness of the grains.

The literature on the subject has been reviewed by Metcalf and Stevenson (1978), as well as West et al. (1979). Early tertiary filters were virtually identical to potable filters. 1-2mm sand became common in Britain up to 1974 when the improved management of the secondary stages following a reorganisation of the industry into larger units able to afford greater expertise made tertiary filters less necessary for several years. Tighter discharge standards are now recreating a market.

The biomass encountering a filter may be regarded as a strong floc compared with hydroxide coagulants. Breakthrough seldom occurs unless it is an unfilterable fraction. The main problem is cleaning of the media. In many older plants the biomass coats the media with a slime that cannot easily be removed by washing alone. This causes the bed volume to grow and media is lost over the wash out weir. Chemical cleaning eg. with caustic soda and hypochlorite, must then be applied. As discussed in Chapter 20, coarse sand eg. 2.0-4.0mm or 2.36-4.74mm allows a greater shear stress to be applied to the surface and combined air and water washing which is virtually essential for such coarse media appears capable of preventing such slime growth, and has been adopted by more than one designer in recent plants. Metcalf and Stevenson (1978) have shown that the filtration efficiency is not greatly affected by media size or filtration rate. The headloss on the other hand is less in coarse beds. The limit on media seems to be that which can be washed without resorting to very high backwash rates.

Tertiary filters operating on treated domestic sewage effluent typically remove about 67-75% of the incoming solids. The removal of the biological oxygen demand (BOD) depends on the ratio of soluble to particulate BOD, the filter has little effect on soluble BOD. Filtration rates are typically 10-15m/hr but higher rates are possible. Bed depths are typically 1.0-1.5m. Chemical treatment is sometimes used to precipitate phosphate as ferric phosphate but this will change the nature of the solids being filtered.

These filters generally follow a conventional layout with some simplification. In the late 1960's and early 1970's several upflow tertiary filters were built. This would seem to be a good application for such filters because one is less concerned about the wash water passing out at the same end as the product. The efficiency demanded is less than that of potable filters. However as discussed in Chapter 25 it is not easy to wash such filters effectively when the deep support gravel is involved in the filtration process and such filters have fallen out of favour. As mentioned in Chapter 26 continuous filters have also been used fairly extensively for tertiary filtration. The travelling bridge variety offers a low headloss which may be an advantage where there is little spare head available between the outlet of a secondary stage and an outfall.

Both upflow and down flow tertiary filters can serve as a vehicle for denitrification. Several papers have appeared which indicate that by dosing a carbon source nitrate is reduced to nitrogen. It is possible also to recirculate part of the effluent from all types of biological treatment process to the inlet and use the nitrate produced in the nitrifying aerobic process as a source of oxygen. If sufficient the BOD in the incoming sewage can be used as the carbon substrate in a plain filter without added air or oxygen. In each case nitrogen is produced within the bed and it must therefore be able to escape. Upflow filters have apparent advantages in this respect but with the coarse media now being used in downflow tertiary filters the nitrogen is able to escape counter to the direction of filtration.

27.3 Biological Aerated Filters

This term has been applied to several devices some of which are not true filters but biological reactors which still need a solids separation stage as with the traditional percolating 'filters'. The term is defined for the present purpose as a granular filter which supports a biomass, allows substantial aerobic oxidation of BOD and also acts as its own solid/liquid separator. In the past very few years a sudden interest in these devices has emerged, spurred on by the tighter effluent standards now being sought and also the considerable space saving provided. The initial cost may be higher than conventional activated sludge or percolating filters treatment but if a tertiary filter has to be added then the difference is less.

Upton and Stephenson (1993) have reviewed biological aerated filters. It appears that the first backwashable version was derived from patent issued to Lowe

(1964) for a 'dry' (non-flooded) downflow filter using 6mm coke (Fig.27.3). This was followed up at WPRL (now WRc) with a filter using small Raschig rings cut from PVC tubing. This technology has developed to the point where several competitors now offer such plants serving populations currently up to 70,000.

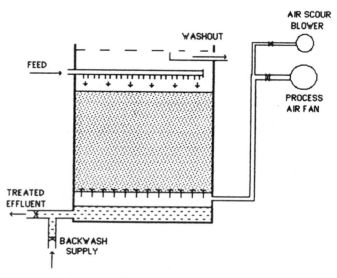

Fig. 27.3 The Lowe percolating filter.

All the existing versions (Fig.27.4) involve submerged granular beds with media within the 2mm to 6mm range. Air is blown through the beds to provide the necessary oxygen. All are fed with settled sewage as they would rapidly block with the suspended solids in raw sewage. There are however several variations. The first to be built at plant size involved a down flow filter using expanded shale as the media. This had a separate process air distribution system (for aeration, nor air scouring) 300mm above the filter floor. Two other designs also employ dense media with downflow filtration but the separate air distribution system which was a feature of the patent has not been found necessary. One version has a single air system that supplies both the process and backwash air which has cost advantages as well as a claimed advantage in blowing out any incipient blockages that might occur with a steady low flow.

Another design is a conventional dense media upflow filter much along the lines described in Chapter 25 although with a deep bed. In this case very considerable care would have to be exercised if the intrinsic problems of such filters, especially the blockage of under drains, are to be avoided. The single coarse media without the finer packing layers of the conventional upflow filter will simplify the situation but careful

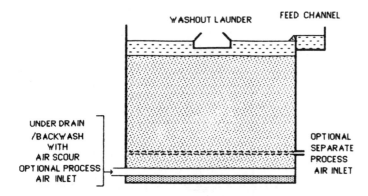

Fig. 27.4 Downflow dense media submerged biological filter.

screening will be necessary if blockage of nozzles is to be avoided.

Two upflow designs use buoyant media. In one (Fig.27.5) expanded polystyrene has been selected. This is held down by an inverted conventional nozzle floor and the media is washed by a downward flow from a header tank above the floor with the discharge at low level, believed to be combined with the inlet.

The second buoyant media design (Fig.27.6) retains a special ribbed polyethylene media under a grid. Cleaning of the media is achieved by increasing the air flow from the process value to a scouring value so that the polyethylene grains which have a specific gravity of about 0.9 are mixed and stirred to loosen the excess biomass. The suspension is then drained off from below. With this design the bed is not well compacted after washing and the wash volume is somewhat limited so that the filter is allowed to recycle after washing to recover the effluent quality before returning to service.

In all cases a rough surface appears to be necessary so that some residual biomass is left after washing. The wash volumes are also less than traditional potable practice, again to leave a seed quantity of biomass.

The design of such filters is based on the ability of the biomass to oxidise BOD and ammonia, and hence the primary design parameter is the loading in kg BOD or $NH_3/m^3/d$. There is also a hydraulic limit, particularly for the downflow versions.

424

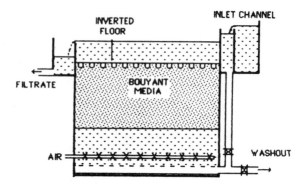

Fig. 27.5 Buoyant media submerged aerated filter, Type 1.

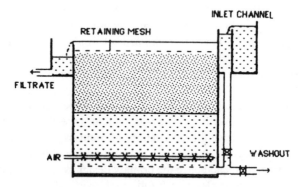

Fig. 27.6 Buoyant media submerged aerated filter, Type 2.

These filters are in fact found to operate with same F/M ratios (daily food/mass ratio) as activated sludge. The biomass concentration held on the media has been estimated at 15,000 mg/l instead of 3500mg/l. Many other parameters such as sludge yield are similar to activated sludge.

Upton and Stephenson's paper summarises the performances as follows:-

For 90% BOD removal   load $= 0.7$-2.8 kg BOD/m^3-d

For 90% NH$_3$ removal   load $= 0.6$-1.0 kg NH$_3$/m^3-d

The nitrification performance improves greatly as the BOD load falls and separate units are preferred by the suppliers of one of the buoyant media versions. The dense media versions appear capable of minimising vertical mixing of the bed so that BOD removal and nitrification can be achieved in a single unit at different levels in the bed.

Filtration rates are limited mainly by the BOD and ammonia. Loading rates and with typical settled domestic sewage the filtration velocities are usually between 1 and 3m/hr, or higher with deeper beds. Most versions use a 2m deep bed but this seems to be a somewhat arbitrary choice resulting from convenience of construction. Data given by Stenson, reported by Upton and Stephenson, indicates air application rates of 10-40m^3/kg BOD and oxygen utilisations of 5-9%. With the 2m depth the power consumption for aeration does not appear to be very different from conventional activated sludge treatment.

Filters are generally backwashed each 24 hours and the backwash water is normally recycled to the head end of the works, the excess biomass is co-settled with the primary sludge. The mixed sludge is discharged for digestion etc. This sludge is reported to settle and thicken well.

The aerated filters provide a final quality equal or better than that of a conventional tertiary filter eg.<10mg/l suspended solids. Indeed an even better quality can be obtained if a high rate tertiary filter is added as in the Thames Water version. Residual ammonia levels depend on loading rates but figures consistently below 1 mg/l have been reported. The vagaries of sludge quality that plague activated sludge plants are absent. No sludge has to be recycled as in activated sludge plants and no control of desludging in involved. No sludge concentration monitoring is necessary. Thus such filters require less vigilance on the part of the operator. On the other hand there is more mechanical plant to maintain.

As far as mechanical design is concerned biological filters generally follow the principles discussed elsewhere in this book. This has indeed been one of the attractions in that the engineering aspects have not required an entirely fresh start. The main difference is that the headloss though such filters with such coarse media is only 200-300mm and most of the headloss through the plant occurs in channels, weirs, penstocks etc. There is therefore a strong incentive to choose the simplest design and minimise components.

Aerated granular biological filters have only appeared on the market in quantity only after 1990 and competitive development is now proceeding rapidly.

27.4 Biological Oxidation of Iron and Manganese

It has long been known that traces of iron and manganese are oxidised biologically in filters. It was reported by Henderson, one time scientist at Mythe Works, Tewkesbury, to occur there if filters were not chlorinated. This observation led to the development of fluidised bed reactors for the removal of ammonia and manganese. The concept was the subject of detailed project at WRc (Gauntlett 1981). However the process has been developed in France as a high rate filtration proceedure and described by Mouchet (1992). Again the hardware is conventional with deep water units being used. It is essential to aerate the water before treatment and some phosphate and presumably nitrate are necessary although the process is inhibited by excess ammonia. A variety of organisms are involved such as *Gallionella ferruginea* and *Leptothrix ochracea*. The interest in the process has been stimulated by very high rates of treatment, up to 40m/hr.

The media also does not appear to be critical with effective sizes of 1.18mm up to 3.25mm being described in Mouchet's paper. One example quoted had a depth of 1.4m of 1.35mm effective size media (ie. 1.18-2.36mm on BS sizes), and a filtration rate of 22-23m/hr. The pH must be close to but below that for physical chemical removal. The process is also capable of removing manganese but the pH must be somewhat higher. Very high rates are still possible.

The main disadvantage of the process is the long maturation time before full efficiency is achieved, perhaps 50-60 days from a clean filter and 5 days from a 2 month shut down.

27.5 References

Anon. (1990), Cryptosporidium in Water Supplies, (First Badenoch Report), H.M.S.O. London

Croll, B. (1993) (Feb), Verbal communication,Scientific Section Meeting, Inst. Water Environmental Management

Gauntlett, R.B. (1981) in Biological Fluidised Bed Treatment of Water and Wastewater, Eds. P.F.Cooper amd B. Atkinson, Ellis Horwood. Chichester, U.K.

Graham, N.J.D. (1988) Slow Sand Filtration, Ellis Horwood, Chichester,U.K.

Graham, N.J.D.and Collins, R Eds. (1996) Advances in Slow Sand and Alternative Biological Filtration, Wiley, Chichester,U.K

Hendricks, D.(Ed), (1991) Manual for the Design of Slow sand Filtration, American Water Works Association, Denver.

Huisman, L. and Wood, W.E. (1974), Slow Sand Filtration, W.H.O. Geneva

Lowe, E.J. (1964), "Method and Apparatus for the Bacterial Disposal of Effluents", British Patent 971,338

Metcalf, S.M. and Stevenson, D.G. (1979) "Tertiary Filtration of Sewage", Public Health Eng. **7,** (3), 134

Mouchet, P. (1992) "From conventional to Biological Removal of Iron and Manganese in France", J. American Water Works Assn. **84,** 158

Upton, J. and Stephenson, T (1993), (Jan) "BAFs - Upflow or Downflow- The Choice Exists", Soc. Chem. Ind. Symposium, Cranfield, U.K.

West, J. Rachwal, A.J. and Cox, G.C. (1979) Experiences with High Rate Tertiary Filtration Treatment of Sewage", J.Inst.Water Eng. Sci.,**33,** (1), 45

CHAPTER 28.

MISCELLANEOUS APPLICATIONS

28.1 Introduction

In addition to solid/liquid separation and biomass support granular media systems are also used extensively for absorption and liquid/solid mass transfer operations. Examples discussed below include absorption on granular carbon and on other materials, ion exchange, and mineralising filters. The term 'filter' is used rather inconsistently and in most of these applications the separation of solids is usually incidental if it occurs at all. Nevertheless the hardware used in these granular absorption reactors is virtually identical to that used in true filtration.

28.2 Absorption and Mass Transfer

The primary mechanisms are diffusion through the fluid phase to or from the solid surface through a boundary layer and also diffusion within the absorbent grains to internal sites possibly accompanied by chemical reactions. In dilute solutions liquid phase diffusion usually controls the rate of reaction. The kinetics which follow the same principles for ions and soluble molecules as well as for submicron particles have been discussed in Chapter 17. The diffusion coefficients for several ions and compounds will be found the literature. In absorption processes at equilibrium there is a defined partition ratio β between the concentration in the liquid phase and that in the solid phase.

Hence the mass transfer equation given in Chapter 17 must be modified as follows:-

$$C - C^* = (C_o - C^*) \exp\left(-\frac{6(1-e)hK_L}{uD}\right)$$

where C^* is the equilibrium concentration in the liquid phase corresponding to the concentration present on the solid phase at that location in the bed.

This will apply as long as there is spare capacity ie. the concentration in the solid phase is below the equilibrium level for the existing liquid phase concentration. It is important to note that if C falls below C^* the desorption will occur. Equations are in fact seldom used because the diffusion coefficients are often not known and must be determined experimentally.

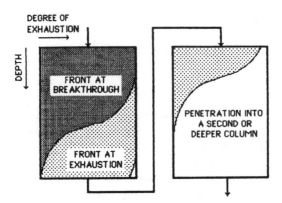

Fig. 28.1 Representation of progressing absorption fronts.

The media bed in an absorption process may be divided into the exhausted material, the 'front' where the absorption is actually occurring, and the clean material, in a manner similar to floc filtration (Fig.28.1). One difference between absorption and filtration is that in the former case any one species can be displaced from the absorbent by a more strongly held adsorbate. Seldom is the process irreversible. On the other hand the material removed is not capable of being dislodged mechanically.

Reference to Chapter 17 will show that the mass transfer coefficient is a function of the square root of velocity. Thus the decay coefficient (corresponding to the filtration coefficient varies inversely with the square root of velocity. The depth of the front therefore increases with the filtration rate and the utilisation of the bed falls as the flow increases for a given breakthrough concentration.

The above characteristics have profound implications for the operation of absorption equipment. An ion exchange column may be backwashed when regenerated but it would not be backwashed during the service run without losing much of its efficiency. Indeed ion exchange beds are only backwashed if absolutely necessary. An activated carbon absorber generally has a long absorption cycle in water treatment applications. Also a small amount of suspended matter is usually present and therefore is filtered out with the resultant increase in headloss. Perhaps more importantly the carbon acts as a support for bacteria and fungi which can grow and cause adhesion between the grains. For this reason regular backwashing is necessary to maintain a free flowing bed which could otherwise clump into mud balls. Washing will inevitably cause some mixing. Any mixing will tend to disperse the absorption front and reduce the efficiency. Taken to extremes, if exhausted material migrates to the bottom of the bed it could release absorbed material and recontaminate the filtrate. This can in fact occur if residual exhausted material is left behind when the bed is replaced with new or regenerated material.

There is an argument for having a wide size range of activated carbon and for applying an extended wash to stratify it and thereby to produce a condition which discourages vertical mixing. In practice it is seldom practical to use prolonged washing. Air scour is used sparingly if at all.

28.3 Granular Carbon Filters

28.3.1 Applications and Processes

The first large scale application of granular activated carbon in water treatment was for the control of tastes and odours. For this purpose an empty bed contact time (EBCT) of 6 min was common (Note some industries use the term EBCT whereas other talk in terms of bed volumes per hour, ie. the reciprocal). The compounds which produce taste and odours, particularly the earthy variety such as geosmin and iso-borneol are biodegradable. The bacteria living on the carbon digest these compounds so that the life of the carbon between regenerations may be three years or so. This behaviour is a common characteristic of activated carbon. It absorbs and concentrates the trace compounds as well as providing accommodation for the bacteria, which prolong the life of the carbon.

Less biodegradable compounds and higher concentrations will saturate the carbon more quickly. The rate of absorption will vary with the molecular weight of the compound and hence its diffusivity. The pore structure of the carbon also influences the capacity for a particular type of compound. The suppliers of activated carbon have experience of many applications and it is wise to consult them. The behaviour can be modelled qualitatively but it is not usually possible to calculate from first principles.

Pesticides have been the subject of trials by several water utilities and EBCTs of 15-30 min are often considered necessary, with a life expectation of perhaps 6 months. The latter will depend on the total amount applied, and the absence of other compounds which absorb more strongly and displace the pesticide. The minimum bed depth in the unexhausted state or EBCT depends on the peak concentration in the feed and the target residue. Additional depth is needed to allow for the progress of the saturated section of the bed (Fig. 25.1). It follows that longer contact times make better use of the absorbent because a larger proportion can be fully exhausted. On the other hand this involves a larger, more expensive, inventory of absorbent and a larger containing vessel. It is therefore possible to calculate a minimum cost situation which balances the saving in overall absorbent use against the increased interest charges on the absorbent inventory and the additional vessel cost. Unfortunately such minimum cost situations are seldom explored and contracts are generally awarded on lowest first cost. Materials such as activated carbon are much more expensive than sand and in contrast being an expendable material bulk contracts are often negotiated by users separately from contracts for the equipment.

To avoid any wastage or under utilisation of the carbon two beds can be operated in series. When the fully exhausted zone reaches the bottom of the leading bed and the absorption front has passed completely into the second (trailing) bed, the leading bed is replaced and the two are changed over so that the new bed becomes the trailing one. The extra cost involved in the doubled number of beds and the additional pipework and valves is considerable and only worthwhile when the rate of exhaustion is high. With gravity filters extra pumps will prbably be necessary.

28.3.2 Engineering Aspects

The design of the contact vessel or filter follows relatively conventional practice for granular media filters. Often activated carbon is used for the dual roles of filtration and absorption simultaneously and many surface water treatment plants in Great Britain have been converted from sand to activated carbon. In this case the design must firstly be satisfactory for filtration. The carbon behaves like anthracite and the wash procedure must follow anthracite practice. Because the full depth of the bed is carbon the bed expansion will be greater than with dual media filters. There has been a tendency for the carbon to be finer than anthracite and where dual media filters have been converted shorter filtration runs have been experienced. A special grade of carbon has been marketed for such conversions to reduce this problem.

If activated carbon is used in a dedicated absorber which is not required to filter solids then there is greater freedom to optimise the vessel. Filter floors and the associated washing equipment are the most expensive component and for a given contact time deeper vessels and rates of filtration higher than for conventional filtration may be chosen. With pressure filters there will be a minimum cost solution which would have to be resolved with the vessel supplier, taking account of the cost of dished ends, vessel side plates and floor cost. The same considerations apply to open gravity filters but pumping costs may have to be considered as well. Dedicated carbon absorbers tend to be at least 2.5m deep on large plant and can be as deep as 4m. The pressure loss through very high rate filters can however be a limiting feature, particularly with open gravity filters.

Continuous activated carbon absorbers have also been used as at Lake Tahoe, Culp and Culp (1971) (Fig.28.2). In this case the flow must be upwards as the carbon is discharged downward. Some of the designs of continuous sand filter and also ion exchange contactors could be used for this purpose but generally to date as with ion exchange continuous moving bed contactors have had limited commercial success. Another combination borrowed from sand filtration is the upflow/down flow combination. In this case the upflow filter is the first stage and when exhausted the carbon is discharged and the charge from the second filter is transferred to the first filter vessel while fresh material is placed in the second (downflow) filter. The principle is similar to the twin downflow bed arrangement discussed above.

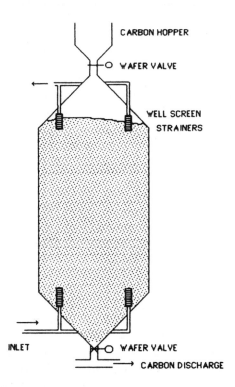

Fig. 28.2 Lake Tahoe continuous absorber.

Many converted potable filters have activated carbon placed on existing gravel packing layers. Where the filter previously had such layers to prevent percolation into the under drain they must be retained. It is some times possible to install a new floor economically but specialist contractors should be consulted. Packing layers are however best avoided on carbon filters because of the inherent difficulties involved when the carbon has to be replaced. Normally about 100mm of carbon has to be left behind because sand and gravel contamination may not be acceptable to the regeneration contractor.

The preferred floor for absorption filters is one without any packing or support media. In this case it is desirable to have strainers with a generous slot area to minimise the local collection headloss around the nozzles. The basic principles of floor design discussed in Chapter 21 must be observed and a sufficient headloss for uniform distribution and fluidisation is even more essential than in filtration where vertical mixing is less important or even beneficial.

While there are arguments in favour of declining rate filtration for solids removal they do not apply to absorption. Constant rate filtration is preferable although great precision is not warranted if the breakthrough of each filter is monitored. Slow start procedures have no relevance as there is no maturation period. The carbon filters on a water treatment plant are usually designed to give the desired contact time with at least two filters out of service, ie. one washing and one out of service for recharging. Washing will be infrequent and will be based on experience, to maintain a clean bed.

28.3.3 Transfer and Regeneration

The removal of carbon from smaller filters is usually a manual operation, however proper access must be provided so that vehicles or skips can get alongside the filter. Suction machines (like vacuum cleaners) are available and these include road vehicles similar to gully emptying vehicles. On large filters special means may be provided for transferring the carbon to a tanker and for replacing it (Fig.28.3). This may be an infrequent operation and therefore a manual one. If the carbon has to be replaced each six months, as may be the case for pesticide removal, a plant with many filters will have fresh carbon delivered for one of them fairly frequently.

It has already been noted that exhausted carbon can release contaminants into clean water and for this reason dedicated absorbers are usually cleaned out carefully to remove all residual material. Such residual material is likely to be exhausted material from the top of the bed if the carbon is discharged at the bottom. Several methods have been developed for this purpose, some patented. It is possible to drain most of the carbon out through a pipe or connection at the base of the filter. It will flow readily when fluidised and the angle of repose under water is low so that fluidisation need be applied only intermittently. The concentrated slurry is diluted to about 25% (settled concentration) as it leaves the filter and it is conveyed by means of an eductor or a pump, designed for solids, to the waiting tanker where it is deposited and excess carrier water allowed to drain away for recovery or disposal.

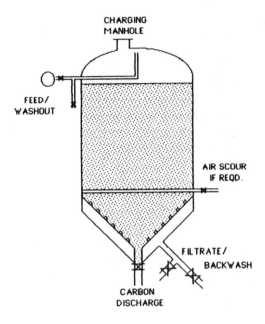

Fig. 28.3 Steel GAC contactor.

The removal of the residual few centimetres presents a greater challenge. It is possible to remove the residue manually with suction tools. One floor design uses recessed nozzles to avoid damage during this operation. A number of designs use sloping floors to facilitate discharge. This has disadvantages in that the bed depth is variable and the filtering flow will vary inversely with the depth. The progress of the front is linked to the flow rate and hence the relative penetration varies inversely with the square of the depth. This can leave a fair proportion of the bed unused when the shallower parts break through. Any inclination of the bed must therefore be small. It is also difficult to achieve a uniform air scour with an inclined floor although not impossible. A separate, level air lateral system may be used. With a sloping floor flushing systems are used to convey the residual carbon to a sump from which it is conveyed to the tanker or a silo. Designs (Fig.28.3) are now available to flush all the carbon away without the need for any manual intervention. Recycling of the flushing water reduces the usage of water to less than that of a single wash. A storage silo may hold the contents of one filter which can then be refilled direct from tankers which are then able to return with a load of exhausted carbon.

Pressure filters are available which have a hoppered floor (Fig.28.4) with nozzles supplied from a separate plenum space below. A centre discharge is fitted for carbon transfer. A standard filter takes one tanker load.

Tankers are discharged generally by flooding and pressurising. The slurry is diluted for easy conveying. To avoid problems of blockage it is advisable to complete the transfer in one operation. Interrupting the transfer is a recipe for blockage. After the solids have passed through the lines are flushed with clean water. Transfer lines should be capable of easy disassembly in case blockage occurs. Each supplier has his own preferred detailed procedure.

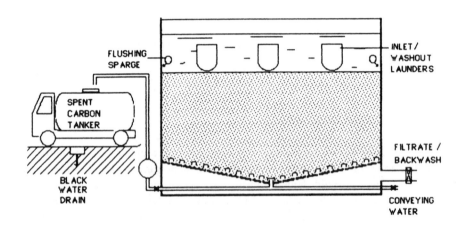

Fig.28.4 Elements of a gravity GAC filter.

The water used to transfer carbon contains suspended fines and is referred to as black water. This can be treated as clarifier sludge or in some cases is run to a separate lagoon for settlement.

Both regenerable and non-regenerable grades of activated carbon are available. The regenerable materials used in potable water treatment are usually dewatered and heated in a controlled 'water gas' type atmosphere with steam present at a high temperature in a multiple hearth furnace or rotary kiln. Few plants have local regeneration facilities. Regeneration is less economical on smaller plants.

For many industrial applications in situ regeneration with steam or chemicals may be sufficient. Organic solvents are driven off with steam. Some industrial absorbers are designed for steam sterilisation. On water treatment plant the intrinsic biological action is usually regarded as beneficial but some of the bacteria present may emerge in the filtrate. These are killed off by subsequent disinfection. Ozone converts many refractory organic compounds present in water into more biodegradable materials and this has caused problems with after growths in the distribution system. One role of activated carbon after ozone is to act as a bioreactor to enable such compounds to be digested in a unit that can be cleaned fairly easily.

28.4   Ion Exchange Contactors

This is an entire technology in its own right and has its own bibliography. Recent examples are by Owens (1985) and Dorfner (1991). Ion exchange for the most part uses insoluble synthetic resin beads with functional groups within the resin matrix so that they are able to act as insoluble but active cations and anions respectively. Thus in the simplest example the sodium form of a cation exchange resin is able to exchange the sodium for calcium in softening units. The calcium form is converted back to the sodium form with strong brine. The purpose of including ion exchange here is to compare the similarities and differences with other granular media processes.

The kinetics of ion exchange are, like granular carbon absorption, governed by diffusion at low concentrations and the equations in Chapter 17 were originally developed in the context of ion exchange. The diffusion coefficients of simple ions are in fact known and reasonable calculations are possible. At higher concentrations, for example, during regeneration internal diffusion within the resin bead limits the rate of reaction.

The exchange equilibrium is again governed by a partition ratio whereby the concentration in the aqueous phase and resin phase are related by a characteristic constant under any one set of ionic conditions. The transition from one form, eg. sodium to the other, eg. calcium, takes place over a front which moves down the bed until breakthrough occurs. Thus mixing of the bed would destroy the front and greatly reduce the efficiency of the process. The length of a run however is much shorter than with activated carbon, possibly only a few hours in recent plant so that washing during a run is seldom necessary. In any case ion exchange resins are seldom required to act as filters for solids. If they do then it is usually by accident rather than intentionally.

Steps are taken in the manufacture of modern resins to obtain the most uniform bead size. Because they are manufactured by the polymerisation of liquid drops it is possible to exercise close control. They do not have to be sieved from a wide size range in the same way as with sand. This great precision assists in the production of sharp fronts both in regeneration and exhaustion, because all the beads present operate at the same rate. The beads are also very smooth and round. They would be less efficient as collectors for solids than most media used for filtration. However there are some applications where the dual role may be exploited. Anion exchangers being positively charged tend to collect naturally negative colloids and macromolecules. As a result they are often easily fouled. Normally pretreatment procedures provide as clean a feed as possible. Certain resin types are more resistant to fouling but as a general rule ion exchange is a process for water free of most organic contaminants and suspended solids. One exception is the use of special resins designed for the absorption of soluble organic matter. Ion exchange for example has been proposed for treating surface coloured water, Odegaard and Torsen (1987).

Ion exchange vessels are almost exclusively pressure vessels similar to pressure filters except that they must be fabricated from or lined with materials that withstand the corrosive regenerant chemicals such as caustic soda and acid. A collecting floor is necessary as well as inlet connections. In contrast to granular media filters more attention is paid to the easy filling and removal of the resin, although this flows more easily than sand. Blank flange connections are usually located near the floor.

The resin beads are less dense than most other media used in filtration. Strong cation resins have a specific gravity of about 1.2, anion resins and weak cation resins may have specific gravities as low as 1.08. Thus they are fairly easily swept away by residual jets from the inlet. Diffusers are commonly fitted to disperse the

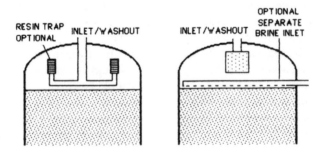

Fig. 28.5 Inlet diffusers for ion exchange vessels.

incoming flow. This is particularly important in applications such as condensate polishing where very high rates are used. One approach is to use a basket rose or upward facing candelabra (Fig.28.5). Without such means serious gouging of the bed with loss of capacity can occur. Also if the resin is regenerated in co-current mode which may often be the case the regenerant solution being denser than the water being treated must also be spread uniformly over the bed. Such inlet fittings are unnecessary in filters containing anthracite or sand although occasionally manufacturers who may be unaware of this may fit distributors. Providing that the surface of the resin is level the headloss through the resin bed ensures an even flow.

Another common feature of ion exchange contactors is the use of screens or resin traps on the backwash outlet to retain any of the expensive resin that might otherwise be lost. Separate inlets are some times used. Such screens can only be used where the feed water is free of debris that could block them. Mistakes are some times made when designers familiar with one technology transfer features to another without understanding the reasons. Granular media filters very rarely include such screens as they would clog, particularly when operating on surface water.

The design of the under drain is identical in principle to that of filters but high velocities in contact with the resin are particularly undesirable and can cause attrition. Uniform flow through the resin right to the base is more desirable than in filtration. The main difference between ion exchange and filtration, as far as the under drain is concerned, is that the regenerant flow is usually much smaller than the service flow whereas the backwash flow in a filter is larger than the service flow. Thus the headloss for distribution which must include a safety factor for the density of the regenerant solution must be a compromise. The service headloss will tend to be higher than for filtration.

Suspended floors with plenum voids below must be avoided in ion exchange because they provide an undesirable dead volume which has to be filled or emptied of chemical solutions and in which back mixing and dilution occurs. The efficiency of regeneration and rinsing would be reduced considerably. The under floor volume should therefore be minimised and header/lateral or radial systems (Fig.28.6) are generally used. Air scour is not normal and this simplifies the design. Where it is used it is normally applied only to cation beds.

The regeneration efficiency of ion exchange is greatly improved if the regenerant solution is applied in counter current mode, ie. upflow if the service flow is downwards. Unless very low flow rates are used the low density of the resin and the density of the regenerant solution make it necessary to have some means of holding the bed in a compact state. Such means are provided by a intermediate collecting system buried in the bed near the top or about one quarter of the way down. Some designers use down flow air to force the water out of the upper section of the

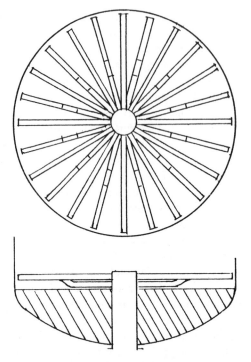

Fig. 28.6 Radial lateral under drain.

bed so that the dead unsubmerged weight of that section of the bed holds the lower part down. The other method is to divide the regenerant flow and feed it from both ends, the so called split flow method. This has the advantage that the upper section is also regenerated at the same time but because this is fully exhausted in service the co-current regeneration of this part of the bed makes little difference to the overall behaviour while the highly regenerated bottom resin provides an excellent final quality.

Such a split flow system is of course similar to the biflow filter described in Chapter 24. When the resin is back washed the intermediate collector may be subjected to a considerable upthrust which can fracture it. Some designs use lance collectors which drop from a lateral system placed above the resin bed. Ion exchange resin has a much lower coefficient of friction than sand and is more able than sand to flow round such collectors. Intermediate collectors also exert a bridging effect and play a part in holding the bed down. Header/lateral systems are more effective than lances in this respect. Such systems must be very strong as they often have to withstand a considerable upthrust. A further complication not experienced in granular media filtration is swelling and contraction of the resin as it changes its chemical form. This can also exert forces that must not be overlooked.

It has been mentioned that mixing due to backwashing is undesirable. Often with split flow systems only the upper part of the bed is backwashed. The lower section which provides the high quality may run for very long periods without washing and remains stratified with respect to its chemical form.

Similar systems are employed when mixed cation/anion resins are used in a single column. The resins are mixed in service eg. by air and restratified for regeneration. An intermediate layer with an inert resin with a density in between that of the anion and cation resins may be included to keep the two active resins well apart and thereby avoid acid coming into contact with the anion resin, and conversely. If this should occur it would completely exhaust any resin involved and if such resin reached the bottom of the bed it would adversely affect the product.

Some designers prefer wrapped laterals or long virtually continuous strainers rather than individual button strainers. Some strainers used in ion exchange have inbuilt non return valves to increase the resistance during upflow regeneration and to direct the reverse flow against the floor. Such moving parts can however be unreliable.

Although they have not been adopted very widely there several continuous ion exchange contactors have been developed. Some may be of interest in granular

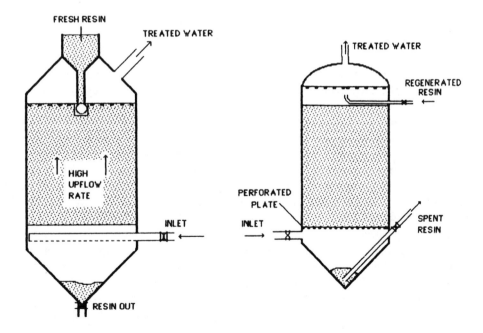

Fig. 28.7 The Asahi contactor for continuous ion exchange.

Fig. 28.8 The Porter contactor.

media filtration. Fluidised bed systems are of no interest in the present context, neither is the Higgins contactor (Higgins and Chopra, 1970) which propels the resin round a circuit between service and regeneration vessels with the aid of an empty chamber and knife valves which would have a limited life against materials other than ion exchange resins. The Asahi contactor (Fig.28.7), reviewed by Solt (1970), uses a high upflow rate to hold the resins against an inverted floor. Cessation of the service flow allows the bed to drop and freshly regenerated resin to be inserted over the existing charge.

The Porter contactor (Fig.28.8) discussed by Clayton (1970),which perhaps is more applicable to filtration, uses a perforated plate near the base of the column. The apertures are large enough to allow the resin or media to flow through, particularly with a co-current flow of water. The open area is small enough to prevent the resin passing downwards when the service flow is upwards. Thus it is possible to discharge the contents below the plate while the upper section remains in service. A short flow reversal allows the upper section to descend and fill the plenum leaving space at the top for a charge of fresh material. Over filling is prevented by strainers in the soffit. This contactor would appear to have applications possibly as a continuous filter.

28.5 Iron and Manganese Removal

At first sight this may appear to be a routine solid/liquid separation process. However, particularly with manganese solid particles may not be involved at all. Certain materials such as magnetite (Fe_3O_4) and manganese dioxide are oxidation catalysts and enable the manganous ion to precipitate on the surface of such materials under conditions which are milder than those for spontaneous nucleation. Thus manganese in solution will precipitate onto manganese dioxide at pH values above about 7.5 in presence of dissolved oxygen. The deposited manganese dioxide is also catalytic and the process needs only a small amount of starter material for the whole sand charge to become catalytic. Eventually after several years the sand becomes coated to the extend that it becomes too large and cannot be washed properly. It must be partially replaced. Like any catalytic process the catalyst can be poisoned and in this case conditions must remain aerobic. Organic matter and sulphides interfere. Manganese dioxide is a catalyst for the oxidation of humic colour but the process is not economical with available forms of manganese dioxide.

A proprietary grade of manganese dioxide called polarite is widely used for the removal of iron and manganese. It would appear from Don 1909) that this was originally a synthetic magnetite.

A variation on the above process uses manganese green sands (zeolites) regenerated with permanganate either intermittently or continuously. Chlorine is an alternative which is usually already available on site. It would seem that after a period the two processes become the same.

28.6 pH Correction and Remineralisation

Aggressive waters containing free carbon dioxide may be neutralised either by dosing with an alkali such as caustic soda, sodium carbonate or lime in which case careful pH control is necessary. Alternatively the water may be passed though a bed of limestone, magnesite or dolomitic limestone which contains both magnesium and calcium. These minerals are dissolved by the carbon dioxide.

While the capital cost of limestone filters may be considerably more than a chemical tank and dosing pump, locally produced limestone may provide a very much lower running cost particularly in areas where chemicals have to be imported.

Demineralisation systems such as distillation or reverse osmosis produce an unpalatable and somewhat corrosive water with a low dissolved solids level. Blending with raw water may introduce too much sodium or chloride. Thus such waters are often remineralised with calcium bicarbonate. This is derived from carbon dioxide, produced either from the combustion of natural gas or from the plant degassing system, which is dissolved in the demineralised water and neutralised by passing through limestone filters.

If the limestone is pure the calcium carbonate equilibrium pH can only be approached but never achieved. Thus small quantities of alkali, usually caustic soda or sodium carbonate are added to trim the final pH. Alternatively excess carbon dioxide may be added initially and the surplus stripped off at the end. If the carbon dioxide is derived from the demineralisation plant there will probably be a considerable excess available and the latter course is attractive. Dolomitic limestone and magnesite will take the pH above the equilibrium level and the final pH can be adjusted by a small amount of by-passing. If such materials have to be shipped from a distance the economic benefit of mineralising filters will be lost.

On a small scale, eg. domestic or private supplies, limestone or dolomitic filters provide a simple foolproof method of pH correction, providing that the water is clean in the first place. If iron is present regular backwashing will be necessary as the limestone can be fouled with ferric hydroxide which will hinder dissolution. Manganese may produce a more resistant coating. Bacterial slimes can also be a problem. In this respect it is better to chlorinate before such filters.

Natural limestone is seldom completely pure, some sources may contain only 80% $CaCO_3$ in which case the filter must be able to retain residual grit and clay. As the limestone dissolves particles eventually become too small to be of use and could even penetrate the under drain. Combined air and water washing is particularly useful because material of a wide size range can be accommodated. Stratification is avoided and furthermore the fines will tend to be washed out of the filters. Because the volume of the grains is a cube function of the diameter grains of 2.0mm starting diameter will have yielded 98.4% if they are washed out at 0.5mm. This is often more than the calcium carbonate content of the limestone.

In contrast to conventional filters, limestone filters show a decreasing headloss as they run. The chains of larger pores within the granular bed, discussed in Chapters 16 and 17 carry a higher flow but instead of blocking they erode preferentially and could ultimately produce short circuit channels. Thus regular back washing is required in order to destroy these channels and restore the integrity of the bed. Washing also removes the insoluble residues and fines. The falling headloss as the run progresses must be taken into account when considering the flow balance in a set of filters.

When the inventory has fallen to the point where the efficiency is no longer adequate the filters must be topped up. This may be after several washes. The incremental charge is a matter of operational convenience. Extra height to reduce the charging frequency is not costly. Typical beds have a depth of up to 2-3m depth and filtration rates of 10-20m/h but much depends on temperature as well as the efficiency required. Studies of the kinetics indicate that at high levels of free carbon dioxide the rate limiting process is ionisation of carbon dioxide to carbonic acid and an increased surface area of media is not beneficial. At low concentrations it appears from Compton and Pritchard (1989) that the surface area is more important.

Combined absorption of carbon dioxide and neutralisation in a single vessel has been attempted but this has obvious disadvantages with impure limestone.

There is no reason why large limestone filters should not be gravity units which can then be recharged by conventional front end loaders. The rate of loss of free carbon dioxide in the supernatant water will not be high providing that there are no weirs and sources of turbulence. The engineering philosophy on reverse osmosis and distillation plants has tended to be directed to steel pressure vessels.

28.7 Other Applications

Granular media filters are sometimes used for the removal of traces of oil after canal tank or lamella separation. The interception processes are similar to those for solid particles. The media acts as a coalescer. The backwashing procedure will depend on the nature of the oil and will be very specific to the application.

Granular media filters find use in several industrial process such as the filtration of brine, and the recovery of fine colloidal precipitates that are difficult to filter on surface cloths and papers. In this situation polyelectrolytes can be very useful and do not contaminate the material with iron or aluminium. It has to be remembered that granular media filtration is suitable for part per million concentrations and can concentrate only to about 1000mg/l. After this point other types of filter, settlement or flotation must take over.

28.8 References

Bratteby, H., Odegaard, H. and Halle, O. (1987) "Ion Exchange of Humic Acids in Water Treatment", Water Research, **21,**1045

Clayton, R.C. (1970) "Continuous and Semi-continuous Ion Exchange Processes".Ion Exchange in the Process Industries, p 98. Soc. Chem. Ind. London .

Compton, R.G.,Pritchard, K.L. and Unwin, P.R. (1989) "The Dissolution of Calcite in Acid Water, Mass Transfer vs Surface Control", Fresh Water Biology, **22,**285

Culp, R.L. and Culp, G.L.(1971) Advanced Waste Water Treatment, Van Nostrand, N.Y.

Don, J. (1909) "The Filtration and Purification of Water for Public Supply", Proc.Inst.Mech.Eng.,London,p7

Dorfner, K. (1991) Ion Exchangers, De Gruyt, New York

Higgins. I.R. and Chopra, R.C. (1970) "The Chem-Seps Continuous Ion Exchange Contactor and Its Application to Demineralisation Processes". Ion Exchange in the Process Industries, p 121. Soc. Chem. Ind. London .

Letterman, R.D., Hadod, M. and Driscoll, C.T. (1991) "Limestone Contactor Steady State Design Relationships", J. Environmental Eng. **117,** (3), 339

Letterman, R.D. (1993) "Pilot Trials on a Limestone Contactor in British Columbia", J.American Water Works Assn. **85,** 91

Owens, D.C. (1985) Practical Principles of Ion Exchange in Water Treatment, Tall Oaks, Voorhees, New Jersey, USA.

Solt, G. (1970) "New Techniques for Carrying out Ion Exchange Processes", Ion Exchange in the Process Industries, p 85. Soc. Chem. Ind. London .

CHAPTER 29.

COMMISSIONING AND PROBLEMS

29.1   Introduction

While it is not possible to list all the problems likely to be encountered with a granular media filter there are certain general points to bear in mind. A distinction must be made between design faults and maloperation or neglect. Owners and operators vary in their tolerance of imperfections. Some faults may be purely cosmetic, some fundamental.

29.2 Commissioning

The correct commissioning of a granular media filter is vital to its subsequent satisfactory service. However constructional programmes often make it difficult to proceed in the ideal way. Often a very limited water supply is available when the media is being placed.

The first step when the structure is complete is to ensure that it is leak tight and that all the pumps and blowers function. Water will be required and if an air scour is installed this will no doubt be tested with the filter empty of media and only sufficient water to cover the nozzles. This itself is a problem because with water at a low depth the under drain will not be subject to the hydrostatic pressure present in service. The difference may be 0.2 bar or more which can affect nozzle behaviour. With a full depth of water it is not be possible to see any inactive nozzles or orifices through the swirling mass of bubbles 1-1.5m deep. Thus it is not possible to reproduce service conditions accurately at this point.

The next step which may seem obvious but which can sometimes cause problems is to ensure that all media being placed is within specification. As discussed in the next chapter it can be difficult sometimes to obtain analytical data on batches of media before delivery and often samples are taken after or during placing. The contractual side needs to be clear because the removal and supply of fresh media costs more than the original material, and arguments are likely to follow any rejection.

A filter must be clean before the media is charged, and constructional debris such as plastic and paper, polystyrene foam, swarf, rubble etc. removed. Ideally the system should be vacuum cleaned. This is particularly important with fine slot strainers.The packing layers, when used, should be placed carefully preferably by hand and certainly not tipped in from the rim of the filter. This could damage nozzles or displace previous layers. Boards should be laid on each layer and the media which will normally be bagged should be lowered into the filter. Every attempt should be made to limit foot traffic on the packing layers and the final interface must be smoothed off with a trammel board. Footprints, as discussed elsewhere, can produce incipient failure points. It is very difficult to rectify any faults which occur at this point and the whole support system may have to be renewed if faulty.

Having laid the support gravel the working media will be placed. However once the first few centimetres have been placed and the support gravel well covered the installation of the remainder is less critical. The surface need not be smoothed particularly carefully because the backwash will do this more efficiently. However it is necessary to have some idea of how much has been added. It is wise to place a board over the area where the working media is to be deposited but of course the board must not become buried. Indeed plastic sacks buried in the media cause dead spots. Tipper trucks or skips should not be discharged into the filter from a height.

After the bulk of the media has been charged, preferably leaving the level about 5% below the final figure, the filter should be filled from below with water at a subfluidising rate. There is a temptation to evaluate the air scour at this time because the system will usually be serviceable but water still scarce. It is recommended that this be delayed until the media has had its first full rate wash with water to flush fines and residual dirt away from the under drain. If no water is available in the clean wash water tank then it must be filled either with a temporary supply or via one filter operating at a low rate. (A compromise is clearly necessary in these circumstances). Again care is necessary to ensure that the upwash pipework is full otherwise surges of air and water may disrupt the support layers. After the first water wash, the full air scour (where used) can then be applied.

Once this point has been reached the filters can be run to produce sufficient filtrate for a programme of washes. It is assumed that a water supply is available, and also that facilities for disposing of dirty wash water have also been installed. This phase of operation is hazardous because many operations which ultimately will be automatic may be manually controlled. Care is necessary to ensure that no damaging operations are initiated such a simultaneous air scour and water wash on a filter not designed for such a system.

The first observation will be a noticeable difference in the height of the bed and some topping up may be necessary or more likely excavation. This is because the voidage of the 'as poured' material differs from that after a wash, especially when the final phase is water on its own at a fluidising rate. Combined air and water washing will reduce the bed depth.

Filter media cannot be produced completely free of dust and undersize material. Indeed specifications usually allow 5% undersize material which with a final water wash tends to migrate to the surface of the filter. At the first wash there may also be other sundry material rising to the surface. The fines must therefore be skimmed carefully, manually. Combined air and water washing tends to prevent segregation of fines unless the procedure ends with a high rate rinse. The air bubbles in the combined wash assist in carrying fines over the wash out weir which is designed to retain the minimum fraction of the media but not the undersize material. Skimming is less essential in this case.

If the filters are to hold dual media then if at all possible they should be run at a low rate with single media only for several cycles until the migration of fines to the surface has subsided. It is not possible to remove such fines after the anthracite layer has been placed. The anthracite will also need to be skimmed, and because of its friability rather more is likely to be removed. Both layers will need to be topped up or trimmed after washing. Skimming of anthracite is best done walking on boards as foot traffic can cause further attrition. (This is even more important with granular activated carbon.) When working in filters it is essential that upwash valves are locked shut because although it is possible to walk safely on a static bed a person will sink abruptly into a fluidised bed (quicksand).

After the media has been fully washed, skimmed, adjusted etc. it is normal practice on potable filters to disinfect the filters before being put into supply. The logic of this practice is perhaps questionable now that prechlorination has virtually ceased because of concern about THMs (trihalomethanes) produced by the action of chlorine on organic matter in the water, especially if traces of bromine are present. Filters do not normally receive a disinfected water and indeed some marginal biological action may be beneficial. Disinfection is a discrete and separate downstream operation. In other situations such as with granular carbon filters the media destroys the chlorine so that disinfection is pointless.

The early days of operation will tend to reveal weak points in the system but the incidence of faults tends to decay with time. There are several applications of granular media filters where the initial filtration efficiency is poor and some days or weeks may be needed before the bed matures fully. This is to some extent true of conventional polishing filtration after clarification. It is more pronounced with the catalytic removal of iron and manganese and even more so with biological processes, where up to 60 days maturation time has been reported by Mouchet (1992) for biological iron removal. This aspect must be considered when acceptance trials are being planned.

29.3 Problems

One potential problem is blockage of the filter floor. It is a wise precaution in design to make all slots and orifices as large as possible. This is a major benefit of support gravel. A filter floor should not be a filter in its own right. It is difficult with a large concrete plant to ensure complete cleanliness as in a dairy with stainless steel plant for example. Concrete slowly etches over the years and particles of sand can be released. Chemicals applied to the water down stream may contain grit, dirt or debris, particularly in territories that lack rigid quality control systems or secure storage. Poor quality lime for final pH correction is a particular hazard. It is wise to add lime after the off take to any clean wash water tank. Lime saturators are sometimes used to avoid such problems. Disinfection is effective at lower pH values and lime is best added after the contact tank..

Some blockage of nozzles with media will be inevitable and the design of the nozzle should be such that this can be tolerated without upsetting the hydraulic characteristics. The application of a backwash as the first step in commissioning a filter helps to move potential blocking particles away from the nozzles. Granular carbon is reputed to produce traces of fine material which can penetrate the floor. These must be trapped by a down stream device, or possibly by a protective sand layer, not by the floor. Upflow filters present a major problem in this respect and orifices in such filters tend to be large (8-12mm).

Blockages may lead to dead zones. Often they can increase the headloss across the floor and possibly cause disruption during a backwash and major failure. Water hammer and other hydraulic transients are particularly dangerous. Some consultants and operators insist on the floor being able to take the closed valve pressure of the upwash pump. In some other cases a standpipe or other safety device is used to protect some floors.

Packing layers can also block. Upflow dense media filters are particularly prone in this respect. In a down flow filter, over running with consequent penetration into the packing layer, or the liberal use of polyelectrolyte can choke the gravel. (This may also occur with no-fines concrete and other porous composition floors.) The backwash is not designed to clean packing layers and if it was the working media would probably be over-fluidised over-expanded and washed out. Chemical cleaning will probably be required but often disruption of the packing layers occurs, in which case the media has to be replaced.

Disruption can be caused by several other factors, the most common being uneven interfaces, the incorrect choice of gravel for the particular wash rate, excessively high wash rates (caused for example by using a standby pump to take the wash rate above the design value), and by an incorrect procedure for starting the air scour or displacing the air afterwards. It cannot be stressed too strongly that air and support gravel are not good partners, although if the design and operation is correct they will live amicably. A common fallacy is that disrupted gravel can be restratified with a high rate wash. It is most likely to make matters worse because the water will channel through zones of gravel.

Mud balling is a term often used in connection with filter washing. It is a form of partially clogged working media and is discussed more fully in Chapter 20. Mudballs behave as extra large particles of media and tend to sink and sit on the support gravel. A good wash procedure should prevent their occurrence. The working media in a filter should remain clean indefinitely, ie. for the life of the filter.

One effective remedy for premature clogging is often to chlorinate the water to be filtered, (after clarification with a coagulant to minimise residual organic matter and thereby reduce the formation of THMs). Residues of several polyacrylamide polyelectrolytes are inactivated by the chlorine in this way. Unfortunately concern about THMs is limiting this once fairly widespread solution but tests to determine the THM level will resolve the matter.

Many filter problems are caused by overdosing polyelectrolyte or bad dosing practice. Such materials must be fully dissolved and mixed uniformly and thoroughly into the main stream. It is most advisable to dilute the material with carrier water so that the solution being blended has a similar viscosity to water and does not behave like lumps of treacle. Residual `fish eyes' must be avoided. A hose without any distributor sparge pipe running into a channel is not sufficient. However even if dosed correctly an excess will still cause the particles to cling to the grains strongly and even cause the grains to agglomerate into mudballs.

Encrustation with calcium carbonate is at times a problem on reservoir waters subject to algal blooms. This can eventually increase the size of the grains and render the wash ineffective. The same thing can occur on filters after precipitation softening. The solution is to correct the pH with acid, carbon dioxide or by blending with a bypassed unsoftened water. Sequestering agents such as sodium hexametaphosphate hinder filtration by stabilizing the particles. Borehole water that has been aerated to remove carbon dioxide and introduce oxygen for the removal of iron and manganese can produce a similar problem and control of the degree of aeration is then the usual answer. Manganese dioxide can similarly coat sand grains and change the diameter. This coating is beneficial as far as furether manganese removal is concerned but when the grains reach the size limit for the installed wash system some the media at least must be replaced. This is usually after several years. A residue should be left to maintain the catalytic activity.

Attrition of media is seldom a problem with sand, or indeed anthracite. Activated carbon however appears to be delicate and air scour must be used with caution. Most losses of media from filters occur as a result of over filling and not taking account of the change in the volume following segregation of the size fractions, or as the result of a faulty wash procedure or a poor design of cill . Bed growth and consequent loss can also occur as the result of gross fouling and inadequate washing.

In some cases accidental air ingress into the suction of an upwash pump (eg. As a result of vortexing in the clean water tank) will produce a plume of air which, particularly with dual media, can cause a steady loss. Leaky air valves may have the same effect. With combined air and water washing the loss of fines may also affect the bed level early in the life of a filter. Media losses in the first few months of operation are usually ignored unless excessive.

The appearance of a filter during washing is generally useful in determining the state of the bed and the uniformity of the wash. One would expect most nozzles to emit air if a filter is air scoured. The air pattern can usually be seen clearly at the surface of the media if the depth of the water is only 100-300mm. However with greater depths the water starts to roll and air plumes coalesce in a manner which is very visible in the aeration lanes of sewage treatment plants. This gives an impression that the air distribution is not as good whereas it may be perfect. A similar situation occurs during fluidisation. Often convection patterns are visible on the surface with apparent shallow boils or spouts. These are evident on larger filters but less so on small columns. Such circulation is beneficial and encourages a degree of particle contact which does not occur in a small column. The spout that is a sign of maldistribution is one that tends to have an empty core and which throws media well above the level of the rest of the bed. Such spouts may be evidence of uneven distribution through the under floor, an insufficient headloss through the floor or disruption of supporting gravel. If air scour is used full distribution acts as a good indicator that the bed is not blocked. The broomstick test referred to below is also useful for checking that the bed is fully mobile. If the filter is working satisfactorily and the dirt content of the media is satisfactory then cosmetic features are best ignored as attempts to replace sections of the support gravel may make matters worse.

Tertiary sewage filters which were discussed in Chapter 27 are particularly difficult to keep clean but the use of coarse media and combined air and water washing removes the resistant slimes. Regular treatment with sodium hypochlorite (eg. injection into the backwash water), caustic soda or both are sometimes used.

29.4 Servicing

Granular filters need little servicing. Components such as valves, controls and instruments must however be maintained. A common source of poor performance is hunting of control systems caused by lack of maintenance, poor design or poor adjustment. Outdoor filters may suffer from wind blown debris. Plastic bags, leaves, particularly large ones such as horse chestnuts can become buried under the media and lead to blind spots that deteriorate into mud balls or dead columns. Filters in areas subject to dust storms can become blinded with fine sand which is too coarse to be removed by the back wash, particularly without simultaneous air scour. Covering may be necessary. The growth of algae can also be a problem and prechlorination was once used primarily to keep the plant clean in this respect. Strings of algae eventually fall onto the media and cause a type of mud ball. Algae growing on walls may fall onto the media causing cracking and clogging. The most acceptable solution from the water scientist's point of view is regular manual cleaning or covering. Overall granular media filters are robust and should last for decades.

29.5 Investigation and Testing

Core samples are sometimes taken from filters to enable the condition below the surface to be assessed. It is not an easy procedure because filter media is seldom cohesive and the grains tend to flow freely from the core tube. On a large filter it is usually best to dig down carefully into damp media with a wall slope of about 45° samples being taken at appropriate depths. Care must be taken to avoid disturbing the packing layers or compacting them by foot traffic and if possible the excavations should stop short of the gravel if present. Excavations readily level themselves again on backwashing and need not be filled in manually. For deeper excavation a weighted caisson box is required. This is sunk into the media during a wash and excavated after the filter has been drained. The media can be examined without the danger of the surrounding media slumping in. (The need to lock upwash valves during such operations has already been mentioned).

A sampler device is illustrated in Fig. 29.1. If a tube is inserted into a bed of sand it will usually slide out when withdrawn.. The device illustrated is a three sided channel with removable fourth side to form a tube 75-100mm square. The bottom is closed by a hinged flap. The device is lowered into a filter bed during washing and the flap closed. The sampler can then be withdrawn and laid on its side and the removable side removed to expose a section of the bed. It can be operated using ropes in the centre of a filter bed. It is not possible to sample an unfluidised bed without an auger (fence post drill). This tends to mix up the media. Digging with a spade is the most satisfactory procedure in this case.

The most useful, common and simple test for the washing phase is the broom stick test. A suitable tube or pole (25-30mm diameter) is inserted into the bed during a wash and moved slowly through the bed. If it moves fairly freely then the bed is regarded as being adequately fluidised. This is applicable also to the subfluidisation combined air and water washes. Any resistant spots indicate clogging or blockage. General difficulty in moving the pole indicates an inadequate wash rate. Such a pole will not readily enter the packing layers and must not be forced into them.

A simple field test for checking the condition of the media is to take a 100cc

454

sample, or any other convenient amount, from a point well away from the walls where movement may be restricted. The sample should be placed in a stoppered 250cc measuring cylinder, water added to fill to the 250cc line and the cylinder shaken vigorously for one minute. The contents are allowed to settle. The settled dirt suspension is quoted as a percentage of the settled sand volume. A very clean filter media will produce very little sediment. 1% would be rated as good, 5% fair and 10% poor. Combined air scour and water washing will usually achieve a very good rating. If the value is only fair or worse, the filter may not be receiving an adequate wash or air scour. It may be overloaded with solids, with polyelectrolyte or sometimes with polysaccharides from an algal bloom or other biological material (see Chapter 27).

For more precise results the suspended solids may be filtered off dried and weighed. An analysis of the solids may sometimes be useful, eg. Al, Fe, Ca and loss on combustion.

A very bad bed can have the appearance of a lunar landscape. It is often possible to recover the situation if the packing layers are intact by digging over with a fork or rotovating the media (avoiding the gravel), by chemical treatment eg. hypochlorite & caustic soda or a proprietary compound. Acid may attack the concrete or steel. The cause of deterioration must be identified and rectified. If the packing layer has disrupted then complete renewal is necessary although the working media can often be salvaged. If it is heavily encrusted then it will be necessary to replace this also.

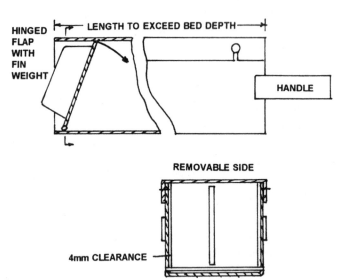

Fig. 29.1 Media sampler

29.6 Reference

Mouchet, P. (1992) "From Conventional to Biological Removal of Iron and Manganese in France", J.American Water Works Assn., **84,** 158

CHAPTER 30.

FILTER MEDIA

30.1 Specification and Testing

A United Kingdom Standard for the Specification, Approval and Testing of Filtering Materials has been issued by the British Effluent and Water Association (BEWA), now British Water, jointly with the Water Services Association (Anon.1993). The American Waterworks Association has also published a Specification (AWWA B100/89)(1989) which also cover some aspects of the installation of materials. The European Standards Organisation (CEN) has under preparation a set of documents covering all of the materials known to be used to treat drinking water, rather than a single generic one. The objective in this case is to remove any barriers to trade between members of the European Union. While there are useful definitions and test methods most users will find these documents too open to be of much use. For example the size of the material may be specified either by the size range as in Britain or by the so called effective size and uniformity coefficient. If any new filtering materials are introduced they will not be covered by CEN standards until the relevant committees can be pursuaded to include them in their programme. The gestation period is lengthy.

There is a danger of specifications being over prescriptive thereby restricting the designer and inhibiting innovation. On the other hand, the requirements must be clearly defined so that the document may be the basis of an unambiguous contract between the supplier and the user. Non-scientific concepts should be avoided. The philosophy behind the BEWA Specification is discussed below. Further background to the testing of filter media has been discussed by Ives (1990).

30.2 Size

As mentioned elsewhere, the most critical aspect of filtration is backwashing. Backwashing behaviour is in turn governed by the hydraulic size of the material, the voidage and the submerged density, all of which appear in the Kozeny and Carman equations which define the hydraulic gradient. This gradient when balanced against the submerged density defines the expansion threshold, as discussed in Chapter 20. The voidage is a function of particle shape. The supplier has little control over the submerged density or shape if he has only one available source. However, the size range is directly under his control, and because the backwash varies with the square of the size, the latter is very critical. Natural sources contain a range of sizes and the supplier must sieve out the band required by the customer. Thus media is normally specified in terms of a size range, the British convention being that the two stated sizes are the 5 percentile and the 95 percentile. Sieving of small grains is a statistical process. Some grains bounce over the screens and pass on with the oversized material. Some find rouge holes or perhaps penetrate edgewise to the production screen, but not to the test screen. One normally finds that with size ranges between 1½-1, to 3-1, the overall size distribution tends to follow a `log-normal' distribution.

Standard sieve series, for example to BS 410:1986, (ISO 565:1990), used for sand and other granular filtering materials have a size step of $2^{0.25}$ ie. ~ 20% which is rounded up to produce convenient numbers where possible. British Standard sieve sizes relevant to filtration are given in Table 30.1.

The sample of material must be dry before sieving and a convenient quantity such as 200 g is placed on the top of a nest of 6 or more standard sieves arranged in decreasing size order and shaken until no further change occurs.

Various mechanical sieving machines are available. These normally use either a orbital motion or a vertical drop, the latter being accompanied by a nutating motion. Machines do not always give the same result as manual sieving and the latter, used with a swirling motion, is therefore regarded as the reference standard.

Shaking is continued until there is no further change in weight (ie, less than 1%) after each further minute of shaking. A standard procedure is evolved and adhered to rigidly. Results are usually plotted. This enables the character of the media to be assessed more easily.

If sand sizes are plotted on a linear scale against the sieve size (which progresses logarithmically) a S-shaped curve is obtained. This is of little use in that any limits also must be drawn in the shape of an S-curve. Much more useful although not yet used very widely is the log/probability plot which is recommended both in the British and American Specifications. The spacing of the probability ordinates is based upon the integral of the equation $Y = 0.3989 \ Exp(-S^2/2)$ where S is the fractional population. An example is given in Fig.30.1. Plotting materials with a log normal distribution on such a paper produces a straight line. Thus it is easy to define 'tram line' limits between adjacent standard sieve sizes and passing through the 5 and 95 percentile points as illustrated in the Fig.30.1. The size band therefore has a width of about 20%. The supplier should aim for a centre line grading. Material falling within the tram lines is regarded as acceptable.

A more quantitative procedure giving a single characteristic figure is the calculation of the *hydraulic* size of the material. This is the parameter to be inserted in the Carman and Kozeny equations and is the *area averaged* size. The surface area of the sphere is 6/D, hence the weight of the fraction retained on each sieve is divided by the size of the sieve and the dividends are summed to obtain six times the total

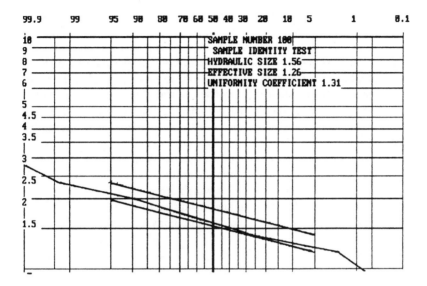

Fig. 30.1 Log/probability plot of media seive analysis with tramline limits.

area of the sample. The reciprocal is then taken to convert this effective area back to an area mean size. The sieve sizes are of course the 'retained' sizes and include grains up to the next largest size. 10% is therefore added to the size to make the figure correspond to the <u>centre</u> line between the adjacent sieves. (See the Appendix at the end of this chapter for a specimen calculation). The hydraulic size is therefore the size that uniform grains would have to be to produce the same hydraulic loss as the disperse mixture being tested, assuming of course that the voidage is the same.

The terms 'uniformity co-efficient' and 'effective size' are widely used, but have little fundamental significance. They were introduced in the USA by Hazen, originally to characterise sands for slow sand filtration before Kozeny's work on flow through porous media. The 'effective size' is the 10 percentile and the 'uniformity co-efficient' the ratio of the sizes of the 60 and 10 percentiles ie. the size that would permit 60% to pass and that which would permit 10% to pass. (No correction is applied to adjust the size from the 'retained' figure to the 'centre line' figure, as with the hydraulic size.) These terms continue to be used and have become entrenched in some countries. However Di Bernardo and Rivera (1990) have demonstrated that these terms have no validity by blending up some sands to produce a set with a constant effective size and varying uniformity coefficient and running them as slow sand filters. The behaviour was certainly not constant but much more in line with the hydraulic or mean sizes. Penetration and run times increased with the uniformity coefficient.

Size ranges expressed in terms of precise limits eg. 0.5 to 1.0 mm, or 0.4 to 1.2 mm, are clear and unambiguous. Traditional terminology has also included various names (torpedo sand), numbers and grades which are confusing, particularly as the same designations mean different things in different countries. The BEWA Specification urges the use of unambiguous terminology based upon BS (ISO) metric sizes.

Mention has also been made of the errors caused by different shaking procedures. Filtration materials are for the most part abrasive and in a laboratory which is testing materials on a routine basis, screens can wear. It is therefore necessary to regard a new set of sieves as reference ones against which sieves in regular use can be cross checked. If a discrepancy becomes apparent, then a replacement sieve must be bought and designated as a reference sieve, and the previous reference sieve brought into routine service. Likewise, it is most desirable that suppliers and customers should examine specimen reference samples from time to time to ensure that their results are in agreement.

Finally, it should be noted that granular material filtration requires a material that is more precisely controlled than that which would be suffice for construction purposes. Few of the BS 812 Specifications for aggregates are relevant. However, some of the test procedures are useful.

Samples obtained for sieving must of course be representative and large samples must be divided by quartering and splitting in rifflers. Continuous transfer from a silo or by hose allows sampling during transfer, although this can lead to situation where the material may be found to be outside the specification after it has been placed, in which case arguments about the cost of its removal may arise.

30.3 Dirt Content

Some filtering materials such as activated carbon and even crushed materials can be dusty. Filters are always washed several times on commissioning before being put to service and it is pointless to be too concerned about dust, providing it is not excessive to the point of the purchaser being short changed (ie, greater than 1% of dust). However, pellets of clay or a general coating of grains by clay can occasionally be encountered. These may disperse only slowly and can prolong commissioning.

30.4 Appearance

Some sands are golden, some grey, some dark. The colour, however, matters little providing that the material is inert. Changes in appearance however can alert an inspector to possible changes in the source and the need for closer scrutiny.

Shape, angularity, sphericity are all subjective terms and some attempts have been made to quantify these. The main concern is that flat, flaky grains tend to orient themselves horizontally and in this way they can increase the resistance to flow. Specifications tend to limit the number of flat particles but in cases of doubt, one would recommend measurement of the hydraulic resistance as a means of resolving disputes.

Many writers have referred to the desirability of using rounded grains. This was discussed in Chapter 30. Providing that the voidage is taken into account in designing the system, shape is merely a matter of ensuring consistency rather than being a fundamental requirement.

30.5 Specific Gravity

The primary importance of this parameter is of course its effect on the fluidisation velocity. Silica sand is a well defined material, with a density of 2650 kg/m³. Most other materials vary. Anthracite is available with densities ranging from 1300 to over 2000 kg/m³. Hence it is recommended that particle densities are always quoted in any offer or purchase order. In Germany some alphabetic designations are also used to denote anthracite grades of various density. This can add further to the confusion.

The measurement of specific gravity using the 'SG' bottle is simple and quick. However, if the material has any internal porosity, the SG bottle will give the figure for the matrix and not the effective density of the particle relevant to its settlement behaviour. Data on activated carbon or expanded materials such as pumice or expanded shale etc. would be meaningless. The only reliable method is to measure the fluidisation threshold, the expansion and headloss in a test column such as that illustrated in Fig.30.2 . 500 mm depth of material, for example, is backwashed to clean it and then allowed to expand to, say, 50% and then to re-settle. Expansion data is related to the final settled bed depth after the upward flow has been switched off,

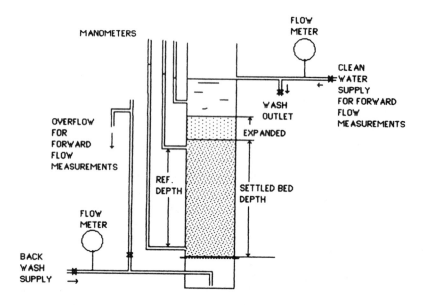

Fig. 30.2 A general purpose test column for clean bed headloss, backwash and media expansion measurements.

not the initial depth. With expanded or porous granular materials care is necessary ensure that all the air which may be present initially within the grains has been driven out. Wetting agents can assist but it may sometimes be necessary to boil the materials in water. Sometimes porous grains are soaked in water and blotted before their density is measured to provide an approximate value of the effective density of the material.

The other density parameter needed by the engineer is the bulk density, defined as the weight of the material at the point of sale that occupies one unit of volume in its working state. Here again the apparatus in Fig.30.2 is required. A known weight of material is charged to the column. The material is expanded with an upward flow of water, allowed to come to rest after switching off the flow and its volume calculated. As mentioned elsewhere in this book, the bulk density can vary depending upon the previous history of the bed. Material which has been washed by combined air and water, and not fluidised with a high rate water wash will tend to have a somewhat higher density than one which has been washed with water only and allowed to settle. Both figures may be lower than the initial poured density of the mixture produced by the supplier.

Finally it is not unknown for filtering materials to comprise a mixture of grains with different densities. This may not be apparent in a mixed sample received in a bag, but in a filter it could cause the bottom layer to remain unwashed while having the appearance of good fluidisation at the top. Again the equipment in Fig. 30.2 may be used to stratify the sample after which fractions from the top and the bottom should be withdrawn from measurement of specific gravity.

30.6   Abrasion Resistance

Specifications of filtering material often refer to the Mho hardness of the material. Although it has been referred to by Ives (1990), the index was intended as a means of characterising minerals and is not a suitable test for bulk materials. It involves attempting to scratch one material with another. Different minerals have been allocated a standard Mho number and unknown materials can be assessed within the arbitrary scale. It is however difficult to scratch small rough grains. In any case, to be statistically significant one would need to apply the test to a large number of grains. The test does not reproduce any action that occurs within a filter. Indeed many soft organic materials such as polyethylene or expanded polystyrene will be exceedingly soft, but are used very successfully in filters.

A far more meaningful test is one for abrasion resistance under the conditions of backwashing (Fig.30.3). It measures the weight loss from a sample which may, if required, comprise several kilograms and is therefore a much more appropriate test procedure for bulk material.

A typical filter bed washed with three bed volumes of water may be subject for abrasion for only five minutes daily during washing thus 100 hours testing represents about three years worth of washing. Such a test can be completed between Monday and Friday in one week.

A previously washed test sample is dried and weighed. It is then charged to the column and washed continuously preferably with the water being recycled. The upflow rate should correspond to that intended for the plant in which the media is to be used although standard conditions are proposed in the BEWA Specification. Combined air and water washing is far more severe and the media must be tested with the appropriate procedure. The test column for the combined wash will require a mesh trap to retain the grains which tend to be carried up by the air bubbles.

Sand should show little abrasion but when anthracite is washed with combined air and water it can cause blackening of the recycled water. Some types of activated carbon are pulverised and are totally unsuitable for combined air and water

462

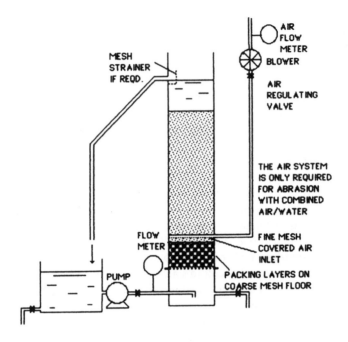

MESH
STRAINER
IF REQD.

AIR
FLOW
METER

BLOWER

AIR
REGULATING
VALVE

THE AIR SYSTEM
IS ONLY REQUIRED
FOR ABRASION
WITH COMBINED
AIR/WATER

FLOW
METER

FINE MESH
COVERED AIR
INLET

PUMP

PACKING LAYERS ON
COARSE MESH FLOOR

Fig. 30.3 Column for media abrasion tests.

washing. Angular materials may show an initial loss as the sharp edges are broken off but may stabilise later. In this case the test should be divided into a number of shorter periods of exposure.

The other related problem encountered in practice is breakdown during handling caused by the friability of the material. If anthracite or activated carbon are handled roughly when bagged, they break down and the size distribution may change. The same is also true when these materials are walked on without protective boards. A test which shears the material under an applied load, to characterise such materials, has been developed by Humby et al. (1996). This has shown significent differences between various types of carbon for example. The test was originally intended to define friable materials which might attract concessions.

30.7 Voidage

This is a most important fundamental parameter. A high voidage indicates a high dirt holding capacity, but also a higher backwash rate and a lower bulk density. As previously mentioned knowledge of this parameter is essential if the fluidisation characteristics of the materials and the hydraulic gradient are to be calculated from first principles. Measurement is simple. With non-porous materials a given weight of known grain density is added to a measuring cylinder containing water. The rise in water level indicates the actual volume of the solids and the difference between this and the apparent volume of the material corresponds to the voidage.

Unfortunately the test is not applicable to grains with open internal porosity although an approximate estimate is possible if grains are soaked and the surface water is removed by blotting. Such internal pores lower the effective density of the grain when filled with water, but do not necessarily provide channels for flow during filtration or backwashing. Parameters such as backwash threshold, headloss during fluidisation etc. must be measured directly if accuracy is required.

Incidently, it seems that the terms *voidage* and *porosity* are used by chemical engineers and civil engineers to describe different things. In this book, voidage is external to the grain, a convention adopted by Coulson and Richardson (1990) in their widely used textbook on chemical engineering. Porosity is internal to the grain, however some authors described the external voidage of the bed as the porosity.

30.8 Acid Solubility and Leaching

In many water treatment plants, water dosed with a coagulant is filtered at a pH well below the equilibrium value and any calcium carbonate present will tend to dissolve. Some sources of filtering material include fragments of shellfish, limestone etc. and loss of this could effect the total inventory of material as well as the size distribution. The presence of a significant amount of acid soluble material is not necessarily a cause for rejection, particularly if the material has a suitably low cost to offset losses in service, providing that such losses do not effect the size distribution significantly.

Support gravel may sometimes include silica pebbles bound by calcium carbonate. These can break down into smaller components which might then come outside the size specification. Clearly, limestone chippings would not be suitable for an aggressive water and the test is a simple way of checking the material.

The AWWA procedure uses 50% hydrochloric acid (ie, 1 to 1 dilution) which provides an accelerated test. It will however dissolve oxides of iron, aluminium and other metals which are unlikely to dissolve in practice, particularly if the purpose of the filter is to remove iron or aluminium floc. There are also many applications where the conditions are non aggressive for example, primary filtration upstream of slow sand filters, slow sand media and filtration after precipitation softening where a limestone material would in fact be suitable. Thus the intended use should be borne in mind when selecting filtering materials.

There are some applications where the material is deliberately selected for its solubility, for example pH correction in mineralising filters. In this case the nature and quantity of the residue may be of interest.

Leaching of potentially toxic elements and compounds from materials of construction in potable water treatment plant is the concern of the UK Water Bylaws Service and the Committee on Chemicals and Materials of Construction used in Potable Water Treatment and Swimming Pools. Filter materials must not contaminate the water to the extent that the parameters listed in the Water Regulations exceed the limits. For this purpose the BEWA Specification and the CEN drafts use a buffered mixture to simulate typical exposure conditions. (The CEN drafts (which are subject to amendment) limit the test to activated carbon, other materials being assumed to have negligible extractable impurities). The sample must of course be clean and free from dust and other extraneous dirt that would be removed during commissioning. Leachable material is most unlikely to be stripped instantaneously, and a dilution to approximately one days worth of filtrate has been proposed.

The possible interaction between disinfectants such as chlorine, chlorine dioxide and ozone and organic materials such as anthracite, must also be considered. Granular activated carbon can produce an extract containing zinc and alkaline salts and recirculation with chlorine produces a brown solution of unknown composition. Ozone oxidises carbon, anthracite and most other organic materials.

The composition of a filtering material in terms of silica, iron, aluminium and calcium content gives little guidance as to its suitability for filtration purposes although draft CEN specifications give composition limits for the purposes of product description.

30.9 Moisture and Volume

It seems rather pointless to insist on dry materials at the point of delivery if the material is the be charged to a filter and submerged in water. Weight is a precise basis for trade, but the designer actually requires a given volume. It so happens that fine materials that originate in the wet state, such as natural sands have to be dried for sieving but the moisture content is purely a matter of weights and measures. If the supplier and purchaser can agree on a basis for trade then there is no fundamental objection to the volume being used.

30.10 Approval and Inspection

It is common for a supplier to submit samples of a material for approval. Such samples will usually be representative of the source and therefore also of shape, quality and composition but not necessarily of a final size grading. The inspector will ensure that the materials supplied comply with the approval sample. Users and suppliers must establish confidence in each other's analytical procedures, otherwise delays in acceptance and delivery become unavoidable. The cost of removing out-of-specification material is very considerable, not the mention the cost of delays in completing and commissioning the plant. Quality assurance procedures are essential.

30.11 References

American Water Works Assn. Specification AWWA B100/89

British Effluent and Water Association, (1993) Standard for the Specification, Approval and Testing of Filtering Materials, British Water, 1, Queen Anne's Gate, London, SW1H 9BT, U.K.

British Standard 410.1986 & ISO 1990 Standard Sieve Sizes

Di Bernado, L. and Rivera, A.E. (1996)"Influence of Sand Uniformity Coefficient on Slow Sand Filtration Performance", Paper 19 in Advances in Slow Sand and Alternative Biological Filtration, Eds. Graham, N, and Collins,R. Wiley, Chichester, UK.

Humby, M.S. (1996), "Development of a Friability Test for Granular Filter Media", J. Chartered Inst. Water and Environmental Management, **10**, 87

Ives,K.J. (1990) "Testing of Filter Media", Aqua, **39,** (7), 144

Table 30.1   B S Standard/ISO Sieve Sizes used for Filter Media

Gravels	Sand
63mm	4.0mm
52	3.3
44.5	2.8
37.5	2.36
31.5	2.0
26	1.7
22.4	1.4
19.5	1.18
16	1.0
13.2	0.85
11.2	0.71
9.5	0.6
8.0	0.5
6.7	0.425
5.6	0.355
4.75	0.3
	0.25
	0.212

Standard Media Sizes based on 2:1 Range	Typical Target Hydraulic size
0.425-0.85mm	0.58mm
0.5-1.0	0.68
0.6-1.18	0.82
0.71-1.4	0.97
0.85-1.7	1.16
1.0-2.0	1.37
1.18-2.36	1.61
1.18-2.8*	1.83 *
1.4-2.8	1.91
1.7-3.3	2.32
2.0-4.0	2.73
2.36-4.75	3.22

* = a non standard but common size.

Note- Other standard seives are available but is is essential that a consistent series with a equal proportional step size be used.

APPENDIX

CALCULATION OF HYDRAULIC SIZE

1. Divide the % retained on each successive sieve by the size of the sieve aperture.
2. Add up the figures thus obtained and divide by 100.
3. Obtain the reciprocal of the above sum.
4. Add 10% to the reciprocal to correct the 'retained' size to the centre size between adjacent sieves (sieves are spaced at 20% increments).

EXAMPLE

	Retained Size mm	Weight %	Calculation
	2.8	0.1	0.04
	2.36	0.5	0.21
	2.0	10.2	5.1
	1.7	22.10	13.0
	1.4	42.3	30.25
	1.18	22.5	19.1
	1.0	1.4	1.4
	0.85	0.3	0.35
Pan	(.71)	0.5	0.7
		100.0	70.1 ÷ 100 =0.701

Reciprocal $1/0.701$ = 1.43

Add 10%   0.14

Hydraulic Size   = 1.57mm

Explanation

Specific Area $S = 6/D$

Contribution to area of each fraction = $(W \times 6)/D$

where w = weight of the fraction and D = diameter

Total Area = Sum $(W \times 6)/D$

   Equivalent Single Size = 6 / Total Area

INDEX

474